Physical Anthropology, Race and Eugenics in Greece (1880s–1970s)

Balkan Studies Library

VOLUME 11

The titles published in this series are listed at brill.com/bsl

Physical Anthropology, Race and Eugenics in Greece (1880s–1970s)

By

Sevasti Trubeta

BRILL

LEIDEN • BOSTON
2013

Cover Illustration: Design by Alexandra Chaitoglou and Ioanna Varia

Library of Congress Cataloging-in-Publication Data

Physical anthropology, race and eugenics in Greece (1880s–1970s) / by Sevasti Trubeta.
pages cm — (Balkan studies library ; 11)
Includes index.
ISBN 978-90-04-25766-5 (hardback : alk. paper) — ISBN 978-90-04-25767-2 (e-book) 1. Physical anthropology—Greece—History. 2. Eugenics—Greece—History. 3. Racism in anthropology—Greece—History. 4. Greece—Race relations. 5. Greece—Social life and customs. I. Title.

GN50.45.G8T76 2013
599.909495—dc23

2013026491

This publication has been typeset in the multilingual "Brill" typeface. With over 5,100 characters covering Latin, IPA, Greek, and Cyrillic, this typeface is especially suitable for use in the humanities. For more information, please see www.brill.com/brill-typeface.

ISSN 1877-6272
ISBN 978-90-04-25766-5 (hardback)
ISBN 978-90-04-25767-2 (e-book)

This book is printed on acid-free paper.

In memory of Georgia Kretsi (1972–2009)

CONTENTS

Preface and Acknowledgements xi
A Note on Greek References and Transliteration xiii
Abbreviations xv

Introduction: Framing the Research: A Medical Discipline with Holistic Claims 1

SECTION I

ON THE EMERGENCE OF MODERN ANTHROPOLOGY IN GREECE

1. Tracing the Intellectual and Epistemic Sources of Greek Anthropology 25
   1.1 The Dispute over Evolutionism and the Alliance of the Natural Sciences and Medicine 26
   1.2 Medical Geography As an Epistemic Source of Greek Anthropology 31

2. The Emergence of Anthropological Institutions and Professional Scholarship 41
   2.1 The Adventurous Establishment of the University Chair for Physical Anthropology 44
   2.2 The Chair-Holder in the Light of His Own Life Narrative 47

Conclusion to Section I 51

SECTION II

CONCEPTS OF ANTHROPOLOGY, INSTITUTIONAL TRAJECTORIES AND SCIENTIFIC CAPITAL

Introduction to Section II: Paths for Passing from the Word to Discourse 55

3. Anthropology at the Museum 57
   3.1 The Museum As a Vehicle for the Mutual Advancement of National and Professional Aims 57

3.2 The Anthropological Museum As National Heterotopia ... 62
3.3 Beyond National Limits: A Public Site to Represent Universal Human Culture ... 66
3.4 Authorization over Ethnology and a Colonial Dream for Athens ... 70

4. Anthropology at the University Chair ... 79
4.1 The Agenda: Teaching Anthropology in the Broad Sense ... 79
4.2 The Implementation: Teaching the Restricted Scope of Physical Anthropology ... 83

5. The Greek Anthropological Society ... 89
5.1 Contours of a Trajectory: Promising Outset, Inglorious End ... 89
5.2 The Internal Debates of the Greek Anthropological Society ... 96
5.3 Quantification of Humans: Criminals, Sound Children and Intelligence Tests ... 99
5.4 Physical Anomaly and Heredity As Nexus to Anthropological Discourse ... 106
5.5 Anthropology As Fetish, Professional Conformity and Opposition ... 110
5.6 Embracing Social Anthropology in the Face of a Noble Anthropologist: Prince Peter ... 115

6. A 'Disinterested Science' in Wartime ... 121
6.1 The Neutrality of Science and the War Narrative ... 121
6.2 The Need for Rehabilitation and Ex Post Facto Patriotism ... 128
6.3 Civil War: Victims, Patriots and Badly Educated Peoples .. 131

Excursus: Anthropological Conceptions and University Capital ... 139

SECTION III

CONCEPTS OF THE GREEK *FILI*: COMMUNALITY IN RACIAL AND EUGENIC TERMS

Introduction to Section III: Terminological Metamorphoses of *Fili*—Diverse Concepts of Communality ... 145

7. Race and Greek Ancestry ... 147
 7.1 From Religious Universalism to Hellenism: A Bridge to Pass from Empire to Nation State ... 149
 7.2 The Arrival of Race via the National Vehicle ... 159
 7.3 Dilemmas of Racial Purity: From Assimilatory Hellenism to Racial Hybridization ... 162
 7.4 To Cope with the Ottoman Legacy in the Shadow of the Nordic Racial Myth ... 171
 7.5 Back to Racial Purity via Mixophobia: In Light of National Socialism ... 177
 7.6 Perceptions of Racial Theories: Absorbed, Absent and Phantom Minorities ... 189
 7.7 Aris Poulianos on the Racial Origins of the Greeks ... 196

8. The Eugenic Concept of *Fili* ... 203
 8.1 Eugenic Representations of Hellenic Antiquity: A Matter of European Self-Consciousness ... 205
 8.2 The Social Question and Overpopulation: Eugenics for Governing Social Stratification ... 207
 8.3 Social Disease, the Promiscuity of the Poor and Eugenically Coloured Hygiene ... 223
 8.4 Scholars on Eugenic Politics, Sterilization and Prenatal Health Examinations ... 233
 8.5 Eugenic Birth Control by State Institutions ... 239
 8.6 Greek Racial Hygienists and the Transient Appearance of Stavros Zurukzoglu ... 246
 8.7 Beyond and against Eugenics: Still a Subjective Decision 258
 8.8 Under the Banner of Reformist Eugenics: The Greek Society of Eugenics ... 263
  8.8.1 Post-War Eugenics in Public Debates: Old Issues—New Controversies ... 267
  8.8.2 Professional Networking and Political Opportunity: A Biographical Note on Nikolaos Louros ... 274

Excursus: *Fili*, Bio-Power and Authoritative Biologism ... 279

9. Concluding Reflections: The Hidden Legacy of Racial Nationalism ... 287

Appendix 293
Sources 295
References 305
Index of Names 331
Index of Subjects 334

PREFACE AND ACKNOWLEDGEMENTS

The initial, though at the time quite vague, idea for this study emerged at the beginning of the 2000s when I was involved in the project "Identities and Alterities" (*Sonderforschungsbereich* 541) at the Albert-Ludwigs-University in Freiburg im Breisgau. As a member of the interdisciplinary sub-project on "Nation-Building Processes and Ethnic-National Contradictions in Southeastern Europe" I was concerned with ascertaining how biologistic and racial arguments were applied in order to buttress national discourses in the Balkans. This was the beginning of a long journey along paths that came to differ from what I had originally imagined. The initial research concept acquired a more concrete form in a subsequent project on "Biologism and National Ideas: The Case of the Greek National Discourse in the First Half of the 20th Century", which was supported by the German Research Foundation (DFG, Deutsche Forschungsgesellschaft) and housed at the Institute of East European Studies in the Free University in Berlin. Crucial impetus to develop the project came from the discussions of preliminary conclusions and hypotheses at the colloquium of the chair for South-East European History and Culture, under the directorship of Professor Holm Sundhaussen, and at the Working Group on the History of Race and Eugenics (based at Oxford Brookes University). A fellowship at the Institute of Modern Greek Studies in Princeton University (2006) opened a new perspective to my approach. In addition to providing an opportunity to exchange ideas with colleagues from different scholarly backgrounds at that institute, my parallel affiliation to the Department of History of Science at the same university broadened my epistemological and comparative perspective. Discussions with individual colleagues, especially from disciplines such as social anthropology, history and biology, also helped me to recognize the multifaceted perspectives this research subject includes and to specify the focus of my project. All these influences have accompanied and acted greatly on the development of this study.

At the risk of disregarding some important colleagues and discussants who motivated me during my researches and the writing of this study, I would like to express my thankfulness first and foremost to DFG for financing my research project and Professor Holm Sundhaussen (Berlin) for his support throughout all of the years in which my project was hosted at his chair at the Free University in Berlin. I also address thanks to the

Centre for Hellenic Studies at Princeton University for the fellowship and especially the director, Dimitri Gondicas, for networking me at the university; the colleagues at the Department of History of Science at Princeton University for including me in their work, and especially Professor Angela Creager for the valuable feedback she and her colleagues provided on my research; to Professor Karl Kaser (Graz) and Professor Miltos Pechlivanos (Berlin) for the careful reading of a first version of the study and especially the latter for the exciting discussions in Berlin; Professor Christian Voß (Berlin) and Dr Christian Promitzer (Graz) for the cooperation and exchanges during all of these years; Dr Nikos Poulopoulos for the inspiring discussions in Cambridge, Boston and Princeton; the members of the Network for the Research of the Civil Wars who responded to my enquiries for additional information or material, especially to Mr Nikos Hatzidimitrakos for providing me with material from his private archive; Professor Béatrice Ziegler (Zurich) for her unpublished manuscript on Stavros Zurukzoglu; Dr Maria Zarifi (Athens) for the material on the Greek university; Dr Katerina Lagos for sending me her unpublished PhD thesis; the staff of the archives and libraries in Athens and Berlin for the support they gave to my research, and especially the staff of the Parliament Library in Athens and Ms Simone Langner at the Bundesarchiv (Berlin). My sincere thanks are addressed to the anonymous reviewers of the manuscript for the enlightening comments and suggestions in the first submitted version of the manuscript; to Brill Academic Publishers, especially the editor-in-chief of the *Balkan Studies Library*, Zoran Milutinovic, for accepting the book for publication; Ivo Romein for his invaluable support and competence from the day of the first submission of the manuscript up to the time when the book was ready for printing; Tim Page for the careful copyediting, the respect with which he treated the manuscript and his huge patience; and the designers Alexandra Chaitoglou and Ioanna Varia for the artistic cover illustration.

This book is dedicated to my colleague and friend, Gogo, who shared in my research adventure for several years and has been recorded in my memory as inextricably linked with the "last period". Every line of the manuscript radiates chapters of my life and the inspiration and encouragement I have received from several friends and colleagues. Last but not least, my gratitude is addressed to Nikos, who accompanies and inspires me with his music and angelic lightness, and who managed to give me back *the vision*. I am grateful for all the support I have enjoyed and the inspiration I received on the long road on which this project has taken me. Of course, I alone am responsible for the views expressed in this study.

Mytilene, May 2013

A NOTE ON GREEK REFERENCES AND TRANSLITERATION

In the list of references, publications in Greek and other languages have been listed together. Only the names of Greek authors have been transliterated whereas titles of the references, place of publication and publishers have been kept in their original forms. However, I have endeavored to give as much information on the Greek references as possible in the text. Thus the titles are often referred to in their translated form either in the main text or in the footnotes. The references which are considered as pertaining to primary sources have been translated into English while simultaneously keeping the original Greek title. With respect to the authors' names, the spelling adopted by themselves or their editors in English, German, Italian, French or other language publications is maintained.

Transliteration from the Greek to the Latin alphabet has been a topic of disagreement among scholars. The system I am adopting here is the following:

Αα:	Aa	Σσ:	Ss
Ββ:	Vv	Ττ:	Tt
Γγ:	Gg	Υυ:	Ii
Δδ:	Dd	Φφ:	Ff
Εε:	Ee	Χχ:	Ch ch
Ζζ:	Zz	Ψψ:	Ps, ps
Ηη:	Ii	Ωω:	Oo
Θθ:	Th, th	αι:	ai
Ιι:	Ii	αυ:	af/av
Κκ:	Kk	ευ:	ef/ev
Λλ:	Ll	ει:	ei
Μμ:	Mm	οι:	oi
Νν:	Nn	ου:	ou
Ξξ:	Xx	γκ:	gk
Οο:	Oo	γγ:	gg
Ππ:	Pp	μπ:	mp
Ρρ:	Rr	ντ:	nt

ABBREVIATIONS

AMPG	Archives of the Max-Plank-Society [Archiv der Max-Plank-Gesellschaft]
GAS	Greek Anthropological Society [Ελληνική Ανθρωπολογική Εταιρεία]
HAUA	Historical Archive of the University of Athens [Ιστορικό Αρχείο του Πανεπιστημίου Αθηνών]
KWG	Kaiser-Wilhelm-Society [Kaiser-Wilhelm-Gesellschaft]
NSDAP	National Socialist German Workers' Party [Nationalsozialistische Deutsche Arbeiterpartei]
PGAS	Proceedings of the Greek Anthropological Society [Πρακτικά της Ελληνικής Ανθρωπολογικής Εταιρείας]

INTRODUCTION

FRAMING THE RESEARCH: A MEDICAL DISCIPLINE WITH HOLISTIC CLAIMS

In an impressive paper presented at the Second International Congress of the History of Science and Technology in London (29 June–4 July 1931), the Soviet physicist Boris Mikhailovich Hessen (1893–1936) argued that Newton's major contribution to mathematical physics had in fact been a response to the demands of England's growing capitalist society in the early 17th century. Hessen's paper on "The Social and Economic Roots of Newton's *Principia*", together with other contributions by the Soviet delegation at that congress, came to be known as *Science at the Crossroads*, and was of particular significance because it suggested that science operates within a wider political economy.[1] Yet if this assertion, which greatly influenced the historiography of science, was a novelty in the 1930s, today it has become the fundament of science studies. Current mainstream historical and sociological studies of science are inclined to assume the interdependency of the contents of scientific knowledge and the societal conditions of its production. Pierre Bourdieu, in his influential concept of 'field', for example, has highlighted the embeddedness of science in societal conditions that extend beyond the narrow settings of scientific terrain (Bourdieu 1998).

In all explanatory models and research programmes developed since then, science studies are still regarded as an insufficiently developed area of research (Weingart 2003). Scholars point to the unsatisfactory state of both the theoretical and methodological instruments used to explore science and scientific knowledge. Some even underline the lack of adequate consideration of the multifaceted and changing relationship between society and science, which means that new paradigms concerning technology in particular still remain 'dark matters' (Yearley 2005).

The topic of this study belongs to the 'old' research paradigms, although the specific case to be explored is still 'dark matter' given the lack of previous research on the issue. The study is concerned with the emergence and

[1] Hessen (1971/1931); Bukharin (1971/1931); cf. Wittich and Poldrack (1990); Freudenthal (2005); Freudenthal and McLaughlin (2009).

trajectory of physical anthropology in Greece, from its first integration into the University of Athens (named Ottonian University until October 1862) in the 19th century by means of the establishment of the Anthropological Laboratory and Museum (1886), the later establishment of a chair for physical anthropology at the same university (1925), up to the founding (1924) and the dissolution of one of the most important anthropological institutions, the Greek Anthropological Society (hereinafter GAS), in the 1970s. The general interest in this topic derives from anthropology's character as a modern science and its claim to status as an institution that influences societal reality. But how far does a discipline that emerged from the realm of the natural sciences and medicine have the authority and capability to provide suggestions for influencing society? And what kind of knowledge of developments in society can the exploration of its outlook and propositions provide?

The answers to these questions should be looked for in the holistic claims made by anthropologists regarding their discipline. Although physical anthropology was originally advanced from the progress of anatomy which effected its close association with medicine in the 19th century, its subsequent development as an academic area of study points rather to its character as a science for exploring humankind in its entirety. But ever since the 19th century, there has been a strong tendency among physical anthropologists to regard the scope of their discipline as not being restricted to the study of human nature and morphology alone. While seeking to reconstruct human evolution, anthropologists raised the global question of 'what is human?' and sought to provide answers that would define the manifold dimensions of the human essence. In doing so, they developed comparative methods and classification patterns that sought to record both commonalities and differences within humankind as a whole. Anthropologists sought to discover the laws of physical, spiritual and cultural forms of life and their impact upon the laws of history and progress. Thus, this discipline could be situated in the field of sciences whose cognitive concern was to trace the dispositive step in history, thinking in terms of a biological law motivated by nature. The subject of anthropological research became the human, not as an individual but as a representative of the genus. Anthropology 'in the broad sense' sought to explore the collective genealogy of humankind, responding, in addition to the natural sciences, to religious and philosophical thought as well (Stocking 1982/1968: 74). From this extensive scope, different schools and traditions were established.

In terms of epistemic contents and methodology, physical anthropology from its very emergence drew on the comparison between peoples on the basis of a wide range of characteristics; these included, apart from physical traits, customs, language and the production of material culture. Crucial to shaping its cognitive realm, therefore, were civilization, culture and the definition of 'human groups'. The latter were specified optionally (or even metonymically) in terms of nations, races, populations or social strata. These optional definitions of 'human groups' frequently appeared to merge or even to exchange semantics. From a holistic anthropological perspective, attempts to localize the driving forces of progress and history led to the interpretation of history as "a racial struggle which produced a constant redistribution of the racial elements in nations according to various laws of social selection" (Stocking 1982/1968: 60). It is therefore not a coincidence that physical anthropology provided the racial discourse with 'hard data' on human races.

The concern of anthropologists was not confined to providing mere explanations for humanity's genealogy and human nature, or even culture. Moreover, in the 19th century and especially in the 20th century, anthropologists sought to use their academic authority to influence political decision-making. This aspiration kept up with that of the natural sciences in general, which claimed to have the requisite authority to explain nature and evolution, and simultaneously to rule over nature, and, not least, to design the world (Goschler 2004). The source from which these aspirations were drawn is not implicit in anthropology, but corresponds to a tendency to apply biological ideas in attempting to understand and shape society. From its emergence, anthropology belonged to those sciences in which the dispute over the boundaries between human nature and human society, biological rules and social conditions of life has rumbled on. The intellectual sources of this dispute might be traced back to the evolutionism of the 19th century (Burrow 1966; Sarasin 2009), but it still remains topical today, although the particular questions of the dispute have been reformulated over the course of time (Macherey 2009), coming to revolve nowadays primarily around new genetic technologies and genetic intervention into human nature.

An influential model in explaining the relationship between the natural and social condition of human life is the one-to-one transmission of biological rules to society, where society is perceived as a living organism. This tendency has been defined in historical and social sciences as 'biologism' or the 'biologization' of society. Biologism is actually a very loosely defined term, although it has increasingly been applied in the sense of biological reductionism, especially with regard to eugenics. As Sören Niemmann-Findeisen (2004: 215) notes, biologism is a term used to describe the transfer of biological thinking to social or political circumstances, while the historian Reinhard Mocek (2002: 37) defines biologism as a notion comprising all attempts to explain social circumstances by means of biological rules or biological and natural-historical analogies.

Earlier sociological attempts to define biologism were made in critical works on biosociology or social biology. They were part of a controversy which was evoked after Edward O. Wilson (1975) published his book on sociobiology. Early critics included Günter Hartfiel (1976) who argued that biologism meant an exclusive implementation of biological explanatory patterns on social conditions of human life. Agreeing with this general assertion, Heinz-Georg Marten stressed the totalitarian character of biologistic approaches, the function of which was to "define integrative principles" (Marten 1983: 15). Marten considered biologism a historical and political phenomenon that communicated biological thinking, facts, imaginations and models to other scientific fields in order to earmark their allegedly 'unitary principles'. Since the late 1990s scholars have suggested that there is a need for a re-examination of the relationship between the biological and social conditions of life on the grounds that:

> the boundaries drawn between such seemingly obvious categories of the "natural" and the "social" are by no means fixed, objective and unequivocal. Rather, it can be shown that they shift in the course of history and are dependent on the context of values and the development of political and scientific discourse (Weingart, Maasen and Segerstråle 1997: 67).

Meanwhile, the humanities and the biological sciences have increasingly been integrated into joint research projects, thereby shaping interdisciplinary approaches that have become instrumental to science studies (Lipphardt and Patel 2008). Nevertheless, the term biologism still remains in use for describing phenomena whose biological outlook, with all its social and political implications, derives from discourses other than biological ones (Sarasin 2009: 274; cf. Butler 1990: 163).

Biologism arose to become an instrumental notion for the study of anthropology, especially in the period prior to the end of World War II, due in the first instance to the holistic claim of anthropologists regarding the scope of their science. Anthropology's attraction to human genetics and even more its involvement in eugenics in the 20th century subsequently increased the topicality of the contradictions in how anthropology defined the boundaries between the 'natural' and the 'social'. The implications of the answers were not exhausted in epistemic issues, given the aspirations of anthropologists to be involved in policy-making processes, or in part, their actual involvement. Case studies on individual paradigms of anthropological schools, national traditions or even individual leading anthropologists can show how anthropologists coped with these questions and the extent to which they drew legitimacy from the scope of their discipline.

Seen in the light of the above considerations, the subject of physical anthropology cannot be taken for granted, not least in the long period of a century considered in the current study. Rather, it has been, if not a contentious point, surely an issue, of shaping and implementing worldviews on the human condition in the interface of the natural and the social. But worldviews are variable and subject to alteration depending on factors which are to a large degree external to science.

On the State of Research on Physical Anthropology

The history of physical anthropology is that of an interdisciplinary field of research in which, in addition to historians and social scientists, physical anthropologists, geneticists and biologists have also made remarkable contributions.[2] The available studies provide illuminating insights into epistemological issues, the histories of scientific institutions and the history of ideas, among many other things. To date, the most profound, complex and multifaceted analyses of the history of European and American anthropology with regard to the interconnection of historical-political developments and production of anthropological knowledge may well have been provided by George W. Stocking Jr, currently professor emeritus in the Department of Anthropology at the University of Chicago. In numerous monographs, edited volumes and articles, Stocking has illustrated the emergence of anthropological thinking and science in the Anglo-American, French and German paradigms, and has described the different traditions and schools in physical, social and cultural anthropology.[3]

---

[2] See two representative works by the geologist, palaeontologist and evolution researcher Stephen Jay Gould (1941–2002) and the physical anthropologist Frank Spencer (1941–99); see Gould (1981) and Spencer (1997).

[3] See Stocking (1982/1968); (1986); (1988); (1995); (1996); (2001).

Several comparative historiographic overviews of leading French, German and US schools of anthropology are also provided in other publications (see e.g. Barth 2005; Calcagno 2003; Caspari 2003). A significant contribution to the exploration of physical anthropological science and thinking has been made by the social anthropologist James Urry (1993). German physical and racial anthropology might be the best-explored case, which is a consequence of the general research interest in the National Socialist era during which anthropologists were involved in the 'Nazi death machinery'.[4] Aside from the National Socialist era, further studies explore the agency of Rudolf Virchow (1821–1902) and the case of imperial German anthropology (Goschler 2003; Zimmermann 2001). The German paradigm, in all historical periods, is a prime example of the merging of scientific, political and social prospects by anthropologists who strove to integrate medicine, a uniform knowledge of humankind and reforms in medicine in order to include hygiene and social politics.

While national physical anthropological traditions have been studied to varying degrees in several Central European countries,[5] anthropology in South-Eastern Europe remains underexposed. The few available studies include some treatises on Romania, such as a chapter on scientific milieus and ethnopolitics in the 1930s and 1940s by Michael Wedekind (2007) that gives new insights into racial/physical anthropology in Romania. Two further articles by Marius Turda (2007a; 2007b) refer to the inter-war period in Romania. A chapter by Rory Yeomans (2007) is dedicated to the Croatian case and an article by Sibel Özbudun Demirer (2011) to Turkish anthropology. The emergence of Serbian and Bulgarian racial and physical anthropology is examined by Christian Promitzer (2003; 2010a; 2010b). Recently, the research interest in anthropology has increasingly addressed Eastern Europe in the post-war period and the socialist era in South-Eastern Europe (Turda and Georgescu 2010; Skalnik 2002; Hann 2005).

Physical anthropology in Greece has not yet become a subject of study in either the historical or the social sciences. The only available publication is a short overview on the emergence and development of this discipline in Greece, which is provided in a work that addresses the educational and research content of physical anthropology (Pitsios 1993). A very early stage of my own research is recorded in two articles published

[4] See, for instance: Massin (1996); Lösch (1997); Lüddecke (2000); Satzinger (2004); Weiss (2004).

[5] On Germany see the references mentioned above; on Austria see, for instance, Koger (2008); Pusman (2008); Berner (2007); Teschler-Nicola (2007); Fuchs (2003).

some years later (Trubeta 2007 and 2010). Hence this present study may in essence represent the very first attempt to approach Greek physical anthropology as a subject of research.

The Search for Appropriate Research Programmes

In terms of methodology, this study is located at the intersection of the history of science, the sociology of science and the social history of medicine. The starting point for specifying the approach of the study is the very programmatic question posed by Georges Canguilhem—"*Whereof* is history of science actually the history?" (Canguilhem 1979: 22). Given its socio-historical character, science can only be examined in the universe that makes science possible, as Canguilhem himself asserted (Marcherey 2009), and when examined in this way the answer to the above-posed question is that the object of the history of sciences is the interpretation of discourses which lay claim to the truth (Canguilhem 1979: 28). In accordance with this suggestion, the subject of this study is actually to trace the transformation of the notion of physical anthropology and its correlation to transforming discourses of truth. The methodological tools for doing so are borrowed from the sociology of science, which examines science as a social institution and focuses especially on the agency of scholars as a category of social actors who are involved in the specification and transformation of epistemic contents, but whose agency is located and examined with regard to other categories of social actors, also beyond the scientific realm. Both the institutional involvement of physical anthropology in the medical sciences as well as its role in discourses on race and eugenics render it a subject of the social history of medicine.

In terms of a discourse that lays claim to the truth, anthropology is considered in this study as providing valid (i.e. legitimized) knowledge on human life and human ontology. A first question arises concerning the grounds on which anthropological assertions have become legitimate knowledge and regimes of truth. The claim to legitimacy and truth corresponds with regimes of power both within and outside the scientific world. Following Michel Foucault (1994: 114), the required legitimacy "it's not so much a matter of knowing what external power imposes itself on science as of what effects of power circulate among scientific statements, what constitutes, as it were, their internal regime of power, and how and why at certain moments that regime undergoes a global modification". A research strategy able to address these questions should allow for the consideration of issues such as the cognitive contents of science, the conditions of their production and transformation, their transfer into society or even the interaction between society and science in producing legitimate knowledge; the agency of the actors in this process. With respect to anthropology, the major questions circulate around the complex resulting from it being embedded in institutional frames of academia and broader society; the conditions of the production, reproduction and transformation of its epistemic contents; the usages of anthropological knowledge in both the academic world and society; the agency of the actors involved in designing the institutional and epistemic profile of anthropology; the role their professional or other interests play in formulating and applying their assumptions; the possible positions of these actors in the societal environment outside academia; and the contemporary scientific or societal questions to which they have attempted to respond with their anthropological proposals. All of the above issues are interrelated without, however, losing their relative autonomy and dynamic.

In order to consider these aspects, and given the presumption that scientific agency is mobilized by the interests of the actors involved (where interests are not a priori to be identified with opportunism), the study avoids making use of a research model based on Actor-Network-Theory.[6] The latter fails to consider the macro-sociological explanatory approach that is embedded in the studies of interests (Weingart 2003: 71–7). Representatives of the Actor-Network-Theory have radicalized the thesis of the unity of nature and society, ascribing the status of actors to material objects, animals, features of the landscape and so on. This approach is inclined to leave underexposed various aspects of power, legitimacy, conflicts, the negotiation of positions and the interplay of politics and science.[7] In contrast to their underestimation in the Actor-Network-Theory, all these issues are considered in this study to be instrumental in approaching science regarded, as Bourdieu suggests, as a 'social field'. Bourdieu locates the production, stabilization, transformation and transfer of knowledge in a 'field' defined as a 'universe' that encompasses scientific and non-scientific actors (Bourdieu 1998: 18–19).

---

[6] Initiators of the Actor-Network-Theory include Michael Callon and especially Bruno Latour. See some of their early works: Callon (1986); Latour (1983); (1988); (1993); (1999); Belliger and Krieger (2006).

[7] Illuminating insights into the interplay of politics and science are provided by Peter Weingart in several of his works published since the 1980s. See, for instance, Weingart (1983) and (2001).

During the past decade, research programmes for science studies have been elaborated on the basis of the Foucauldian concept of discourse (Sarasin 2003; Keller et al. 2001). The usefulness of this concept derives from the understanding of discourse to extend linguistic frames and hermeneutic settings in which social reality is 'read' as a 'text' (cf. Chalaby 1996), and to reveal instead both its symbolic and material dimensions. The latter model of discourse analysis[8] suggests linking epistemic content with social and institutional practices and the agency of social actors. As Foucault has specified, discourses are dispositions of themes, disciplines and fields of meanings that are institutionalized to different degrees and are produced, reproduced and transformed in sets of practices (Foucault 1992/1972). Starting from this definition, Reiner Keller (2001) suggests a discourse analytical approach to knowledge (both general and scientific) as a 'genuine research programme' intended for practical implementation. Keller perceives discourse as going beyond the bare aggregation of statements and as including patterns of behaviour regulated by societal practices; in other words, patterns of legitimate utterances and actions. The focal point of his research programme is the action that produces knowledge able to constitute, disseminate, reproduce or transform regimes of reality and their symbolism. Practices (institutional practices in particular) are the means by which the regimes of realities and their symbolism are designed, reproduced, disseminated and transformed. Agency is considered against the institutional background of the actors, who appear as spokespersons and representatives of discourses. However, their positions are not totally determined by the conditions of the discourse; they are not passive recipients but can decisively influence its conditions (Keller 2001: 129, 131).

Obviously, this research model is strongly oriented towards institutional sites in which knowledge is produced, although it leaves some leeway for considering the relative independence of individual action. For all its constrictions, it could be used as a basis for examining the transfer of symbolic into material capital, albeit on condition that the approach will also focus on further aspects. First, the agency of the actors deserves to be seen more broadly. Actors are not only collective agency, nor mere spokespersons of discourses, but should also be considered with regard to their subjectivity. Apart from external conditions and professional or other interests, social agency results from the ways actors cope with given opportunities.

[8] Especially by Chalaby (1996) and Keller et al. (2001).

They usually face multiple options and their choice is optional. Moreover, in making decisions, agents are mobilized not only by rational but also by irrational choice. A further aspect to enrich the discourse analytical approach is the mutual influence between knowledge that is produced in institutional sites, such as academia, and that diffused throughout society, and which, in turn, comes to flow into academic knowledge as well.

The Research Questions of this Study

This study covers a period of almost one century, starting from the 1880s up to the 1970s. This time frame has been chosen on the grounds of the creation of distinct anthropological institutions in Greek academia, namely, the establishment of an anthropological museum in 1886 and a chair for physical anthropology at the University of Athens in 1925, which followed the founding of the Greek Anthropological Society in 1924. However, the intention of the study is not to provide a simple reconstruction of the history of the anthropological institutions in Greece, nor to describe a mere series of events that are associated with the emergence and trajectory of the anthropological discipline. Rather, it should be read in light of the disposition of anthropological discourse in Greek academia and the proclaimed intention of its representatives to influence societal developments. Accordingly, this study traces the aspirations, the voice, the echo and the silence of those involved in the anthropological discourse in academia and society. The organization of the chapters is based around conceptual questions rather than historical periods so that a pendular movement between historical periods is not excluded, which ultimately serves to reveal the conceptual aspects of the study.

An initial difficulty in framing this study was to find an appropriate term for its subject: physical anthropology or simply anthropology? The former refers to the official institutional denomination of this discipline at the University of Athens up to 1950, and the latter to the term that the protagonists in the Greek anthropological scene applied themselves. In the following I will use the term 'anthropology' with reference to both, underlining in advance that this is a conventional term since the scope of this discipline and its lines of demarcation from other fields in the anthropological sciences were a point at issue during the entire period considered by this study. Actually, the matter is anything but normative. Due to the discursive character of science, the scientific object is transformable and indeed becomes transformed by means of the reformulation of the research questions (Canguilhem 1979: 152). This very transformation along with the conditions in which it happened is inevitably at the centre of my study not only for methodological reasons, but also because the epistemic content and the specification of the scope of anthropology was a major concern for the actors involved in the Greek anthropological discourse and even interpenetrate the sources used in my research. I will argue that the inconsistency in the denomination of this discipline corresponds with epistemological issues related to the specification of the scope of anthropology which was a matter of dispute generally from the initial emergence of anthropology as a modern scientific discipline, and especially in the period with which this study is concerned.

Embracing Foucault's suggestion, as outlined above (Foucault 1978: 26), the central question arises as from which sources anthropology drew its legitimacy in the particular Greek case, and what effects of power circulate in scientific assumptions formulated by those involved in the Greek anthropological discourse. The main pillars of the analysis are the agency of the actors involved in the anthropological discourse, the production of anthropological knowledge, its contents and embedment in regimes of truth both outside and inside the academic terrain. The agency of actors will be examined with respect to their positions in the academic field; their possible involvement in political or other, non-scientific fields; the ways in which anthropological knowledge was professionalized; their career ambitions; the individual and professional interests that influenced the epistemological paradigms and contents of Greek anthropology; and, finally, the strategies these actors pursued in realizing their scientific and political-societal prospects and interests. This approach aims also at revealing the ways in which symbolic capital was transferred into material capital, and the factors which impacted upon this.

Outline of the Book

The book is divided into three main sections which reflect the central research questions as outlined above. The FIRST SECTION addresses the origins of modern anthropology in 19th-century Greece with respect to its intellectual sources and its emergence as a scientific subject, up to its establishment as an autonomous discipline in Greek academia. The central argument of the **first chapter** is that anthropology emerged as a contested notion of reference to the cosmological views related to the Darwinian theory of evolution and its implementation by the natural sciences. In this phase, the epistemic scope of anthropology was determined by its alliance with a group of sciences (i.e. the natural sciences and medicine). A particular focus here is addressed to medical geography as an epistemic source of anthropology. This preference is not arbitrary but is grounded on an occurrence which proved to be crucial to the institutionalization of anthropology in Greece as a distinctive academic field, and which also left a mark on its contents for a long period: namely, a French treatise that made Greece its subject and approached it by means of medical geography and its allied natural sciences. This treatise was important as it united in a sole approach all sub-disciplines which constituted the subject of medical anthropology and by this means approached Greece as a diachronic entity buttressing the dominant national doctrine by 'hard data'. The treatise entitled the author, Clon Stéphanos (1854–1915), to direct the first anthropological institution in Greece, the Anthropological Laboratory and Museum, which was founded in 1886 at the medical faculty of the University of Athens. The processes that led to the emergence of the anthropological institutions in Greece are the subject of the **second chapter**, which examines this issue in relation to specialization in the epistemic scope of the sciences and the accompanying creation of professional academic scholarship. The concluding argument, which also leads to the next section, is that the institutionalization of anthropology in Greece preceded the creation of a professional group of trained anthropologists. This argument is advanced by the illustration of the circumstances under which Clon Stéphanos's successor, Ioannis Koumaris (1879–1970), occupied the directorship of the Anthropological Museum after Stéphanos's death (1915), and later also the first chair of physical anthropology.

The conclusion that the physical anthropological discipline emerged in Greek academia without a professional group of trained anthropologists forms the starting position of the SECOND SECTION, which is dedicated to conceptions of anthropology after its integration into academic institutions. The leading questions raised in this section concern how the epistemic contents of Greek anthropology were shaped given the initial lack of trained anthropologists, the integration of the discipline into the medical faculty and the claim of anthropologists to be relevant for society. In which ways were the epistemic contents of anthropology interrelated with its institutional trajectory and the university capital that its representatives drew from their academic positions? I argue that the scope and contents of anthropology varied with respect, first, to each institutional representative, namely, Clon Stéphanos and Ioannis Koumaris, and, further, to each of the three anthropological institutions. The **third chapter** addresses the transformation of the conceptualization of anthropology as reflected at the Anthropological Laboratory and Museum during the eras of its two directors. On the one hand, Stéphanos's anthropology bore a national outlook and, in epistemic terms, was coined by the alliance of physical anthropology with archaeology, linguistics and ethnography. During his directorship, anthropology at the museum corroborated the antiquity cult and acted as a national heterotopia from which he himself drew his scientific capital. Koumaris's conception of anthropology at the museum, in contrast, exceeded these narrow national limits and additionally included universal human culture and colonial racial anthropology. The realization of his agenda was challenged by the competition with other sciences relative to anthropology, first of all ethnology, on their respective authority over human culture and its material objects. The argument of this chapter is that by means of Koumaris's agenda Greek anthropology exceeded the frames of the medical faculty, at which it was institutionally positioned, but also the narrow settings of the antiquity cult that was served by the first director of the Anthropological Museum. The broadening of the epistemic scope of anthropology was reflected also in the agenda of the university chair, which is the subject of the **fourth chapter**. This chapter argues that there was a basic contradiction between the initial agenda of the chair-holder, on the one hand, and its realization, on the other: while anthropology as a university subject was programmatically inaugurated by the chair-holder as a holistic science intended also to provide objective knowledge for national concerns and topical social questions, and to popularize anthropological tenets in order to reach broader society, in reality the teaching of this subject was restricted to the very limited scope of physical anthropology. Finally, the university chair failed to train a new generation of anthropologists and was therefore abolished as soon as the chair-holder retired (1950). I argue that the failure of the university chair meant that the university capital of the chair-holder decreased, especially since, with his retirement, he also abandoned the directorship of the museum. The **fifth chapter** examines the conceptions of anthropology at the Greek Anthropological Society (founded in 1924) in light of the decreasing university capital of its president, Ioannis Koumaris, who started his career in the 1920s as the holder of all academic positions connected to anthropological science in Greece. This chapter argues that the more his university capital was decreasing and the failure of the university chair to train a group of professional anthropologists was becoming obvious, the more the scope of anthropology elaborated at the GAS was broadening, coming finally to attach to anthropology an idiosyncratic, fetish character that justified the interpretation of any scientific and human activity as anthropological. This development is ascertained on the grounds of the internal works of the GAS and its internal organization, the trends in membership and its internal debates. The internal debates suggest that the GAS was hardly an institution for producing anthropological knowledge; rather, it became a forum for presenting expertise from diverse research contexts that often moved around the interface of medicine, the natural sciences and the humanities. The GAS acted as an umbrella institution that gathered biologistic and medicalized approaches of human action under the common denominator of pathology, either bodily and genetic or mental and social. In the 1960s, however, social anthropology gained a preference in the conception of anthropology by the (now sole and nominal) president of the GAS, Ioannis Koumaris, insofar as it was epitomized in the person of a noble anthropologist, namely, Prince Peter, who in that very period gave a series of lectures in Athens. Reflecting on the downward trajectory of the Greek anthropological agencies, which became obvious after World War II, the **sixth chapter** argues that this outcome was anything but independent from the political profile of those involved in the Greek anthropological discourse. This argument is brought forward after tracing the activities and voices of the GAS's members both inside and outside the anthropological discourse during World War II and the Civil War. I argue that the pretence of political neutrality, all too often proclaimed by those involved in the anthropological discourse even during World War II, actually masked the fact that the scholarly community in Greece was deeply divided into hostile fronts. Given that its internal debates portray its profile as conforming to prevalent power relations, involvement in the GAS was a political position. Moreover, the agency of its president was strongly embossed by his sympathy for German anthropology, if not Germany generally, even in a period in which German anthropologists collaborated in the Nazi crimes. A central issue that marked the profile of the GAS and its president was the emergence of the racial question. I argue that the anthropological discourse became the most important terrain in Greek academia in which racial, and racist, thoughts were voiced and became 'normalized' through the use of scientific objectivity and patriotic aspirations to defend Greece (perceived in terms of racial communality) as a vehicle.

The concepts of Greek communality that are epitomized in the notion *fili* are the subject of the THIRD SECTION of the book, which addresses the diverse usages of this notion in racial anthropological and in eugenic discourses. The starting argument is that although *fili* is usually translated into race, its semantics has been subject to alterations coming to provide it with polysemy. The conceptual analysis of *fili* undertaken in this section aims to provide evidence for the representations of Greek communality and the ways in which the members of the Greek community are thought to be tied together. It thus brings to light transformations in Greek society and, in turn, the ways in which these transformations irrupt into the perception of Greek commonality and shape the semantics of *fili*. The **seventh chapter** addresses the racial anthropological conceptions of *fili* and argues that they were racial anthropological in character and national in outlook. They drew arguments on pre-existing Greek national ideas, notably the 19th-century idea of Hellenism, and also reflected the state of affairs in the racial sciences in Europe, being strongly linked to the German racial sciences, and particularly those associated with National Socialism. I raise the argument that the racial connotations of *fili* were transmitted and diffused in Greece using as a vehicle the 19th-century controversy on the contested continuity of Hellenic ancestry and the asserted decay of Hellenic civilization because of miscegenation with alien races. By this means, the question of continuity came to be at the core of the racial representations of Greek communality—however, in racial terms, it was conditionally interpreted either as racial constancy, or as homogeneity, or/and purity. I pose the argument that the more the distance from the Ottoman period increased and the national states in the Balkans stabilized, the more the discourse was shifted from racial constancy to racial homogeneity and purity. This trend was obvious in the ideas of Clon Stéphanos, but it obtained its more demonstrative and elaborated form in the racial conceptions of Ioannis Koumaris. Since the latter was concerned with theorizing the Greek race, most of this chapter is dedicated to his agency and especially to his theory on the 'strong nucleus' of the Greek race. I argue that this theory drew on the 19th-century idea of Hellenism and the supposed ability of the Greek *fili* to acculturate and absorb foreign peoples or nations without losing their proper Hellenic qualities. Racial constancy, rather than purity, was the central element in Koumaris's theory of the 'strong racial nucleus' which, I argue, was elaborated in the shadow of the myth of the Nordic race and sought to cope with the Orientalistic prejudices inherent to this. The argument of the next part of the chapter is that the rejection of racial purity was conditional, insofar as it was a defensive reaction to the National Socialist attack against Greece, but he was also a fervent advocate of racial purity when the matter concerned the persistence of the Greek race and its diachronic characteristics. Then, the "racial purity of the Greek blood" appeared at the top of his eugenic agenda which, in a contradictory way, drew on the National Socialist model of racial hygiene. The subchapter addresses Koumaris's racial hygienic agenda in greater detail and raises the argument that it was marked by singularity, given that his concerns varied remarkably from the usual tenets of racial hygiene. His anxieties concerned the possible infection of Greek blood caused by miscegenation (i.e. marriages between Greeks and racial aliens in emigration). The singularity of his racial hygienic model comes to the fore when one considers the mainstream eugenic agendas in Greece (which form the subject of the following chapter), and the perceptions of his theory in the anthropological discourse. The latter is outlined in the two final subchapters in which the argument arises that the idea of the racial homogeneity of the Greek population was implied in the anthropological discourse almost exclusively in correspondence with the minority question, which was dictated by the transforming relations between Greece and its neighbouring states. This focus is evident also in the perceptions of the Greek race by Aris Poulianos, an anthropologist by training, who entered the Greek anthropological scene from the Soviet scholarly tradition in the aftermath of the Koumaris era.

The **eighth chapter** addresses the particular character of eugenics in Greece and raises the question as to which ways eugenicists conceptualized the Greek collectivity which they termed *fili*. How is the eugenic conception of *fili* related to the prevalent national ideology, and which semantics do heredity and ancestry obtain in this? Which questions rising in society did eugenic proposals seek to correspond with and how were these questions and the provided answers imprinted in the eugenic usages of *fili*? The central argument developed in this chapter is that eugenic thinking in Greece was embedded in socio-political reformist projects that advanced the establishment of welfare institutions and the modernization of society. Adherents of racial hygiene were exceptional among Greek eugenicists and sympathizers of eugenic ideology. In the post-war period, eugenics became closely interwoven with genetics, conforming to an international reformist eugenic movement to which Greek racial anthropological discourse did not find any link. This chapter argues that the antiquity cult played only a marginal, if any, role in the eugenic programmes, while at their core was the intention to manage social inequality in modern society through control over human reproduction. This argument is advanced by an analysis of eugenics as a mode to govern social stratification and to provide answers to two issues that arose during the modernization of Western societies—the social question and overpopulation—which were connected to the problem of governing social stratification and arranging the distribution of resources in a modernizing society during a period in which social welfare institutions were founded or developed. Essentially, eugenicists proposed the redistribution of societal resources, suggesting the reduction of the percentage of the needy in the Greek population by restricting their biological existence rather than by changing the conditions of inequality. These arguments are developed on the basis of the Greek discourse on eugenics by considering its important representatives. In a following subchapter the eugenic proposals are analysed with respect to the concept of the 'medicalization of society', which refers to an interpretation of poverty and precarious life conditions in medical and moral terms that is imprinted in the terms of 'social disease' and 'social pathology'. In the former, the manifestations of social inequality are medically defined as disease and morally evaluated as promiscuity. I argue that this idea is not an exclusively eugenic one, and that it draws its origins from the discourse on hygiene that arose in 19th-century Western societies. I explore this phenomenon in the particular Greek case with respect to the transforming connotations of both disease and poverty during the first wave of industrialization in the 19th century. In this context, the manifestations of social inequality were described as a pathological condition not only by those demanding repressive measures against the poor, but also by their lobbyists, who pursued, however, a different line of argument, while they accused state policies of ignorance towards the people and those at the lower end of the social strata. These views were also shared by members of the Saint-Simonist circle in 19th-century Greece, as well as by socialists, communists and (to some extent) liberals in the 20th century. Given that the boundaries between hygiene and eugenics had often been blurred, I suggest a crucial difference between them regarding how each sought to cope with these questions in practice. Eugenicists demanded state intervention not to improve the living conditions of the poorest parts of the population, but to restrict their biological reproduction. A part of this chapter presents proposals and attempts by Greek actors (including scholars, technocrats, physicians, feminists, etc.) to impose eugenic measures in Greece, especially sterilization and medical examinations prior to marriage. A particular subchapter is dedicated to state attempts to apply such measures in Greece, arguing that they were more or less successful. The analysis of the Greek eugenic discourse reveals that the most fervent proponents of applying eugenic measures in Greece were liberals. In the subsequent part of this chapter, the racial hygienic projects are presented with respect to their main representatives and their positions in academia. I argue that the German version of eugenics, to whose conceptualization anthropology and race theory were instrumental, was exceptional in Greece and had hardly any influence either on mainstream Greek eugenics or on state politics. Nor did Greek racial hygienists manage to establish their agendas in Greek academia, although there were occasions to do so. The main adherents of racial hygiene were, next to Ioannis Koumouris, the professor of hygiene Konstantinos Moutousis and Stavros Zurukzoglu. The latter was a theoretician of racial hygiene with an international reputation whose appearance in the Greek scholarly scene was ephemeral, since he made his career in Switzerland. The attempt of Moutousis, again, to introduce racial hygiene into the Greek university failed because of professional competition with other professors of hygiene. Given the idiosyncratic project of racial hygiene by Ioannis Koumouris and the absence of collaboration between the proponents of racial hygiene in Greece, this chapter argues for the very limited influence of racial hygienic ideas over Greek academia. Keeping up with the analytical agenda in this whole chapter (i.e. to situate eugenic projects in the broad landscape of different views and approaches to society), the following subchapter argues that even if eugenics had the status of a mainstream ideology in Greece, it still existed parallel to other considerations of society which were either beyond or opposed to eugenics. The third subchapter addresses the reformist current of eugenics that appeared in Greece after World War II and which was incorporated in the Greek Eugenic Society, founded in 1953 by the gynaecologist Nikolaos Louros (1898–1986). I argue that the activities and the debates of this agency betray a conformity with the reformist post-war eugenics which has come close to embracing the concerns, methodology and narrative of the new genetics, in which it eventually became integrated. The eugenic questions were now reformulated in a new context which, after the experience of the recent world war, was embossed by a new sense of political correctness. I suggest that a quite important difference from the 'old eugenics' was that the individual and his/her rights now were not disadvantaged in the favour of communality and that the discussion was shifted to intervention into human reproduction to individual luck and favour. In the concluding chapter of this section, I suggest that the meanings of *fili* in the racial anthropological discourse and in eugenics differed significantly. The former was a concept that sought to buttress the prevalent Greek national doctrine by providing racial arguments and anthropological hard data in representing Greece as a diachronic entity.

In the fashion designed in the anthropological discourse and especially by Ioannis Koumaris, the racial anthropological conceptions of *fili* failed to address the topical questions of society. In contrast, *fili* in mainstream eugenics was defined in bio-political terms and sought to develop proposals for current social developments. In doing so, it faced the future, rather than the past. I suggest that *fili* is close to the Foucauldian notion of 'race' in referring to a type of modern society that is distinctive for the internal differentiation and inequality that take place along socio-biological lines and is transmitted by discourses of sexuality and reproduction.

The short concluding chapter of the book reflects on the heritage of Greek anthropology against the background of the decaying trajectory of its institutions and poses the question as to whether this should be understood as a general failure of the Greek anthropological discourse to find any connection to Greek society. This chapter suggests that, for all the institutional failure, the heritage of the anthropological discourse is present and effective in the racial nationalism that is still operative in Greece. This becomes possible because of the lack of criticism of racial and racist views that still provide support for nationalist aspirations. Indicative enough of this circumstance is that the anthropological discourse and the racist views of its main representative have not yet been subject to research nor to any criticism. Finally, this chapter formulates advanced questions on the interplay of scientific and popular knowledge in the context of the society of knowledge.

A Note on the Sources

The most important bodies of source material considered in this study include university archives, working materials of scientific associations and the publications of the central actors, as well as contemporary statements made in the press (either scholarly or popular) relating to the questions raised by this study. Archival sources include the Historical Archives of the University of Athens, especially the portfolios of the Anthropological Museum and those relevant to the chair of physical anthropology, notably, the minutes of the meetings of professors at the medical faculty, the sessions of the university senate and yearbooks of the University of Athens with the course calendars (1925–70). The university course calendars contain exact details of the curricula and were printed regularly, with a brief interruption during World War II and the Greek Civil War (i.e. from 1941 until 1949).

Auxiliary data was found in the General State Archives (Athens). Apart from Greek archives, research was also conducted in Germany in order to ascertain the involvement of Greek actors in the anthropological, racial and eugenic discourses in German academia and the scholarly scene. The decision to consult the German archives took into account the intensive collaboration of the most important Greek actors in this study with German academia and the strong German influence, especially upon Greek medical scholarship. The archives that provided valuable material were the Max-Planck-Society in Berlin (Kaiser-Wilhelm-Institut für Anthropologie, menschliche Erblehre und Eugenik), the Humboldt University (Institute for Anthropology) and the Federal Archive (Bundesarchive, Berlin).

For the most part this study is based on published sources which have not yet been considered systematically (if at all) in the research, or have not been read from the perspective of the questions this study raises. The most important source material includes the *Proceedings of the Greek Anthropological Society* (hereafter *PGAS*, published from 1924 to 1970) and the Historical and Ethnological Society (*Bulletin de la Societé Historique et Ethnique de la Grèce*, during the period 1882–1970), as well as the three volumes of public debates hosted by the Greek Society of Eugenics and Human Genetics (1953–78). The most informative and invaluable source for drawing on the work of the Greek Anthropological Society is its proceedings, which have been printed without any interruption[9] and provide illuminating insights into the pathway of this anthropological body over a period of 46 years—from 1924 to 1970.[10] The *PGAS*, in which the internal debates of the Greek Anthropological Society are recorded without omissions and in detail, are a major source for this study, especially as the GAS's members are registered by name, profession and date of joining the society. They also include documents, correspondence and information from which the networks, activities and attitudes of the actors in other contexts can be interpreted, transcending the narrow anthropological framework.

---

[9] A regular income from membership fees and occasional donations facilitated the printing of the *PGAS*. Occasional advertising (from 1945 onwards) brought a supplementary financial resource, while effective cooperation with the printing houses appeared to assist the publication as well. For instance, the 1947 edition was even printed for free by the publisher of the journal *Helios [Ο Ήλιος]*. *Proceedings of the Greek Anthropological Society* (1 May 1948): 9.

[10] Extracts from the *PGAS* were reprinted in the Greek journal *Medical Annals [Ιατρικά Χρονικά]*. The contents index and summaries of the papers presented at the Greek Anthropological Society were published in French in the *Revue Anthropologique*, the journal of the Paris-based International Institute of Anthropology. See Jean Koumaris (1931) and (1934).

Along with the *PGAS*, the writings of the two leading anthropologists, Clon Stéphanos and Ioannis Koumaris, form the major primary material on which this study is based. The three-volume autobiography (1951, 1961 and 1970) of Koumaris has been a particularly valuable source. It is an excellent reference work for examining the institutional trajectory and content of the anthropological discourse in Greece given that he was its only protagonist in the Greek academia. Apart from the detailed catalogue of his publications,[11] his autobiography also encompasses references to his works by other scholars, providing in this way insights into debates Koumaris had been involved in and the reactions provoked by his views. Information of such references to his work is valuable for researching his networking and attitudes during the long period in which he led all anthropological institutions in Greece. A large part of his autobiography also includes his own reflections on the trajectory of the anthropological institutions and estimations on their outcome. His conception of anthropology is also outlined with respect to each individual anthropological project (notably the Anthropological Museum, the Greek Anthropological Society and the university chair for physical anthropology), and he delivers comments on the implementation of each one. Reading the autobiography alongside the *PGAS* is especially insightful in that these two bodies of material reflect different levels of perception: the former is contemporary to the events whereas the latter provides information and reflections on the same events, mostly retrospectively. Koumaris's publications in Greek and foreign periodicals are considered among the primary sources, but I have listed only those articles relevant to the questions of this study.

On matters relating to eugenics and racial issues the most important sources considered were, next to the above-mentioned volumes of the Eugenic Society, the social and scientific press in which the then contemporary debates were carried out. Along with some monographs, on eugenics in particular, such articles to a great extent make-up the body of the source material on which this aspect of the study is based. Here, the Historical Archives of the University of Athens were also considered with respect to debates on eugenics and hygiene in academia.

[11] However, the list of references does require verification due to inaccuracies and mistakes.

SECTION ONE

ON THE EMERGENCE OF MODERN ANTHROPOLOGY IN GREECE

CHAPTER ONE

TRACING THE INTELLECTUAL AND EPISTEMIC SOURCES OF GREEK ANTHROPOLOGY

The epistemological sources of anthropology go back to cosmological views on humanity and its place in the universe; views which are encompassed in discourses produced before the anthropological discipline had been established in the academic terrain. What does it mean, for a research programme tracing the origins of a science, if the history of a science is not identical with the history of a scientific discipline, as Georges Canguilhem (1979) argues? Certainly, what this does *not* mean is that there is a linear series of developments in the form of a succession of causally interconnected ideas or/and events that advanced anthropological thought and anthropological sciences. Rather than following such a linear approach, this study is interested in ascertaining the processes through which anthropological science became embedded in academia as a scientific discipline and drew its epistemic sources from universal anthropological knowledge.

Starting from these preliminary explanations, the first section of this book aims to portray the intellectual and scholarly environment in which anthropology started to become an issue of discussion (and a point at issue) among scholars in Greece, either as an intellectual problem, or as a subject integrated into the natural sciences and medicine. In the first instance, this section of the book can be read as a short introductory historiography of the emergence of anthropological science in Greece; the focus addresses both its epistemic sources and the institutional frames which had proven crucial to designing anthropology's contents up to the establishment of the first anthropological institutions in Greek academia in the 19th century. As a point of departure, consideration is also made of the broader context in which Greek scholarly elites in the then young state (founded in 1832) sought to find their way in the world of modern states by creating appropriately modern institutions. The main questions raised include: how was anthropology perceived by Greek intellectuals and scholars as a matter correlated to the turbulence in cosmological views that was caused by the Darwinian theory of evolution? How did these perceptions change in the course of establishing academic institutions?

In which way did developments in the natural sciences and medicine and their increasing specialization influence the alliance of anthropology with other sciences, and how did the emancipation of the former come into being in the particular Greek case? Dealing with these questions requires an exploration of the paths through which anthropology had been introduced into the Greek academic terrain, leading to the creation of a professional anthropological scholarship; or, at least, the conditions for creating such scholarship.

1.1 *The Dispute over Evolutionism and the Alliance of the Natural Sciences and Medicine*

Following the publication of Charles Darwin's *Origins of Species* (1859), a set of developments led to the emergence of anthropology 'in the broad sense' as the central intellectual problem of the 1860s (Stocking 1982/1968: 74). The intellectual problem arose as soon as, on the grounds of Darwin's theory, humanity was becoming repositioned in the macrocosm, thereby challenging cosmological views which identified humankind as the centre of the universe. As George Canguilhem (1979: 135) argues, Darwin's theory implied no less than the defeat of anthropocentrism on the grounds of which humanity was faced with the challenge of yielding anew its position in the animal kingdom, whereas it had previously been believed to have been above and able to rule over it "by virtue of the divine right of kings". Thus it is not surprising that anthropology became the terrain for a contest between religious cosmology on the one hand and the scientific rationale for human evolution on the other. The fronts of this intellectual dispute included on the one hand representatives of theological worldviews and, on the other, advocates of the new evolutionary theories. Developments in the sciences came to provide the rational world outlook of the latter with hard data. The intensification of research in prehistoric archaeology, archaeological investigations of ancient civilizations and progress in comparative philology all supplied basic material for "the study of the physical types of mankind, the sociological and historical theorizing of writers like Comte and Buckle" (Stocking 1982/1968: 74). In this context, anthropology acted as a mode for understanding and explaining the world and humanity beyond religious cosmology; it became instrumental to what Max Weber termed the "disenchantment of the world" [*Entzauberung der Welt*] (Weber 2006/1919).

This intellectual dispute is evidenced also in 19th-century Greece where, keeping up with the general trend, anthropology formed part of a cluster of worldviews and as such became a subordinate issue to the controversy between Darwinists and anti-evolutionists. The former included, to a great extent, natural scientists and physicians who advanced the establishment of natural sciences in Greece as a field of knowledge, of education and as professional specialization. The anti-evolutionist front, on the other hand, involved religious-minded scholars, predominately theologians and jurists who interpreted Darwin's theory as a belief in science that competed with Christianity. The public polemic between the fronts is recorded in periodicals which were addressed to a wide audience with multiple interests. Proponents of the new ideas on human evolution (labelled by their opponents as adherents of 'scientific materialism') were provided with a forum for expressing their thoughts in several periodicals, the most prominent among which was *Hearth* [*Εστία*, founded in 1876].[1] The views of the anti-Darwinists were mainly presented in theological journals such as the *Redeemer* [*Σωτήρ*][2] and more importantly *Recreation* [*Ανάπλασις*][3] (Sotiriadou 1990: 142). In most of these publications, anthropology was only one point at issue among a number of others; it was integrated into a complex of arguments which endorsed evolutionary ideas or opposed them. One exception to this was a serialized article by the anti-evolutionist jurist Ioannis Skaltsounis (1821–1905), which polemically addressed anthropology as a specific issue and which was published in *Recreation* between 1888 and 1889.[4] Here, like in several of his other articles and monographs,[5] the author was essentially concerned with disproving 'materialism' from a religious standpoint.

---

1 Other publications that promoted evolutionary ideas in Greece included the rather ephemeral journal *Prometheus* [*Προμηθεύς*, published from 1890 to 1892], *Attic Diary* [*Αττικόν Ημερολόγιον*, 1867–96] and several others. For a detailed account see Sotiropoulou (1990) and Krimpas (1993).

2 See some writings in this journal by Papadiamantopoulos: for example, Papadiamantopoulos (1877a); (1877b); (1878). For an extensive presentation of the disciples and opponents of Darwinism in Greece, see the PhD thesis by Sotiriadou (1990).

3 *Recreation* [*Ανάπλασις*] first appeared in 1887 and was published by a Christian organization of the same name.

4 The serialized article by Skaltsounis, entitled "Anthropology" appeared in the journal *Recreation* [*Ανάπλασις*] during 1888 and 1889; Skaltsounis (1888a); (1888b); (1889a); (1889b).

5 See his monograph on *Religion and Science* in which he devotes a chapter exclusively to Darwin's theory: Skaltsounis (1884); (1893); (1898). Cf. a review of Skaltsounis's book in *Bulletin of Hearth* [*Δελτίο Εστίας*], no. 391 (1884).

On late receptions of Skaltsounis's ideas, see a publication by the Christian Union of Scientists on the occasion of the 50th anniversary of his death: Christian Union of Scientists (1957).

Certainly, the challenge that the theories on human evolution posed for theological cosmology remained topical in the following period and continued to preoccupy scholars and intellectuals even in the 20th century. But, over the course of time and while natural sciences and anthropology were increasingly becoming established as fields of legitimate knowledge and professional scholarship, the dispute tended to dampen. Religious beliefs could coexist with scientific rationality as soon as the positions of individuals, and scholars in particular, could justify different views in diverse societal contexts. The 1880s was a period of transition for the discourse on anthropology in Greece in which this intellectual dispute took place alongside the simultaneous specification of anthropology's epistemic scope and an increase in its legitimacy as a science, the latter of which preceded and facilitated the entrance of anthropology into mainstream educational institutions.

Before anthropology rose to form a distinct academic area, anthropological knowledge in the young Greek state was integrated into the natural sciences in the process of specifying the objects of research and the emergence of sub-disciplines in the sciences and medicine. More specifically, the interest in anthropology, at least as a field of knowledge, was initially aroused on the grounds of the alliance of medicine with the natural sciences and in making the Greek readership familiar with Darwinian theories. Particular scholars played a leading part in this process by communicating the intellectual and scientific spirit of their time to the Greek scientific institutions in making. Their professional biographies and scholarly achievements bear testimony to the initial alliance of natural sciences and medicine, and then to the gradual emancipation of medicine during the process of specifying research objects and establishing sub-disciplines, including anthropology. Among these scholars, many occupied distinguished positions in Greek academia.[6]

---

[6] For instance, the professor of physiology at the University of Athens, Ioannis Zochios, who published in the periodical *Attic Diary* [*Αττικόν Ημερολόγιον*]. On his scientific achievements and his views see Sotiriadou (1990): 124f.

Others include the biologist, physician and founder of the first laboratory for physiology at the Athenian university, Rigas Nikolaidis (1856–1928), and the physician Theodoros Tsikopoulos (or Theodor Zikopoulos in his German publications). Rigas Nikolaidis, after finalizing his studies in Athens, continued his postgraduate studies in German universities in Tübingen, Berlin and Leipzig. Afterwards, he practised in Paris at the laboratory Ranvier of the Collège de France. From 1883 to 1926 he made a career as professor of anatomy and physiology at the University of Athens where he founded the first physiology laboratory. For biographical data see Krimpas (1993): 94f.

Anthropology became a specific objective in the work of the physician and botanist Spiridon Miliarakis (1852–1919), the son of a famous Cretan family of scholars. His monograph on anthropology (Miliarakis 1889) was preceded by a large number of publications on Darwin and Darwinism, from which Miliarakis's work grew. Miliarakis studied medicine in Athens and practised in Skopje from 1876 to 1881. He then went on to study in Würzburg, in Germany, where he earned his PhD in botany (Miliarakis 1884). Returning to Athens, he was engaged as curator in the Botanical Museum, which was part of the university, thus succeeding Theodor von Heldreich (1822–1902) in this role who subsequently kept only his position as the curator of the Royal Botanical Garden in Athens. Miliarakis also later became a professor of botany at the University of Athens. Aside from his pioneering work in the field of botany (in which he collaborated closely with Heldreich), he was also engaged in familiarizing Greek scholars with Darwin's work and its perceptions. He translated a respectable number of texts, mostly from German editions, into Greek,[7] including, among others, Darwin's biography written by William T. Preyer (1841–97), the German professor for physiology and director of the Physiological Institute at the University of Jena.[8] Darwin himself might have been informed about these translating activities through his correspondence with Theodor von Heldreich.[9] It appears that Miliarakis's intention was not merely to make the Greek readership familiar with Darwinian thinking. As he announced in the foreword to his pamphlet on Darwin's biography, he considered his translating work as a preparatory stage for realizing another project, his own monograph on anthropology (Miliarakis 1880: 5). Indeed, nine years after disclosing this intention, his book, entitled *Handbook for*

The physician Theodoros Tsikopoulos/Theodor Zikopoulos earned his PhD in Germany: Zikopoulos (1886). He became a target of Skaltsounis's anti-evolutionist polemic. See Skaltounis's polemic against Tsikopoulos/Zikopoulos: Skaltsounis (1894).

7 Some of Miliarakis's texts were published in the periodical *Hearth* [*Εστία*] (cf. Krimpas 1993: 93). Apart from publications on Darwin, his writings included handbooks for natural history, zoology and botany, addressed to schoolchildren. On his scientific achievements see Sotiropoulou (1990): 110–13.

8 See the translation by Miliarakis (1880). Miliarakis made no reference to the German text he translated. William T. Preyer later wrote another biography of Darwin: Preyer (1896).

9 In a letter (dated 27 January/8 February), Heldreich informed Darwin about this publication. The correspondence between Heldreich and Darwin was published by Krimpas (1993: 86–91), who notes that the correspondence and the respective translated article by Miliarakis are retained in the Darwin Archive (Darwin's Papers) at Cambridge University.

Zoology Including a Complete Anthropology for the Secondary School, was published (Miliarakis 1889).

With this handbook Miliarakis provided the Greek audience—and schoolchildren in particular—with an important work which linked Darwinian theory to anthropology. But he was not the first to do so. Indeed, a few years earlier anthropology had been the subject of another monograph, written by a physician and printed in Smyrna (Diamantopoulos 1880). This book, which makes no reference to Darwinian theory nor to the dispute regarding evolutionism more generally, provides clear evidence of the integration of anthropology into medicine, anatomy and physiology in particular. As was betrayed by the title, and as the author himself disclosed in the foreword to his book, this publication was intended to be used as a kind of handbook targeted at schoolchildren at the secondary level of education in the Greek Kingdom and the Greek cities of the Ottoman Empire. The author stressed the importance of anthropology first and foremost for a general education that was necessary for everyone; in terms of anthropology's content, he suggested that this discipline formed the basis for hygiene and regarded "somatological anthropology as physical science" (Diamantopoulos 1880: δ'). The book was organized into two main sections: in the first, the organs of the human body were described in detail, while the second part referred to 'physiology', or the function of each human organ. An appendix addressed issues related to dietetics and health in different phases of human life and proposed possible therapies for each body organ in case of illness. The author noted that his study was based on a "reliable bibliography on physiology", which was primarily drawn from German sources. He indicated his intention to compose a study that, although not necessarily popular, would still be understandable and easy to read for the target group. His book therefore included numerous (90) illustrations and a detailed anatomical map borrowed from the German anatomist Hubert von Luschka (1820–75).

In 19th-century Greece, the epistemic subject of anthropology was being shaped on the basis of the natural sciences and medicine, and especially at the intersection between anatomy and physiology. In that period anthropology was integrated into a group of sciences from which it was becoming gradually emancipated in the course of the specification of academic fields within the sciences and medicine. Medical geography was a wide scientific terrain which was comprised by a group of sciences, with anthropology among them. For my study, medical geography is important for a further reason related to the emergence of anthropology as an academic discipline in Greece, namely, a treatise on medical geography that accredited its author with all necessary symbolic capital for heading the first anthropological institution at the Greek university.

1.2 *Medical Geography As an Epistemic Source of Greek Anthropology*

A French monograph with the title *La Grèce au Point de Vue Naturel, Ethnologique, Anthropoloque, Démographique et Médical* appeared in 1884 in Paris. The author, Clon Stéphanos (1854–1915), was of Greek origins and, after having studied medicine in Athens, continued his postgraduate education in Paris, where he earned his doctorate at the beginning of the 1880s. Stéphanos's treatise was important because of the way that it reflected the contemporary stage of developments in the sciences of humankind and its environment, and the way in which he associated these sciences with Greece and, not least, with the Greek national idea. The treatise was based on a multifaceted epistemic approach which at that time was still exceptional for studies on Greece. The author gathered and systematized previous studies in disparate disciplines, such as geography, ethnology and anthropology, and provided the possibility of corroborating Greek national ideology by means of 'hard data' based on medicine and sciences.

In this medical geographical study, anthropology was integrated into the archipelagos of allied sciences by means of which contemporary scientists, and medical geographers in particular, sought to approach humankind in its natural environment. Current Greek scholars honour Stéphanos's monograph as the "first systematic effort to establish a 'medical geography' of Greece providing current researchers of medical geography in our country with valuable information" (Stavropoulos and Marketos 1986). Such an assumption is well grounded considering the scope of Stéphanos's study, which was organized into chapters that included climate, geology, flora, ethnology, anthropology, demography, hygiene and pathology. With this broad scope, the study reflected the state of the art in medical geography (and maybe also medical topography),[10] which in that period "had been articulated as a discipline encompassing elements of geography, meteorology, medicine, cartography, and geology" (Valenčius 2000: 14). Stéphanos

[10] Medical topography discusses the relationship between health and the environment in a certain area, unlike medical geography which has a more global scope. See Valenčius (2000): 5. Cf. also the discussion of the problems in defining medical geography by the same author (p. 14, note 52) as well as *passim* in the same volume.

followed the epistemological tradition that explored human beings as a species within their environment and the milieu in which they lived. Consistent with the prevailing contemporary approaches in medical geography, his treatise interconnected issues of the natural environment with those of human health. It further correlated meteorological and climatic phenomena as well as the diffusion of diseases by way of climatic conditions. In terms of methodology his treatise suggested the quantification of environmental conditions and the conditions of human life at large. In expanding on his approach, Stéphanos called upon the assistance of medicine and its allied disciplines, which together formed the subject of medical geography. Certainly, there was not a united epistemic school of this discipline that Stéphanos could follow. In his time, the specification of medical geography's subject was becoming more and more disputable. The more the internal distinctions in the sciences and scientific disciplines concerned with nature and humanity proceeded, the more the disagreement increased among scholars as to the weight of each of these disciplines in shaping the scope of medical geography. And this dispute was not independent of the dynamics of the medical profession and the boundaries that were drawn so as to separate scientific disciplines, and hence also professional groups. The major point at issue was the differences in the estimated share of hygiene, medicine and anthropology in shaping the subject of medical geography.

In France, the hygienist Leon Poincaré (1828–92) criticized his contemporary medical geographers for overrating climatic factors to the disadvantage of hygiene. The controversy between Poincaré and other geographers is illustrated by contemporary scholars as follows:

> Wishing to emphasize the aspects of medical geography with hygienic value and produce as accessible work, Poincaré categorized the work of Hirsch and Lombard as overly climatological, and that of Bordier as too anthropological for his task (Osborne 2000: 44).

On the other hand, the professor of medical geography at the Parisian School of Anthropology, Arthur Bordier (1841–1910), broadened the scope of medical geography and placed its subject at the intersection of anthropology and medical microbiology; he furthermore pointed towards the political utility of this discipline (Osborne 2000: 44).

In Germany, medical geography was related to the 'history of medicine' by the brothers Heinrich Rohlfs (1827–98) and Gerhard Rohlfs (1831–96), who were editors of the journal *German Archive of History of Medicine and Medical Geography* [*Deutsches Archiv für Geschichte der Medicin und*

Medicinische Geographie], which was published from 1878 to 1885.[11] This correlation was heralded programmatically in their journal; the issues addressed therein confirm that medical geography was thought of as a field subordinated to the history of medicine.[12] Drawing their understanding of medical geography from Leonhard Ludwig Finke (1747–1837), the brothers Rohlfs advocated a universal history of humankind in terms of 'natural scientific medicine'. In a programmatic introduction to their journal, the editors argued that medical geography helped the reader to understand:

> how human behaves in his various forms of birth, education, life style, nutrition and climate, how his health and physical structure is influenced under all these conditions, which diseases and ailments he has the most because he lives here and not elsewhere, because he breaths this and no other air, consumes this and no other food, drinks this and no other water, lives this and no other lifestyle, to see and illustrate what man can withstand and suffer... to finally see where chance, instinct and spirit will drive a man made transparent through science so that his bodily ailments might be put to an end (Rohlfs and Rohlfs 1971/1878: 5).

The brothers Rohlfs argued that medical geography should form the basis for an "international hygiene" (Rohlfs and Rohlfs 1971/1878: 5). Gerhard Rohlfs in particular strove to apply this understanding of medical geography in several of his treatises on climate and diseases in North Africa (Rohlfs 1971/1879; Rohlfs 1971/1880).

Despite all of the differences between the diverse schools, medical geography drew its sources from a global outlook that exceeded the limits of an individual country. This is anything but surprising considering that the rise of this discipline was inextricably connected with colonialism. But, if this is so, how far can such a global approach be adequately implemented in a study which addresses an individual South-East European country and its inhabitants, as Stéphanos did in his own treatise?

The global scope of medical geography was evident first of all in Stéphanos's methodological approach, and the bibliographical references disclose

---

[11] *Deutsches Archiv für Geschichte der Medicin und medicinische Geographie*, Heinrich Rohlfs and Gerhard Rohlfs, eds, Leipzig: Hirschfeld Verlag (1878–85) [reprint: Georg Olms Verlag: Hildesheim and New York 1971]. The brothers Rohlfs jointly published the first three volumes of the journal. From the fourth volume (1881) on, Heinrich Rohlfs alone was the publisher.

[12] On the failure to treat this subject see Edith Heischkel-Artel, "Vorwort", *Deutsches Archiv für Geschichte der Medicin und medicinische Geographie*, reprinted edition of 1971: v–xiii.

that the author embraced both French and German scholarship. This preference stands to reason in view of the fact that, while he may have been studying in France, it was German scholars who dealt extensively with Greece, addressing archaeological, anthropological, ethnological, ethnographic and historical issues, not to mention issues concerning the influence of the environment upon the idiosyncrasies of its inhabitants.[13] Assembling the most important studies on Greece in all disciplines subsumed under the common denominator of medical geography, including physical anthropology, his study functioned as a handbook for his contemporaries. Indeed, it became a reference in the works of prominent researchers in the field of anthropology such as Eugen Petersen (1836–1919) and Felix von Luschan (1854–1924) in their 1889 volume *Travels in South Western Asia Minor* (Petersen and Luschan 1889).

Aside from its significance in advancing contemporary scientific knowledge, Stéphanos's study was also important because it suggested a model to approach 'Greece' as an entity in diachronic spatio-temporal terms. Yet while this was significant, it was not the first treatise to link medical geographical approaches with Greek national aspirations. Almost a century earlier, in 1800, another PhD thesis had been submitted in Paris by the physician and intellectual Adamantios Coray (1748–1833),[14] whose two-volume study on Hippocrates and his natural philosophy has been acknowledged by both his contemporaries and later scholars as an important contribution to medical geography in general (Coray 1800 and 1801). Coray is still thought to be the most prominent representative of the so-called 'Greek Enlightenment' (1774–1821), an intellectual current which is perceived in the dominant Greek national ideology as having expanded the ideological foundation for the subsequent establishment of the Greek nation state (Dimaras 1989). His scientific achievements were also not unnoticed. The historian Genevieve Miller characterized Coray's two-volume work as the "most extensively annotated 'Air, Water and Place'

[13] See, for instance, Bernhard Ornstein, "Über die Farbe der Augen, Haare und Haut der heutigen Bewohner Griechenlands", *Verhandlungen der Berliner Gesellschaft für Anthropologie*, Berlin: 1870; Bursian, "Über den Einfluß der Natur der griechischen Länder auf den Charakter seiner Bewohner", *VI u. VII Jahresbericht der geographischen Gesellschaft in München*, Munich: 1877; both cited in Stéphanos (1884): 440. Stéphanos's monograph contains an extensive bibliography of German and French anthropological treatises on Greece and its population.

[14] His name appeared in the bibliography with diverse spellings, Adamantius Coray(s), Adamantios Korae or Koraēs (Greek: *Αδαμάντιος Κοραής*). The present study uses the form Adamantios Coray, unless referring to the bibliography, when the original form will be kept.

which has ever appeared".[15] The acknowledgement this work received in France was not accidental if one agrees with Michael Osborne that France, "perhaps more than any other country of Europe, has long celebrated the works of Hippocrates" (Osborne 2000: 32). This treatise has been portrayed as "a junction of his [Coray's] medical studies and his philological knowledge" (Dimaras 1989: 335). While analysing Hippocrates's thinking, Coray provided, in addition to historical and philosophical content, also extensive medical commentaries. From his perspective, the major question dealt with by Hippocrates in his certain works was how it was possible that humans in diverse parts of the world were different (Coray 1800: i–ii). Even though Coray has been largely perceived (and to some extent he indeed operated) as an ambassador of Greek national aspirations in Europe, above all he was a cosmopolitan intellectual whose spirit and activities reflected the turbulent period in which he lived. As a witness of the French Revolution, he was deeply influenced by rising intellectual and ideological currents, and was himself a far from passive recipient of these contemporary developments since he is considered to have influenced scholarly circles in France, the anthropological ones in particular. According to George Stocking, in order to make known the state of Greece, in 1803 Coray addressed the Société des Observateurs de l'Homme (founded in 1799) where he presented "the celebrated memoir on *The present state of civilisation in Greece*".[16] Yet Stocking argues that the influence of Coray's memoir upon the Société was so strong that this institution "soon became the resort of the Philhellenes, and accordingly lost its scientific character" (Stocking 1982/1968: 21).

This influence and philosophical dimensions of Coray's work might make a comparison between him and Clon Stéphanos appear disproportional. What is certain, however, is that while the former contributed to the creation of the Greek national doctrine using, among other means, the sciences as a vehicle, the latter corroborated the rising nationalism by drawing on arguments regarding medical geography. In spatio-temporal terms, his subject spanned Hellenic antiquity, even though the focus addresses, for the most part, the recent period and the regions encompassed within the borders of his contemporary Greek state. Unfortunately, it is not evident whether the specific proportion of antiquity and present in addressing his subject was a preference of the author himself or was forced by

[15] Miller (1962): 138. On the perception of Coray's volumes, see Osborne (2000): 32.
[16] Stocking (1982/1968): 20; Stocking means Coray (1979/1803).

his publisher. (A footnote on the first page suggests that the published treatise might have been only a fragment of a larger study that had been shortened by the publisher.) This restriction notwithstanding, the idea of the timelessness and continuity of the Greeks from Hellenic antiquity to the present emanates from the content. In terms of methodology, the way in which he addresses the issue of hygiene indicatively shows that he embraced the macro-environmental approach that was typical of medical geography. In the chapter entitled "Hygiene" Stéphanos addressed the "living conditions" and "customs" of Greece's inhabitants, for the most part peasants and villagers. Matters of housing were treated with regard to the construction of houses and indoor furnishings as well as the cleanliness of the habitants. (Stéphanos observed that the mountaineers neglected cleanliness, especially those of Albanian origins. Exceptions were the islanders of Spetses and Hydra, which he mentioned as exceptional instances of extreme affection for cleanliness by the Albanian inhabitants.) Further issues discussed under the rubric of 'hygiene' included heating systems and material used in the houses, illumination, alimentation, eating habits and natural resources, potable water, consumption of tobaccos, costumes and agrarian and pastoral life. In all, he painted a precarious picture of the hygienic conditions in which the Greek population lived, which he attributed to the 'enormous hardship' the country had suffered for centuries under Ottoman rule.

The national outlook of Stéphanos's treatise is evident not only in his judgemental view of the Ottoman period, but first and foremost the specification of Greece in spatio-temporal terms according to national simultaneousness; he stretched the focus of his research back over past centuries, thereby presupposing the historical continuity of Greece. The treatment of the individual issues bears witness to the projection of his research over Greece as a diachronic entity. Questions related to hygiene were also addressed under the rubric "Pathology" (in a chapter with the subtitle "Nosological Geography and Epidemiology") in which he outlined some widespread views on hygiene in antiquity, the Middle Ages and in the recent period prior to the founding of the Greek state. In doing so he was keeping up with a trend in contemporary medical geography in which, under the term 'pathology', the history and the current state of the spread of diseases in Europe and in the colonies was subsumed.[17] The

[17] See for example, Hirsch (1860) and (1862–4).
Especially in the first volume (1860), Hirsch explored the dissemination of diseases, particularly Beriberi, in India, as correlated to 'nationality and race', and concluded that Beriberi was equally disseminated among native populations and Europeans.

major topic of this chapter on 'pathology' dealt with the most relevant diseases in various Greek regions (including Crete, which at the time still belonged to the Ottoman Empire) in both antiquity and the present time. He depicted the most prominent views about the combating of diseases in Hellenic antiquity as suggested by Thucydides, Theophrastus, Aristotle, Strabo and Plutarch. The dissemination of diseases was also treated in correlation with rates of population growth in a separate chapter on "Demography". The glance here was not restricted to the present, but went back to past periods, providing a rough historical insight into quantitative demographic trends in contemporary Greek territories during the previous centuries, regardless of whether or not they had in fact belonged to Greece in the past. The contemporary Greek population was recorded in extensive statistical data on fertility and mortality rates, age and sex groups, family status, profession, religion, population movement in terms of marriages, natality (considering both 'legal' and 'illegal' births) and mortality. 'Migrants and aliens', meaning individuals of Albanian origin or migrants from territories that were still Ottoman, were examined separately.

Anthropological issues were addressed along with ethnological ones in a chapter that, under the headings "Ethnology" and "Anthropology", explored the origins and racial composition of the Greek population, applying the same model of periodization. Now more than ever the look back at the past was well justified, since the time past was of importance for defining the epistemic scope of these two disciplines. In the ethnological section the author separated the historical periods of Hellenism into a prehistoric (or 'primitive') epoch, antiquity, the Greek and Roman ages and the medieval and modern periods. In this time frame he specified the peoples who inhabited the different Greek regions with respect to their origins and their linguistic and racial features. Racial issues were addressed at length in the anthropological chapter. Based mainly on secondary literature, this treatise provides a comprehensive conspectus of hitherto physical anthropological research on the Greek inhabitants conducted either by Greek or by foreign scholars in the 19th century. With this division of the chapters, Stéphanos embraced a concept of anthropology that was reduced to physical anthropology and which used predominately anthropometric methods.

Indeed, in Stéphanos's time a good deal of physical anthropological data on skulls, the colour of skin, hair, eyes and other features were collected by intense studies, measurements and observations. The inhabitants of Greece and the Balkans had generally become the object of physical and racial anthropological examinations conducted by mostly foreign scholars, long before this discipline had been established at the University of Athens. This was the very time in which measuring somatic traits was seen as a major means for exploring human evolution in the paradigm of diverse population groups. Such research was clearly marked by the idea of ancestry which, again, was interpreted in very topical terms. And Greece was particularly attractive for anthropometric and racial anthropological investigations in the 19th century due to its association with two legacies, stemming from Hellenic antiquity and the Ottoman Empire, both of which were closely linked with European self-consciences; the former in an inclusive and the latter in an excluding way. The interest of 19th-century researchers was related, on the one hand, to the possible affinities of current Greeks to ancient Hellenes and, on the other, to the ethnic or racial diversity that was thought to be a typical property of the Balkans. As a Balkan country, Greece was part of a historic geographical space that was established in Western European thinking as the place *par excellence* of human, and even racial, diversity, with all the implications of diversity inherent in the ideology of Balkanism (Todorova 1999; Sundhaussen 2002 and 2003). Mobilized by their interest in human diversity, anthropologists and, more often, physicians with or without specific anthropological training, conducted anthropometrical investigations on living individuals and human remains in the Balkans. Such investigations were facilitated by the military operations in the Balkans, from the 19th century and even up to World War I, where local military forces and those of the Great Powers stationed in the Balkans were largely utilized as samples for anthropological investigations. In the course of the decline of the Ottoman Empire and during the nation-building process in the region, the diversity was reinterpreted by the local Balkan elites (scientific, political and others), and often applied in order to substantiate the uniqueness of their emerging nations. A modern science, as anthropology was thought to be, was instrumental in providing solid knowledge for buttressing national aspirations. Moreover, it was in this fashion that anthropological approaches became embedded in academic institutions, shaping research programmes and schedules at university departments in Western Europe at which Balkan elites also studied.

Clon Stéphanos's treatise witnessed this state of affairs in that it encompassed a great number of anthropometric studies. He drew arguments from his contemporary famous anthropologists and physicians such as Augustin Weisbach (1837–1914) (Weisbach 1881 and 1882) and Bernhard Ornstein (1806–96). Researchers of Greek origins were therein largely absent, although a notable exception to this was the physician N. Apostolidès, whose anthropometric research was presented by Joseph Jegorowitsch Deniker (1852–1918) in a French journal of anthropology (Apostolidès 1883a).[18] Ornstein's numerous publications also linked medical geography with demography and physical anthropology. He was chief surgeon in the Greek army and this professional position had facilitated his research activities, which included anthropological measurements and observations of the colour of eyes, hair and skin. His conclusions were released in anthropological and ethnological journals, and the periodicals of the related ethnological and anthropological societies in Berlin.[19] Ornstein became particularly famous due to an 1875 report in which he depicted the case of a Greek recruit with an unusually hairy patch at the base of his back. This case evoked a controversy between Ornstein and Rudolf Ludwig Karl Virchow (1821–1902) who considered it as nothing but an exceptional pathological deformation, in contrast to Ornstein who represented the case as a typical phenomenon of atavism (Zimmerman 2001: 74f.). This controversy was indicative of the early debates on Darwinism and atavism in Germany where shows involving 'human monstrosities' (including, among others, hairy and caudate individuals) were common in the 1870s.[20] Certainly, Stéphanos was least of all interested in discovering cases of atavism in the Greek army. In his treatise he was concerned with portraying Greece as a diachronic entity which drew its origins from antiquity.

This very treatise qualified him finally to head the first anthropological institution in Greek academia, the Anthropological Laboratory and

[18] The article is written by Deniker, member of the Paris Anthropological Society [Société d'Anthropologie de Paris], who presented Apostolidès as "docteur ès sciences de la faculté de Paris". Deniker wrote that the data were collected by Apostolidès during several travels in Greece and Turkey (Asia Minor).

N. Apostolidès also published in the *Bulletin of the Historical and Ethnological Society of Greece* [*Δελτίον της Ιστορικής και Εθνολογικής Εταιρίας της Ελλάδος*] (Apostolidès 1883b) a three-page Greek edition in which he demonstrated in a rather descriptive way the cranial measurements of living Greeks. His sample was chosen to include different groups according to age, place of origin and profession.

This short article was descriptive and the demonstration of the data was not accompanied by any analysis or comments; these, as Apostolidès announced, were reserved for a future publication. The publication of such a subject in *Bulletin of the Historical and Ethnological Society of Greece* might be exceptional, since physical anthropological issues were rather rare in the columns of this bulletin.

[19] Ornstein was a corresponding member of the Berlin Society for Anthropology [Berliner Gesellschaft für Anthropologie]. See Ornstein (1881); cf. Mason (1887). On Ornstein, see Zimmerman (2001): 77; Hunter and Macalpine (1960): 296.

[20] See, for instance, Goschler (2003): 183. Germany was certainly not exceptional in this. Cf. Oldenburg (1996).

Museum (hereinafter Anthropological Museum). In epistemic and academic terms, the process of its establishment is illustrative of the initial alliance of anthropology, the natural sciences and medicine and their gradual separation after the increasing specialization in the subjects of each discipline that was connected also with professional specialization.

CHAPTER TWO

THE EMERGENCE OF ANTHROPOLOGICAL INSTITUTIONS AND PROFESSIONAL SCHOLARSHIP

The institution in which a scientific discipline is hosted is indicative of its epistemological origins and is also essential for (although it does not determine) the latitude for shaping its scope and drawing its methodological tools. Physical anthropology emerged from the large group of human sciences and, in the 19th century, was to a large extent integrated into medicine, and anatomy in particular, even though the discussion on its epistemic scope was still ongoing. This state of affairs was reflected in the establishment of the anthropological discipline in Greek academia with the founding of the Anthropological Museum in 1886 and several years later of a chair of physical anthropology, both at the University of Athens. These two institutions were integrated into the medical faculty, and were intended to be headed by Clon Stéphanos. Their establishment was the outcome of a series of developments that lasted several years and kept pace with the increasing specialization in sciences and medicine and the accompanying creation of respective scientific institutions, within and outside the university.

Before the establishment of university departments, such institutions included scientific associations for the natural sciences which were concerned not merely with advancing evolutionary ideas but, moreover, with establishing sub-disciplines and promoting the concerns of the rising scientific professions.[1] Thus the first institutions to advance the natural sciences and medicine in Greece were created even before the establishment of the first Greek university, the Ottonian University in Athens, which was founded in 1837 and named after the then ruler, Bavarian King Otto (1815–67). A pioneering role in this process was played by the Natural Historical Society of Athens [Φυσικο-Ιστορική Εταιρεία Αθηνών], which was established in April 1835. On the initiative of some of its members, the

[1] Physicians in particular had to remove traditional medicine and folk healers in order to establish modern medicine and assert their professional claims. On the competition between healers and physicians in 19th-century Greece, see Korasidou (2004/1995); Ruisinger (1997).

Medical Society of Athens [Ιατρική Εταιρεία Αθηνών] was created only a few weeks later.[2] Scholars of German origins[3] and education were often active in such associations, as for instance the physician and botanist Nikolaus Karl Fraas (1810–75), who contributed a great deal to the founding of the above-mentioned Natural Historical Society of Athens. After the founding of the university in Athens, many of them also staffed the university faculties. The Ottonian University of Athens consisted initially of four faculties: theology, law, medicine and philosophy.[4] The medical faculty was unique in South-Eastern Europe in that it did not belong to the military.[5] The Faculty of Philosophy included Greek and Latin philology, archaeology, history, mathematics and physics. With this composition, the newly founded Greek university provided the younger generation of the upper classes (and later that of the middle class as well) with the opportunity to earn their first university degrees in Athens. Many continued their academic careers abroad, in Germany and France in particular, as these were the preferred countries for studies, at least in the natural sciences and medicine. After graduating abroad, many of them returned and partially staffed the Greek university.

In this university structure in transformation, the first academic institution in Greece with an anthropological focus was established: the Anthropological Museum was founded in 1886 at the medical faculty of the University of Athens. With this, one of the world's earliest anthropological institutions (anthropological museums in particular) to be incorporated into a university was created. Indeed, the establishment of anthropological institutions at the universities began worldwide in the middle of the 19th century and these were, as a general rule, university departments or chairs: in France a chair of anthropology was established in Paris in 1855 and was held by Jean Louis Armand De Quatrefages de Bréau (1810–92); in Russia a department of anthropology was created in 1864 at the Moscow

---

[2] For more about the founding of these associations see Ruisinger (1997): 79, 81.

[3] The involvement of German scholars in the Greek scientific scene was anything but unusual, especially during the rule of the Bavarian king, Otto (1833–62). In that period, German natural scientists and doctors often staffed state institutions such as the Royal Garden in Athens or even the Greek army (e.g. the above-mentioned Theodor von Heldreich and Bernhard Ornstein, respectively).

[4] "On the Constitution of the University", *Governmental Gazette of Kingdom of Greece*, no. 17, 27 April 1837. ["Περί συστάσεως του Πανεπιστημίου", *Εφημερίς της Κυβερνήσεως του Βασιλείου της Ελλάδος*, Αρ. 17, 27 Απριλίου 1837].

[5] For an account of the establishment of medical universities and faculties in South-Eastern Europe see Promitzer, Trubeta and Turda (2011).

University, where Anatoly Petrovich Bogdanov (1834–96) became professor of anthropology.[6] In Budapest the Department of Anthropology was created in 1881 at the Pázmány Péter University, followed three years later by the Anthropological Museum (Turda 2007b: 362). In North America the first university department of anthropology was founded in 1887 by Franz Boas (1858–1942) at the Clark University in Worcester (Leaf 1974: 2–9). In Greece, the Anthropological Museum was unique in that its establishment preceded the creation of any other anthropological institution, either within or outside the university. Obviously, both the timing of its founding as well as its denomination betrays conformity with a general trend in international academia to establish laboratories that collaborated with autonomous museums. This strategy facilitated teaching and research within the universities. By this means, scientific associations were incorporated into university museums which benefited from the collections of the former. The creation of the museum and laboratory for anthropology in the Greek university occurred relatively early, given that the University Museum for Palaeontology and Geology emerged as late as 1906 from the collaboration of the university with the Physiographical Museum [Φυσιογραφικό Μουσείο]. The latter had been founded in 1835 by the Physiographical Society [Φυσιογραφική Εταιρεία].[7]

The establishment of the first anthropological institution marked the beginning of a period of professionalized anthropology in Greece. This museum was the initial step in integrating the anthropological discipline into Greek universities, followed by the decision of the Faculty of Medicine in 1915 to advance not only anthropological research but also the teaching of anthropology by creating a university chair as well, which was designed to be occupied by the museum's director. But the establishment of the chair proved to be adventurous, since Stéphanos died suddenly in the same year (1915) before having taken up this new position. The following chapters of this section illustrate the circumstances under which the university chair had finally been established, giving rise to anthropology as a field of studies in Greece.

---

6 On the professionalization of anthropology in Europe, see Hoßfeld (2005).

7 See the museum website: http://arch-museum.uoa.gr/~emusweb/paleontology/index.php?lang=en (last accessed October 2012).

2.1 *The Adventurous Establishment of the University Chair for Physical Anthropology*

The medical faculty decided to establish a chair of physical anthropology in a period in which many organizational changes took place at the University of Athens. Several academic disciplines, which up to that point had been collected under the roof of a single faculty, were separated to form autonomous departments; faculties were split and new institutes or chairs emerged. This process had intensified from 1904 onwards when several disciplines, including physics, mathematics and pharmacy, departed from the original Faculty of Philosophy and were reassembled in a new Faculty of Sciences. Later on, in 1919, an institute of chemistry was also incorporated into this new faculty.

After Clon Stéphanos's sudden death before he was able to take up the professorship, the issue of retaining the chair and nominating a new candidate, and, more urgently, a successor for the directorship of the Anthropological Museum, appeared on the agenda of the medical faculty.[8] The problem the faculty now faced was the lack of eligible candidates not only for one but for two positions. In the 29-year period between the founding of the museum and the creation of the chair, no anthropologists had been trained. In effect, Stéphanos's failure to train successors threatened the entire future of physical anthropology at the University of Athens. This, in turn, represented a tremendous disadvantage for the medical faculty, which saw itself threatened with the loss of an already integrated institution, the Anthropological Museum, and the curtailment of the desired expansion through losing an additional chair.

Despite the lack of trained anthropologists, five candidates, all of them anatomists, applied for the chair. However, none complied with the requirements for taking up the position since they had neither studied anthropology nor had any other qualifications in this field. The question now arose as to whether the chair should be occupied or abolished. The majority of the professors who participated in that session of the medical faculty (12 against five) decided in favour of keeping the chair and choosing one of the five applicants, regardless of their lack of appropriate qualifications. The professor of anatomy and distinct authority in the Faculty

[8] It was discussed at a meeting of the professors on 30 May 1915. See Historical Archives of the National University of Athens, *Proceedings of the Meetings of the Medical Faculty*, vol. 8 (17 April 1915–1 June 1918): 118–29.

of Medicine, Georgios Sklavounos (1869–1954), began assessing the candidates and suggested that a qualification in anatomy should be used as a criterion, arguing that "anatomy is the basis and the beginning of any anthropological study".[9] Sklavounos knew all five candidates well, since he had trained each of them in the anatomy laboratory that he directed. Excluding four candidates on the grounds that they either lacked competence or had more career chances elsewhere, he finally proposed Ioannis Koumaris as a candidate for holding the chair. Sklavounos praised the commitment of this candidate to medical sciences and his outstanding professional responsibility, and, as Sklavounos noticed, he had not yet managed to gain a permanent academic position. An additional reason for proposing him was his PhD thesis, which Sklavounos (who also was Koumaris's mentor) characterized as "an original study close to field of anthropology". His suggestion was embraced by the other professors of the medical faculty who argued that the lack of an anthropological qualification was a problem that could easily be solved if the future chair-holder upgraded his competence by means of additional education in anthropology.

Despite the positive decision of the faculty,[10] the chair remained vacant since the university senate did not confirm the nomination, probably for administrative reasons related to the proliferation of irregular chairs at the university and the lack of funding to cover the additional expenses. The members of the university senate argued that the university had to justify to the ministry not only the necessity for an additional irregular chair but also the nomination for professor and chair-holder of a physician who was not trained in this scientific subject.[11] The senate did conclude that the proposed candidate was suited for taking over the direction of the Anthropological Museum which had temporarily passed to its own

---

[9] Historical Archives of the National University of Athens, *Proceedings of the Meetings of the Medical Faculty*, vol. 8 (17 April 1915–1 June 1918): 125.

[10] Finally, 12 of the 17 participants at this meeting of the medical faculty voted for Ioannis Koumaris.

[11] This scepticism was expressed in reference to further irregular chairs, in the medical faculty as well as in other faculties: see Regular Meeting of the University Senate, 10 October 1915: Historical Archives of the National University of Athens, *Proceedings of the Meetings of Senate* (1915–16), vol. 28: 73–5; Regular Meeting of the University Senate, 28 November 1915: Historical Archives of the National University of Athens, *Proceedings of the Meetings of Senate* (1915–16), vol. 28: 90–4.

With respect to the chair for anthropology, Koumaris noted in his autobiography that his nomination did not come into effect "for political reasons". Koumaris (1951): 12.

jurisdiction.[12] Therefore, at that point in time, Ioannis Koumaris became director of the Anthropological Museum only, while the chair for physical anthropology was left for a later date.

The fact that the medical faculty and the senate of the University of Athens decided anew ten years later to establish a chair for physical anthropology[13] might not be independent from the achievements of the director of the Anthropological Museum who, from the year of his appointment (1915) onwards, was committed to the anthropological sciences. He published in Greece and abroad, and established significant connections with anthropological institutions across Europe. All his activities culminated in the founding of the Greek Anthropological Society in 1924, one year before the creation of the university chair and his election as professor for physical anthropology.

Once the creation of an irregular chair for physical anthropology was decided again in 1925 by the medical faculty, the director of the Anthropological Museum was the only candidate, having in this period of ten years monopolized the field of this scientific discipline. The lack of interest in physical anthropology among physicians in Greek academia worked to his advantage and meant that, as director of the Anthropological Museum, he headed the only anthropological institution in Greece. Although his nomination for the university chair did not pass without scepticism regarding how far the establishment of a chair for physical anthropology was really necessary at that time, as well as criticism of the competence of the candidate himself,[14] he was finally nominated for the chair for physical anthropology by 19 votes to one. The university senate confirmed the proposal at a meeting on 16 May 1925.[15] Koumaris became the first professor of

[12] Regular Meeting of the University Senate, 10 October 1915: Historical Archives of the National University of Athens, *Proceedings of the Meetings of Senate* (1915–16), vol. 28: 73–5.

[13] Historical Archives of the National University of Athens, *Proceedings of the Meetings of the Medical Faculty*, vol. 15 (2 June 2–14 January 1927): 72–91.

[14] The major point at issue arose when the professor for surgery, Konstantinos Mermigkas (1874–1942), questioned the scientific quality of one of his publications, namely the *Surgical Handbook for the Doctor on Campaign* (Koumaris 1915). This handbook had been written during the First World War for surgeons in the field and included practical advice for urgent operations. The critic saw in the treatise numerous and grave errors which, in his opinion, disqualified the author from teaching at the university, all the more so since this handbook was his sole didactic work. However, this criticism was not shared by the other faculty members.

[15] 30th Regulatory Meeting of the University Senate, 16 May 1925: Historical Archives of the National University of Athens, *Proceedings of the Meetings of Senate* (1924–5), vol. 36: 224.

physical anthropology in Greece and, as he used to remark, throughout the whole of the Balkans.[16] He held this position until 1950 when he retired, upon which the chair was abolished because of the lack of successors. Up until today there has not been another chair for physical anthropology in Greece.

Because of his singularity as an institutional representative of the anthropological discipline in Greece for a period of almost half a century, his impact on the design of anthropological discourse in Greece was considerable. A cursory glimpse at his life history, as suggested by his autobiography, might serve as a meaningful reference for understanding his agency, or at least as a useful accompanying note for reading the rest of this book.

2.2 *The Chair-Holder in the Light of His Own Life Narrative*

Born in 1879 in Athens as the eldest son of a doctor, Ioannis Koumaris completed his primary and secondary education at Athenian private schools[17] and started his studies in medicine, anatomy and surgery at the University of Athens in 1896. In the period from 1899 to 1902 he was an assistant to Professor Georgios Sklavounos in the anatomical laboratory. He completed his PhD there in 1901 with a study in anatomy which remained unpublished,[18] although a summary appeared in a German medical periodical in a joint publication with his mentor (Kumaris and Sclavunos 1903). Koumaris earned his licence to practise the medical profession (*approbation*) in 1903, after which he obtained professional experience at the university, in a military hospital and in some clinics, first at the university surgical clinic 'Aretaieion' [*Αρεταίειον*] in Athens, and later (1904–6) at a private surgical and gynaecological clinic 'The Redeemer' [*Ο Σωτήρ*]. From 1906 to 1908 he was abroad undergoing advanced surgical training, in Berlin with Professor August Bier (1861–1949), and in Paris. From 1913 to 1915 he lectured in surgical and topographical anatomy at the University of Athens. Koumaris also gained professional experience

---

Koumaris was officially appointed irregular professor for physical anthropology on 20 May 1925.

16 Koumaris (1951): 13–14.

17 On private schools in Athens in the 19th and 20th centuries see the edition of the Athenian University: *University of Athens (Museum of the History of Education)* [*Πανεπιστήμιο Αθηνών. Μουσείο Ιστορίας της Παιδείας*], 1999.

18 See Koumaris (1951): 49.

in the army. During the First Balkan War he served on the front in Macedonia as a volunteer in the mobile surgical service of the Red Cross.[19] In the last months of World War I he served as military surgeon. For these services he was awarded medals which, as he stated, were the only ones he received in his whole life.[20]

This biographical information is included in his three-volume autobiography,[21] which he started writing at the age of 71, following his retirement in 1950. However, these volumes are an account of professional achievements, rather than a more conventional autobiography. They contain details of his professional career, his time in the field of surgery, his subsequent commitment to anthropology and his engagement at the Anthropological Museum, as chair-holder of physical anthropology and president of the Greek Anthropological Society. They also provide information about his training in physical anthropology after his nomination to become director of the Anthropological Museum.[22] According to this, during the summer semesters he had attended university lectures, visited excavations and laboratories, institutes and museums of physical anthropology, prehistoric anthropology and ethnology in Paris, Brussels, Vienna, Munich, Berlin and Rome.[23]

For the most part this autobiography is a detailed register of his publications, including possible feedback, both praise as well as criticism. His publications were divided according to their subject into the categories "Surgery" and "Anthropology"; and according to their appearance in either the 'scientific press' or the 'social press'. A final section lists some early, non-scientific writing, including essays, poems and literary texts published under his *nom de plume* 'Ion Mistras'. All these latter writings were addressed to his personified Hellas and to a national ideal. The choice of his pseudonym was allegorical in a double sense: on the one hand it revealed nationalistic aspirations, while, on the other, it evoked associations with his family's descent: 'Ion' is a short form of his first name Ioannis, but it also refers to ancient Ionia; 'Mistras' (also Mistra) is the Byzantine town close to Sparta in which his father was born. His autobiography contains little information about his private life or his family, with the exception

19 Koumaris (1951): 18. Cf. Stathaki-Koumari (1993: 10).
20 Koumaris (1951): 17.
21 The first and main volume was published in 1951; the two much shorter supplementary volumes appeared in 1961 and 1970. Koumaris (1951); (1961); (1970).
22 Koumaris (1951): 11–12.
23 See also an additional note on his anthropological training in Agelakis (1997).

of a brief note at the beginning of the first chapter where he introduces himself in a laconic and rather military style:

> Ioannis Koumaris, father's name Georgios; regular Professor at the University of Athens, regular Chair of Anthropology (Faculty of Medicine); Director of the Anthropological Museum; Permanent President of the "Greek Anthropological Society"; Born in Athens, on November 12/24[24] 1879; Son of a doctor from Sparta resident in Athens; unmarried.[25]

Some additional personal information can be found in a brief essay written by his niece, the folklorist Rodoula Stathaki-Koumari (1993), who portrays her father's oldest brother as a man "who all his life pursued a vision that he has never abandoned" (Stathaki-Koumari 1993: 13). She describes her uncle's frugal life, with his university salary as his only income, and asserts that his absolute commitment to the anthropological sciences prevented him from enjoying private life, starting a family of his own or acquiring property. This absolute and selfless devotion to science was asserted by Koumaris himself on several occasions. For sure, the assertion of such a commitment was not unusual in academia generally. Rather, devotion to science with abandon is thought of as a specific quality that 'sublimates' scholars, as Pierre Bourdieu argues: it evokes, and simultaneously is based on, a specific form of illusion that science itself is unselfish (Bourdieu 1998: 27). Indeed, dutifulness and the conviction of a lifelong mission permeated the autobiography of the chair-holder, by means of which he intended to provide his Greek readership with "a modest note by a more modest scientist".[26] He explained his belated commitment to anthropology (the 'abrupt alteration' in his career)[27] in part due to his "repeated unfortunate attempts"[28] to start a career in the field of surgery and his failure to find an appropriate professional position elsewhere. The main reason for deciding to study surgery had been the tradition that the first son of a doctor should carry on his father's profession regardless of whether or not this was consistent with his own interests.

The title of his autobiography, *50th Anniversary*, suggests that he considered his life a kind of institution. Seen in this light, his frequently mentioned 'vision' acquired a military nuance. He portrayed himself as a

---

[24] He means 12 November according to the Julian calendar and 24 November according to the Gregorian calendar.

[25] Koumaris (1951): 10.

[26] Koumaris (1951): 7.

[27] Koumaris (1951): 12.

[28] Koumaris (1951): 12.

"volunteer soldier who served the scientific army of the homeland starting in the academic year 1896–97 and formally finishing in 1950 when retired from active service".[29] The decision to write his autobiography originated, he said, in his conviction that any biography was useful and important for educating future generations. Coming generations would "benefit colossally" from the biographies of "top-flight and prominent scientists as well as of simple workers in the sciences".[30] A review of his works and activities persuaded him that such a collection might fulfil an obligation towards the Greek and international scientific communities by facilitating the studies of young students and anyone else concerned with the "Science of Human". Given that Ioannis Koumaris became the leading anthropologist in Greek academia for more than half a century, his autobiography is indeed an important source for studying the history of anthropology.

[29] Koumaris (1951): 7.
[30] Koumaris (1951): 9.

CONCLUSION TO SECTION ONE

The short historiography of the emergence of Greek anthropology undertaken in the previous chapters casts light on anthropology's different meanings and connotations in the 19th century and the path that led to its establishment as a scientific discipline in Greek academia. This process bears the sign of the specialization of scientific disciplines, the emancipation of groups of sciences and the creation of appropriate departments at the university. Following a strong tendency in the 19th-century sciences, anthropology in Greece became embedded at the medical faculty. This affiliation is illustrative of its epistemic sources and its institutional specification as *physical* anthropology. However, with its integration into the medical faculty, the question of its epistemic scope was still far from being answered; in fact, it was only as a result of this process that the question first emerged. This is, first of all, since on the international scale, physical anthropologists claimed to have authority over anthropological sciences at large, expanding the scope of their discipline beyond the limits of physical anthropology; not least, the contents of anthropology were transforming while the sciences of human evolution were leaping forward. Furthermore, the institutional affiliation of a scientific discipline is crucial to its academic trajectory in terms of competition or alliance with other disciplines that can prove instrumental to shaping the boundaries, and thus the scope, of scientific disciplines. And, not least, a factor with a decisive impact upon the epistemic scope and the trajectory of a scientific discipline is the subjective agency of its representatives.

A condition that had been crucial to anthropological discourse in the Greek case was that the establishment of anthropological institutions, such as the museum, but foremost the university chair, paved the way for the creation of a professional group of anthropologists, rather than the reverse. The head of the academic institutions obviously began his anthropological career by coincidence; it was a concurrence of fortunate events that gave the impetus to his university post. Becoming an anthropologist and training a first generation of anthropologists then became a target to be reached. This is the point of departure for examining how anthropology was conceptualized in Greek academia, given its integration into the medical faculty and its institutional designation as physical anthropology.

SECTION TWO

CONCEPTS OF ANTHROPOLOGY, INSTITUTIONAL TRAJECTORIES AND SCIENTIFIC CAPITAL

INTRODUCTION TO SECTION TWO: PATHS FOR PASSING FROM THE WORD TO DISCOURSE

The main aim of this section is to ascertain how anthropology, even though institutionally designated as physical anthropology, was perceived and conceptualized in the Greek anthropological discourse following its establishment at the Greek university as a field of research and teaching. How were the elaboration of anthropological conceptions, their possible transformations and the scientific capital anchored in these conceptions related to each other? Instead of contextualizing the elaboration of conceptions and anthropological statements, this question will be treated with regard to the organization of the field of anthropological statements, as suggested by Michel Foucault: "Rather than wishing to replace concepts in a virtual deductive edifice, one would have to describe the organization of the field of statements where they appeared and circulated" (Foucault 2005: 62). Taking the field of statements, again, as a place in which scientific capital (Bourdieu 1998) is produced and reproduced, the question arises as to which way the anthropological discipline in Greece acted as a resource of symbolic capital in terms of providing recognition and acknowledgement to both professional anthropologists and those involved in the anthropological discourse, even though they were not necessarily anthropologists by training. And how did anthropology's performance as a potential resource of symbolic capital, in turn, influence its epistemic scope?

In treating these questions, anthropological concepts will be examined in this section with respect to the positions occupied by the main actors who designed the Greek anthropological discourse (i.e. their leading position in research institutions and scientific departments which provided them with power over scientific reproduction); in other words, the positions that gave them the power to influence career chances, or to support career development. The academic anthropological institutions will be taken as the frames for the production and reproduction of scientific capital and as a starting position for examining the shape of anthropology's scope. I will argue that in the period under consideration the Greek anthropological discourse is shaped by two eras, each of them associated with the professional anthropologists Clon Stéphanos and Ioannis Koumaris. The former was associated with the emergence of professional physical anthropology, whereas the latter shaped the development and content of anthropology up to 1970, the year in which he died at the venerable age of 91. Although Koumaris's professional career apparently started as a result of a coincidence, he turned this coincidence into commitment and even broke with Stéphanos's tradition. In doing so, he used as a vehicle three institutions that he himself led: the Anthropological Museum (from 1915 to 1950), the university chair of physical anthropology (1925–50) and the Greek Anthropological Society, which he founded in 1924 and directed up to his death.

In accordance with the proposed approach, this section is divided into four chapters. The first three examine the concepts of anthropology at the three anthropological institutions in relation to each other but also to the scientific and university capital derived from these institutions that came to be imprinted in these concepts. I will trace the trajectories of these institutions along with the anthropological statements made in their frameworks in seeking to identify the relationship between these trajectories and the epistemic conceptions of anthropology. The question arises as from which epistemic and societal contexts did their speakers draw, or seek to draw, legitimacy for representing their statements as secure scientific knowledge? In the fourth chapter of this section I will illustrate the political profile of the anthropological institutions and also their head, Ioannis Koumaris. The argument is that, for all proclamations of the scientific objectivity of anthropology as a disinterested and politically neutral science that transmits objective truth, this political profile became highly apparent during World War II and the Greek Civil War, in conditions in which the scientific community was deeply divided into hostile fronts.

CHAPTER THREE

ANTHROPOLOGY AT THE MUSEUM

This chapter addresses the concepts of anthropology as they were designed in the first anthropological institution founded in Greece, the Anthropological Museum and Laboratory at the University of Athens. It shows how the transformations these concepts experienced are closely associated with the personal change in the museum's directorship. Expounding these conceptual transformations brings to light the different understandings of the anthropological sciences that were held by their representatives, and the differences between them are so deep that there is good reason to assert the existence of diverse eras. Furthermore, the chapter reveals the rise in antagonistic claims among scientific disciplines to hold authority over the terrain of the representation of human culture.

3.1 *The Museum As a Vehicle for the Mutual Advancement of National and Professional Aims*

Museums have been designed as sites for the representation of material culture in the public, and anthropology belongs under the first of the 'exhibitionary disciplines' hosted in museums (Bennett 1995: 75). The Anthropological Museum and Laboratory at the University of Athens, at least under the directorship of Clon Stéphanos (1886–1915), differed from this paradigm in that it was a research institute, a laboratory, rather than a public location for the representation of material culture. This restriction was likely due to circumstances endogenous to the university that were related to a shortage of resources, rather than to an intentional programme by the museum's director. Nevertheless, the fact remains that this first university institution for anthropology in Greece under Stéphanos's directorship was a research institute, a place for gathering primary material from excavations and anthropometric measurements, as well as ethnographic data. This is the character the museum attained under Stéphanos, since the first material he received from the medical faculty was appropriate for establishing collections on 'human normality' and 'abnormality'; this in turn meant that the museum became specialized in the area of anatomical pathology (a branch of pathology developed in the 1840s), which used to be integrated into anatomy in 19th-century medical science. Indeed, the 'legacy' that the Anthropological Museum inherited from the medical faculty in the early days of its establishment was material that belonged to the subject of anatomical pathology. This included stillborn, malformed babies or embryos that were sent by doctors to the medical faculty's Museum for Anatomy and Pathology, prior to the establishment of the Anthropological Museum. The Historical Archive of the University of Athens documents at least two such cases from the 1850s.[1] The file includes the correspondence between the Greek ministries of education and the interior, the dean of the University of Athens and the pharmacist who conserved 'conjoined twins' for the purpose of research. (In the correspondence the pressing issue was whether the ministries would reimburse the pharmacist the expenses for the conservation.) A similar case of a 'monster' (as it was literally referred to) was reported two years later.[2]

The Anthropological Museum was, however, not concerned with cases of anatomical pathology. Once founded, it acted as a site for collecting objects of anthropological significance that earlier had been exported to anthropological institutions abroad. For instance, archaeological artefacts, such as ancient Greek craniums and an antique skeleton, were sent to the Society for Anthropology, Ethnology and Prehistory in Berlin [Berliner Gesellschaft für Anthropologie, Ethnologie und Urgeschichte] by the curator of the Royal Botanical Garden in Athens, Theodor von Heldreich.[3] Other such objects were transported abroad by foreign archaeologists

---

[1] One of them concerned 'conjoined twins' born (apparently stillborn) in 1855 in a village on the island of Euboea. The archive contains the correspondence between the ministries: see Historical Archives of the University of Athens, File 3/5–21: Anthropological Museum.

[2] The eparch of the island of Hydra informed the Ministry of the Interior of the birth of a 'monster' in his eparchy. This letter (dated 4 July 1857) was attached to a report written by the doctor who conserved the 'monster', and both documents accompanied a container with the conserved corpse. After correspondence between the Ministry of the Interior, the dean of the University of Athens and the Faculty of Medicine, the package was finally accepted by the Museum of Anatomy and Pathology. See correspondence in the Historical Archives of the National University of Athens, Protocol, File 3/5–21, Anthropological Museum.

[3] Theodor von Heldreich was a corresponding member of the society. The donation was announced at the society's meeting on 20 February 1875, see *Verhandlungen der Berliner Gesellschaft für Anthropologie, Ethnologie und Urgeschichte* 7 (1875): 26. Heldreich was also a corresponding member of the anthropological society in Vienna; see annual meeting of the anthropological society in Vienna, 14 February 1882. *Mitteilungen der Anthropologischen Gesellschaft* (Vienna) 11 [new series 1 (1)] (1882): 211.

and anthropologists who conducted excavations in Greece. Indeed, with the creation of the Anthropological Museum, a site to host material culture of national significance was created. Soon, the collections were enriched occasionally by means of donations (e.g. by the Greek Archaeological Society) but mostly by new material gathered by researchers. Extensive information on the activities of the Anthropological Museum is contained in the reports regularly submitted by Stéphanos to the dean of the university which give detailed accounts regarding the progress of the investigations.[4] These reports reveal that excavated human remains of archaeological importance were anthropologically examined[5] and that anthropometric measurements of living individuals were also conducted. The sample for the latter measurements was, for the most part, taken from school pupils and soldiers in Greek military institutions[6] and from among foreign forces, such as the Italian, Russian, French and British crew of the foreign armadas docked in Greek harbours and in Piraeus.[7]

Apart from craniometrical research, ancillary ethnographic studies were included as research activities during Stéphanos's directorship in that Greek toponyms were recorded and compared with toponyms of other Balkan areas. This was not a novelty in Stéphanos's research interests; his early publications were concerned with the issues of local dialects and local history (cf. Stéphanos 1879 and 1875). The annual reports to the dean of the university provide information about genealogical studies in progress; the registration of families had begun, and in some regions these had already been completed. This activity justifies the specification of the investigations at the Anthropological Museum by its director himself as

---

[4] Letter from 24 July 1887 addressed to the dean of the National University of Athens. Historical Archives of the University of Athens, Protocol, File 3/5–21, Anthropological Museum.

[5] In a report dated 24 July 1887, he reported having conducted research in Lavrio (Attica), taking as a sample 400 workers 'of diverse descent'; he had also taken cephalic measurements of 200 children in Attican schools and the islands of Naxos and Syros. Furthermore, he had excavated and examined antique craniums, four of which were identified as craniums of slaves and were discovered in the mine of Lavrio, while three others were found in an ancient tomb in Megara (both in Attica). Letter from 24 July 1887 addressed to the dean of the National University of Athens. Historical Archives of the University of Athens, Protocol, File 3/5–21, Anthropological Museum.

[6] The letters were undated but are included in a sub-file dated "1891–1893": Historical Archives of the National University of Athens, Protocol, File 3/5–21, Anthropological Museum.

[7] Historical Archives of the National University of Athens, Protocol, File 3/5–21, 1891–2, Anthropological Museum.

both "anthropological in a narrow sense" as well as "ethnological", as he argued at the outset of the very first annual report (dated 24 July 1887).

Indeed, Stéphanos's research was anthropological in a "narrow sense", being descriptive and lacking any attempt at theorizing. He embraced the comparative method by which his contemporary anthropology approached 'human groups' from a 'comparative standpoint', and positioned the object of his research in the interface of disciplines, such as anatomy, archaeology, demography and linguistics. His main concern seems to have been the association of scientific hard data with Greek national ends. And yet he underlined the national importance of the progress hitherto made in his research, as well as his aspiration that it would advance national science to reach international scientific standards in the future, which, reciprocally, would corroborate national aspirations by scientific means.[8]

In the course of his research, the collections of objects and data grew significantly to include useful documents and diverse material that had in the meantime been acquired from private persons and by donation. Stéphanos asserted that the collection had become "unique world wide" and that the museum's profile had reached international standards. In his progress reports during the 1890s he announced plans for new projects, including the development of a map showing variations in stature among the Greek male population. On the basis of the anthropological material being collected, the museum's director hoped in the near future to be able to provide anthropological maps with the cephalic index of the Greek population taking into account the diverse regions of the country. Indeed, the conclusions of his research were published in 1911 as a Greek-language monograph (his only one after the French treatise) on cephalic indexes (Stéphanos 1911). Yet he made the successful realization of projects contingent upon increases in funding, which continued to be insufficient and hampered the progress of the research. He attached detailed catalogues of the technical equipment that was necessary for the continuation of the research to his reports,[9] and he also complained of a shortage of space, which was an additional problem that had emerged as a result of

[8] Letter dated 24 July 1887 addressed to the dean of the University of Athens: Historical Archives of the National University of Athens, Protocol, File 3/5–21, Anthropological Museum.

See also: letters which were undated but included in a sub-file dated "1891–1893": Historical Archives of the National University of Athens, Protocol, File 3/5–21, Anthropological Museum; and the Historical Archives of the National University of Athens, Protocol, File 3/5–21, 1891–2, Anthropological Museum.

[9] This catalogue was not included in the files of the archives.

the flourishing activities; while new objects and data had been collected, they could not be utilized due to the lack of appropriate facilities. This was a major reason why the Anthropological Museum could not work as a museum, and shortages of space and resources would continue to pose a serious obstacle to its work. The repeatedly emphasized national importance of the projects in the annual reports might have aimed to persuade the dean of the university to increase the budget for the Anthropological Museum.[10]

Indeed, in all of the submitted reports the museum's director emphasized that the increasingly enriched collections were of national importance in that they documented the preservation of ancient Hellenic elements in various Greek regions and the racial anthropological unity of the Greek population. Corroboration of the prevailing Greek national doctrine through 'hard data' and anthropological arguments marked the content of Stéphanos's work and simultaneously operated as the main resource for his professional position. Certainly, the national idea he supported was elitist in outlook, and was restricted to a cult of antiquity and issues of Hellenic ancestry. His approach neglected the question of its relevance to contemporary society and made no reference to topical issues. The Greeks, who were the object of his anthropological, ethnographic and linguistic investigations, were barely considered as a subject within their contemporary societal context. Rather, they were positioned in a historical vacuum in which Hellas was imagined to exist diachronically, and this very diachronic existence was allegedly recorded in the bones, customs and the language of the inhabitants in the geographic territory of the modern Greek state. In this project, the Anthropological Museum acted not only as heterotopia, as any museum ultimately acts, but as *national* heterotopia in particular. In the following chapter I will argue that restricting anthropology to narrow national frames was clearly a subjective decision

10 He himself had applied for an amount twice as high as the ministry had conceded to the museum in the previous academic year. Stéphanos justified his claim on the grounds of the high price of the necessary technical equipment and the costs of books, expeditions, excavations, preparing photographs of the findings in order to exchange them with foreign museums, etc.

The first year after the establishment of the Anthropological Laboratory and Museum, Stéphanos applied for an amount of 6,000 drachms. However, the ministry granted him only 3,000 drachms. For the academic year 1887–8 he again applied for 6,000 drachms. Letter dated 24 July 1887 addressed to the dean of the University of Athens: Historical Archives of the National University of Athens, Protocol, File 3/5–21, Anthropological Museum.

of the museum's director, but his contemporary anthropological discourse provided him with legitimacy and arguments.

3.2 *The Anthropological Museum As National Heterotopia*

The alliance of anthropology and archaeology (and partly also linguistics and local history) characterized the anthropological discourse during the period of Stéphanos's directorship of the Anthropological Museum. Not only was research material borrowed from archaeology (ancient bones to be anthropologically examined), but anthropology also shared archaeology's power to affirm the Greek national doctrine (Hamilakis 2007; Damaskos and Plantzos 2008). In other words, his anthropological understanding shared with archaeology the property of being a "device fundamental to the national imagination; it is a device that produces facts on the ground, the experiential and physical national truths. It creates regimes of truths for the nation" (Hamilakis 2007: 293).

With a clear focus on national imagination and using archaeology as a primary ancillary science, the Anthropological Museum under Stéphanos's directorship acted as the "heterotopia Hellas", to borrow the term used by Dimitris Plantzos with respect to archaeology (Plantzos 2008).[11] Heterotopias (among which museums are a generic paradigm) are defined by Michel Foucault as the reflection of a "counter-site" by a mirror that:

> exerts a sort of counterreaction on the position that I [the object of the reflection] occupy. From the standpoint of the mirror I discover my absence from the place where I am since I see myself over there. Starting from this gaze that is, as it were, directed toward me, from the ground of this virtual space that is on the other side of the glass, I come back toward myself; I begin again to direct my eyes toward myself and to reconstitute myself there where I am. The mirror functions as a heterotopia in this respect: it makes this place that I occupy at the moment when I look at myself in the glass at once absolutely real, connected with all the space that surrounds it, and absolutely unreal, since in order to be perceived it has to pass through this virtual point which is over there (Foucault 1986: 24).

The concept on which the Anthropological Museum was based, at least until 1915, was national and heterotopic in that it reflected the primordialistic principle of Greek nationalism and produced simultaneity, making

[11] On further approaches to Greek nationalism based on the Foucauldian notion of heterotopia, see Leontis (1995) and Hamilakis (2007).

representations of the past meaningful for the present. The research conducted at the museum was informed by the doctrine of the ancient Hellenic ancestry of the contemporary Greek population, and it sought to buttress this doctrine by providing scientific 'hard data'. Anthropological arguments acted in the way that Foucault termed a "heterotopology": "a sort of simultaneously mythic and real contestation of the space in which people live" (Foucault 1986: 24). Human remains in anthropology, like classical ruins in archaeology, act as material signs of a heterotopia. As such, they function like the 'material landmarks' of the archaeological heterotopias which, as Yannis Hamilakis discerns, operate:

> not simply as the iconography of the national dream (however important that role is), but also as the essential (in both senses of the word), physical, natural, and real, and thus beyond any dispute, proof of the continuity of the nation, a key device for its naturalization (Hamilakis 2007: 17).

Certainly, neither the use of anthropology in the service of the nation nor its alliance with archaeology was invented by Clon Stéphanos. In his time, the advancement of the national idea by drawing arguments on the basis of anthropology, archaeology and their allied disciplines was justified by the inclusive character of the anthropological sciences and the embracement of any discipline related to humans and human life. This idea, which was programmatic to anthropological sciences at the end of the 19th century, formed part of the presidential address given by the American surgeon and professor of archaeology and ethnology, Daniel Garrison Brinton (1837–99), at the International Congress of Anthropology held in Chicago (28 August to 2 September 1893). On this occasion Brinton talked about "The Nation as an Element in Anthropology" (Brinton 1894), a subject that illustrated how multitudinous the mutual influences between archaeology, ethnology, folklore, religious studies, linguistics and anthropology were; how closely these disciplines were woven together; and how each bore upon the whole of human nature. Seeking to explain these complex interrelations, Brinton made the following statement which betrays, if anything, how closely the idea of the nation with that of humanity was associated in the conscience of the scholars:

> nationality has ever been and is to-day an agent more powerful in modifying both the physical and the psychical elements of man than either race, climate, religion or culture; and therefore that it must constantly occupy the attention of the anthropologist, whether his researches are in the purely physical or in the intellectual fields (Brinton 1894: 20).

The basis for the alliance of anthropology, archaeology and the human sciences was the universal perception of humanity as a totality. In turn, the reflection of this worldview in designing the scope of the anthropological sciences led to a holistic epistemic conception of anthropology that emerged at the turn of the 19th century and which was concerned with:

> the Science of Man, the study of the nature of man, the search for the correct expansion of those laws, and all the laws, which govern the birth, growth, development and decay of all his traits, powers and faculties.... It embraces everything and excludes nothing which pertains to humanity, whether in the individual or in his various aggregations (Brinton 1894: 19–36).

This statement is of programmatic significance because it too was made by Brinton at the International Congress of Anthropology in 1893. On the grounds of this holistic overview of anthropological science, research in prehistoric archaeology, supplied by comparative philology, constituted the basic material for all humanities.[12]

In epistemic terms, archaeology and anthropology were thought to overlap, at least with respect to exploring the very earliest periods of anthropogenesis that were the research subject of palaeoanthropology. But the distance between these disciplines began to grow as anthropology continued to move closer to the present. Then, according to the American social and cultural anthropologist Murray Leaf (1974: 20), the "distinction between cultural dynamics and physiology and between physical anthropological studies and archaeology becomes more pronounced". At the turn of the 20th century, the widening of this already broad scope of anthropological sciences led to an increase in the distance between anthropology and archaeology, not due to epistemic discrepancies, but rather because the former came closer to a large spectrum of other disciplines. If at the end of the 19th century anthropology was perceived as "a group of sciences", as Edward William Brabrook (1839–1930) stated in his presidential address to the British Association for the Advancement of Science (Brabrook 1898), in the period that followed anthropologists asserted "all-inclusiveness" for their science (Fallaize 1921). Indicative of anthropology's "all-inclusiveness" is a classification scheme for anthropology that was published in the *International Catalogue of Scientific Litera-*

[12] Professor Daniel Folkmar thought that sociology ought to be based on anthropology. In a lecture given in the 1890s at the University of Chicago, he depicted this relation as being determined, like other studies, by the relative importance of sociology in human life and action: Folkmar (1893–9). Cf. Stocking (1982/1968): 74.

ture at the beginning of the 20th century (between 1901 and 1915), and which is depicted by James Urry (1993: 10) as follows:

> Two major headings were devised: "Man's Place in Nature" and "Specifically Human Activities = Cultural Anthropology". In the first category there were three sub-sections: "Man as Organism = Anthropography" which included anatomy, physiology, the senses and related psycho-physical and psychological aspects and demography; "The Distribution of Man in Space" including ethnographic and ethnological description, comparative ethnology, racial distributions, migrations and diffusion; [and] "The Distribution of Man in Time" or palaeoanthropology, origins, descent and stages of development. Cultural anthropology was divided into two sub-sections: "Man as Individual" including the gratification of the senses, art, music, etc., and gratification of the intellect, science, philosophy, cosmology, etc.; [and] "Man in Relation" either to others, i.e., "sociology" which was further divided into communication of meaning, (language, signs, literacy, etc.) and aspects of social structure (internal and external), or to the environment, "to the seen" (technology, subsistence, etc.) or "the unseen", (religion, magic, folklore, etc.).

Thus, in contrast to the international trend to conceptualize anthropology as an all-inclusive science, the Greek anthropologist applied a concept of anthropology that was restricted to the alliance with archaeology and ethnography. In his ethnographic research, Stéphanos collected written testimonies and, in doing so, he preferred to build up the idea of national continuity and diachronicity by means of 'more hard data', rather than by folklore and oral testimonies. This is a remarkable preference given that in his time folklore was regarded as a discipline allied to anthropology (Brinton 1894), and that, in particular in the Greek case, folklore (*laography*) arose very early to become a discipline of national importance (Herzfeld 1982: 97f.). Yet *laography* sought to provide evidence for the timelessness of Greek culture and the persistence of Hellenic antiquity in the present, arguing on the grounds of the folk tradition. Clon Stéphanos favoured a more positivist approach. Instead of recording the customs and oral testimonies of the folk tradition, he collected written sources, material objects (inscriptions or documents) and pedigrees. The evidence of the consistency and continuity of the Greeks was, in his approach, registered in the bones and material testimonies of history, rather than in oral traditions. A further property of his era was the lack of interest in popularizing anthropological knowledge and disseminating it to society, as well as in formulating assumptions for current social issues that were independent of anthropology by making use of the authority that academic positions lend to their holders. In a period during which anthropologists in other countries sought to gain influence upon policy-making and to have a voice in contemporary developments in society, Greek anthropology was strictly limited to the academic sphere and was concerned with substantiating Hellenicity and advancing the antiquity cult. The present was not a matter for anthropology as applied by the first director of the Anthropological Museum, despite the fact that the discipline as a whole was becoming increasingly concerned with the present from the turn of the 20th century onwards. In conceptual terms, Stéphanos's approach remained descriptive; he did not reflect theoretically on the findings of his research nor on the notions he used. Indicatively, while he drew on his contemporary racial anthropological discourse, he took the notion of race for granted. The succession at the directorship that came after Stephanos's sudden death in 1915 brought a substantial change in the anthropological discourse in Greece, starting from a novel concept of the museum and the efforts of the new director to implement it into practice.

3.3 *Beyond National Limits: A Public Site to Represent Universal Human Culture*

Keeping pace with the most current stage in anthropology as exhibitory science demanded a revision of the concept of the Anthropological Museum in terms of re-conceptualizing anthropology to extend the hitherto restricted national focus, including human culture in its global dimension, and creating a public exhibition site open to society. These targets were indeed taken up as a cause by the new director's concept of the Anthropological Museum, Ioannis Koumaris. He had the vision to create a site accessible to society for representing human culture and, not least, encompassing colonial heritage. The Anthropological Museum reformed in this fashion was thought to act as a vehicle for the popularization of anthropological knowledge.[13]

The reformist project did not, however, omit the previous character of the museum as a research laboratory. On the contrary, along with its new design, the intention was that the museum would continue its research activities, especially after the establishment of the chair for anthropology, which would further support teaching. An example of research pertaining to the museum's activities is that of the 'anthropological examinations' of

---

[13] Koumaris (1951): 33.

the remains of Princess Alexandra (1870–91),[14] which the museum's director himself conducted when Princess Alexandra's remains arrived from Leningrad before being entombed in the royal cemetery of Tatoi in Attica.[15]

For all the continuing research activity, the major concern of the new director was the expansion of the collections to reflect adequately the dual-disposition of the anthropological sciences (i.e. their universal and national profile). The intended organization of the museum[16] reveals a more advanced stage of specialization and classification in the anthropological sciences than Stéphanos undertook during his directorship. The new concept included both natural and human history but, at the same time, these were conceptually disconnected from each other. This approach was reflected in the organization of the museum, which was designed to be divided into three main sections. The first would be a section of 'physical anthropology' to include collections of human remains from different Greek regions, and probably coincided with Clon Stéphanos's concept of representing ancient Hellenic heritage. A further section would be related to 'prehistory and human palaeontology', which was intended to host collections of items from prehistoric periods of human evolution, such as stone tools and weapons, which did not necessarily originate in Greece. The third section would be dedicated to 'ethnology', which was a true novelty for Greek science and reflected Koumaris's vision of establishing Athens as an equal to European colonial capitals.

Whether this concept was indeed implemented in the internal organization of the museum is open to question since a series of problems hampered its function as an institution open to the public. Nevertheless, the concept was relevant insofar as the collections were indeed enriched by items consistent with the overall project, as I will show in the following.

---

14 Princess Alexandra was sister of the sister of King Constantine and the eldest daughter of King George I (1845–1913) and Queen Olga Konstantinovna Romanova (1851–1926). She was married to the Russian Duke Pavel Alexandrovich Romanov (1860–1919) and lived with him in St Petersburg until her death at the age of 21, shortly after giving birth to her second child, Dmitri Pavlovich Romanov (1891–1941). Her son is best known for his involvement in the murder of Grigori Yefimovich Rasputin (1869–1916).

15 According to Koumaris's autobiography, these 'anthropological examinations' took place (on 24 November 1939) in the church of the Royal Residence in Tatoi (Attica), in the presence of King George II (1890–1947), his heir to the throne and other members of the royal family. This occurred during the second era of King George II (from 1935 to 1947). See Koumaris (1951): 74.

See Stamatopoulos (2004); Van der Kiste (1994); Zeepvat (2004).

16 In his autobiography he illustrated the conceptual reform he had intended to realize. Koumaris (1951): 27–8.

A serious impediment to opening the Anthropological Museum to the public was the permanent lack of an appropriate building to accommodate the museum and its collections.[17] This was the same problem about which Clon Stéphanos had complained in his annual reports to the university dean, and the situation does not seem to have changed greatly from his era onwards. Indeed, in its founding year (1886), the Anthropological Museum was housed in the building of the university in Athens and stayed there for the next ten years. It was then repeatedly uprooted as the university grew, and its permanent displacement indicates that it enjoyed only a secondary importance in comparison to other institutions and museums. It moved for the first time in 1896 into the building where the Academy of Athens is housed today. In 1923 this building was given up to the Byzantine Museum, and three years later (in 1926), when the Academy of Athens was founded and took up residence there, the Anthropological Museum moved back to the university building and shared the rooms of the anatomical laboratory of the medical faculty. Following plans for a new university building complex to host the Faculty of Medicine and the Anthropological Museum, the cornerstone of the university building was laid in 1929 in the district of Goudi in Athens, an event that the press described as the first step in "creating a proper university".[18] Following the medical faculty, the Anthropological Museum moved into these buildings in the early 1930s, where it remains today, still working as a laboratory rather than as a public site for exhibiting its collections.

During Koumaris's directorship, the collections had been considerably enlarged. Acquisitions included artificial models of palaeontological skulls, protanthropus and telanthropus skeletons and "facial models of diverse racial types".[19] This resulted from donations by other museums or private persons, but it was also due to a long and insistent campaign by the museum's director who solicited state institutions in Greece and diplomatic representative authorities abroad, as well as international scientific institutions and private persons, to donate materials or funds to the museum.[20] Heads of foreign institutions did occasionally present the museum with objects or human remains of archaeological significance that had originally

---

[17] For an account of this problem see Koumaris's autobiography: Koumaris (1951): 27–8.

[18] See for instance the publication of the opening speeches in the Athenian newspaper *Free Forum* [*Ελεύθερον Βήμα*], 30 April 1929: "In order to acquire a true university".

[19] Koumaris (1951): 25.

[20] Koumaris (1951): 25.

been excavated in Greek territories and then transported abroad.[21] A large collection of palaeontological objects was donated by the chairman of the Austrian Speleological Society, Adalbert Markovits/Markovich (1897–1941) who lived and worked in Greece from 1927 until his death in 1941 owing to a plane accident. Markovits bequeathed his entire collection to the Anthropological Museum under impressive circumstances only shortly before he died.[22] For this donation Markovits is still honoured as the main benefactor, even as a co-founder, of the Anthropological Museum, next to Clon Stéphanos and Ioannis Koumaris.[23] His collection formed an essential basis for a Museum of Prehistory that would emerge with the reorganization into three main departments.[24] Additional gifts to enrich the collections of the section related to 'prehistory and human palaeontology' came from other museums that were in part related to the University of Athens. One example was the Museum for Palaeontology and Geology (founded in 1906) that possessed collections of prehistoric tools and weapons discovered in Greek and Dutch territories.[25]

However, the relationship with other museums was not always easy or based on partnership, and in some cases conflicts about exhibits caused or increased competition. This occurred with the Archaeological Museum when both museums laid claim to the same exhibits, for example, some

[21] For instance, the Swedish anatomist Carl Magnus Fürst (1854–1935) is recorded as having donated ancient Greek skeletons to the museum, and in 1923 August Rutot, director of the Natural History Museum in Brussels, donated prehistoric stone tools and weapons. Further donations from abroad are mentioned to have been made by Salomon Reinach from the Museum Saint-Germain en Laye in France, by Dimitris Kitikas, a Greek who lived in Egypt, and others. Koumaris (1951): 25.

[22] Markovits/Markovich, an amateur pilot, was killed when his private plane crashed in an accident near Thessaloniki on 28 October 1941. Only a few months before his death he bequeathed his entire collection to the Anthropological Museum. However, at the time, the collections could not be accepted due to the permanent shortage of space in the building of the medical faculty and also because the building was occupied by the military. Markovits had therefore stored his collection elsewhere temporarily, but had given two testamentary letters to the director of the Anthropological Museum (dated 10 March and 23 March 1941), containing detailed catalogues of the items and declaring his wish to leave his entire collection either to the museum or the Greek Anthropological Society in the event of his death. Reported in *PGAS* (27 December 1941): 37–9.

[23] See the website of the Anthropological Museum at the Medical Faculty of the University of Athens: http://anthropology-museum.med.uoa.gr/3d.htm (last accessed August 2012).

[24] On these collections see Koumaris (1946e).

[25] *PGAS* (20 December 1939): 37. This museum also contained large collections of vertebrates, invertebrates, plant fossils, micro-palaeontological collections and fossils from several Greek regions and elsewhere. See the website of the museum: http://arch-museum.uoa.gr/paleontology/ (last accessed May 2013).

Mexican objects donated by the German ambassador in Greece. The Greek Ministry of Education decided to present these artefacts to the Archaeological Museum, causing the director of the Anthropological Museum to protest.[26] In his view, while there could be several local archaeological museums countrywide, only one anthropological museum should exist and this should be under the aegis of the university. The epistemic alliance of archaeology and anthropology appears not to have prevented competition between their representative institutions. But, if the competition with the Archaeological Museum concerned single points, the conflict with Greek ethnological institutions extended to more essential questions concerning the line of demarcation between anthropology and ethnology. In fact, it was the echo of an international dispute over the authority of anthropology over ethnology that led to the uneasy relationship between these two disciplines from their very emergence. In the Greek case, this conflict arose as soon as the realization of the reformist project of the Anthropological Museum was at stake. At the heart of this conflict were resources, both symbolic and material. The most troublesome part of the designed anthropological museum proved to be the third section to represent universal human culture, wherein there were two points at issue: ensuring anthropology's authority over ethnology, and acquiring the appropriate artefacts for the collections.

3.4 *Authorization over Ethnology and a Colonial Dream for Athens*

> Athens and its university lack an Ethnological Museum, that is a museum to house all objects which bear evidence for the development of civilization of various peoples worldwide. (The museum that resides in the Polytechnic University… only bears the name "ethnological" by mistake and will be soon renamed) (Koumaris 1931d).

This statement was made in 1931 by the director of the Anthropological Museum in the Athenian daily press, and is only one instance of his long campaign to prompt the renaming of the Ethnological Museum [Εθνολογικό Μουσείο] and its parent agency, the Historical and Ethnological Society [Ιστορική και Εθνολογική Εταιρεία]. The argument behind the campaign was that the latter museum was not authorized to call itself

[26] The minister defended the decision, responding that the Anthropological Museum already possessed a similar collection, but the Archaeological Museum did not. Cf. report in *PGAS* (8 October 1946): 28–9.

'ethnological'. Considering the nature of its exhibits, it could more properly be classified as a national museum, and as this had not occurred, "after the founding of the Greek Anthropological Society, the country had two Ethnological Societies, one in name only and another true" (Koumaris 1946b). On these grounds, renaming the 'Ethnological Museum' 'National and Historical' became a central concern which the director of the Anthropological Museum eagerly pursued for over 30 years. From the 1920s to the late 1950s he addressed countless letters to the public, to state institutions and even to governments (including the dictator Ioannis Metaxas),[27] but also to the directorate of the Historical and Ethnological Society itself, demanding the renaming of both this society and its museum.[28]

But to what extent was this demand reasonable? And how did this society identify itself and its museum? The character of the Historical and Ethnological Society was stipulated in its journal as serving "modern and contemporary Greek history and historical research".[29] From 1882 its objectives were specified as being:

> to collect historical and ethnological material and any kinds of objects that could contribute to the elucidation of recent and contemporary Greek history and philology, the subsistence and the language of the Greek people.[30]

The Historical and Ethnological Society possessed and administered its own historical archives, library and museum,[31] the latter having been established for "encompassing memorials of the national entity".[32] The initiative for the creation of the museum pre-dated the founding of this

[27] In the *Proceedings of the Greek Anthropological Society* he referred to a letter of 8 December 1939 addressed to Ioannis Metaxas: cf. *PGAS* (20 December 1939): 38.

[28] Early publications in the popular press include Koumaris (1921) and (1929a).
The actions he undertook were reported in the *PGAS*, in which he regularly gave an account of his initiatives and their outcome. See a first account of this correspondence in the *PGAS* (3 December 1929): 67–72. See also his letter in the Athenian newspaper: Koumaris (1936).

[29] "The Revival of the Historical and Ethnological Society" ["Η αναβίωσις της Ιστορικής και Εθνολογικής Εταιρείας"], *Δελτίον της Ιστορικής και Εθνολογικής Εταιρείας* (*Bulletin de la Societé Historique et Ethnique de la Grèce*) 11, (1956): α–η.

[30] "Constitution of the Historical and Ethnological Society of Greece" ["Καταστατικόν της Ιστορικής και Εθνολογικής Εταιρείας της Ελλάδος"], Εν Αθήναις: Εκ του Τυπογραφείου του Αιώνος, 1882: 1.

[31] "The Revival of the Historical and Ethnological Society" ["Η αναβίωσις της Ιστορικής και Εθνολογικής Εταιρείας"], *Δελτίον της Ιστορικής και Εθνολογικής Εταιρείας* (*Bulletin de la Societé Historique et Ethnique de la Grèce*), 11 (1956): α–η.

[32] "Constitution of the Historical and Ethnological Society of Greece" ["Καταστατικόν της Ιστορικής και Εθνολογικής Εταιρείας της Ελλάδος"], Εν Αθήναις: Εκ του Τυπογραφείου του Αιώνος, 1882: 1.

society when various scholars, most of them members of the philological club 'Parnassus' [Παρνασσός] (which also attracted a wide range of intellectuals and scholars), decided to collect "national heirlooms of the struggle for national independence",[33] starting from 1821 (the emblematic year symbolizing Greece's emancipation from the Ottoman Empire) and extending to the establishment of the Greek national state in 1832. Thus, when the Historical and Ethnological Society was founded in 1882, the collections already existed and the presence of a museum could be encompassed within the society's statutes:

> The Society possesses a Historical and Ethnological Museum that is divided into two sections: one section, named "archive", includes written historical documents; another, called "museum", comprises all other documents concerning the existence of the nation.[34]

The Historical and Ethnological Society faced the usual problem of finding a building for preserving and exhibiting its collections. Until 1940 its historical archives, its library and the collections all resided in the building of the Polytechnic University (founded in 1836). The rapidly growing interest in the exhibits led to the establishment of a permanent exhibition, but it also put the acquisition of its own building urgently on the agenda. Actually, the director of the Anthropological Museum fully realized the significance of this museum in representing *national* material culture and supported the demand to possess its own building. He even argued that the best place for accommodating collections of such vital national importance was the Old Parliament Building, located in the centre of Athens. This building, he said, was worthy of such valuable collections and would be well positioned to raise the status of a Greek 'National Museum' to that of the German Zeughaus or the French Palais des Invalides (Koumaris, 7 January 1936). His objections were not in fact addressed to the Historical and Ethnological Museum as such, but concerned only the denomination 'ethnological' in its name.[35]

---

[33] "The Revival of the Historical and Ethnological Society" ["Η αναβίωσις της Ιστορικής και Εθνολογικής Εταιρείας"], *Δελτίον της Ιστορικής και Εθνολογικής Εταιρείας* (*Bulletin de la Societé Historique et Ethnique de la Grèce*), 11 (1956): α–η.

[34] "Statute of the Historical and Ethnological Society of Greece" ["Καταστατικόν της Ιστορικής και Εθνολογικής Εταιρείας της Ελλάδος"], Εν Αθήναις: Εκ του Τυπογραφείου του Αιώνος, 1882: 1.

[35] For instance, when a new directorate was elected in 1931 at the Historical and Ethnological Society, Ioannis Koumaris addressed once again his claim to the society, using the same arguments. The Historical and Ethnological Society responded immediately to this letter, assuring Koumaris that they would certainly consider his concerns; however, any

The Ethnological Museum was indeed renamed the National Historical Museum [Εθνικό Ιστορικό Μουσείο] in July 1942;[36] it also took up residence in the Old Parliament in the centre of Athens, even if it did take until 1962 to make the move. For the director of the Anthropological Museum, however, this was not sufficient as long as the Ethnological Society still carried that name. He thus continued his crusade, addressing letters to the directorate of the society and to the press (Koumaris 1946b), but without success; the Historical and Ethnological Society still carries that name today.

The initial question as to how reasonable the claim to rename these institutions was can now be partially answered. Doubtless, neither the collections nor the overall character of these institutions could really justify their denomination as 'ethnological'; at least not in the sense of cultural anthropology. Indeed, 'ethnological' was used with different meanings in this dispute. The Historical and Ethnological Society used 'ethnological' in terms of popular culture and with reference to ethno-romantic local history, according to the German tradition. On the eve of the establishment of the Greek state and in the immediate aftermath, this agency's collections and overall activity revealed a tendency among Greek intellectuals to perceive the 'people' [*λαός*] in rational terms, as being the only authentic creators of national culture, representative of the political sovereign body in the making; and as a rational moment of national history (Kirakidou-Nestoros 1993; Herzfeld 1982). This concept differed essentially from the meaning ascribed to the term 'ethnological' by the head of the Anthropological Museum, who defined 'ethnology' by drawing on the Anglo-American and French context of colonialism.

But *what's in a name?* to borrow the Shakespearean question George Stocking posed with regard to a similar controversy that occurred in 19th-century Great Britain (1971: 376). The dispute that broke out in the 1920s in Greece reflected debates that had been fought out elsewhere since the first anthropological societies were established in the 19th century over

decision would be postponed in the immediate future. The correspondence was published in the *PGAS* (30 December 1931): 55–6.

The historian Faidon Koukoules (1881–1956), a member of the anthropological society and the then general secretary of the Historical and Ethnological Society, also received a personal appeal to promote the concern. Faidon Koukoules only repeated the decision of the directorate and promised to examine the matter in the near future. *PGAS* (30 December 1931): 57.

36 "The Revival of the Historical and Ethnological Society" ["Η αναβίωσις της Ιστορικής και Εθνολογικής Εταιρείας"], *Δελτίον της Ιστορικής και Εθνολογικής Εταιρείας* (*Bulletin de la Societé Historique et Ethnique de la Grèce*), 11 (1956): α–η. See also *PGAS* (25 September 1942): 29–30.

the epistemological boundaries between these two disciplines and their authority in symbolic and material matters. The controversy was about defining the scope of anthropology relationally, which effectively means that it was about drawing the demarcation line between scientific disciplines and their agencies, and the claim each could make to exclusivity in the country. This had practical implications insofar as resources were in play. The campaign for removing the contentious word 'ethnological' from the name of a society and its museum was clearly related to the project of the Anthropological Museum, which aimed to encompass an ethnological section from whence an autonomous ethnological museum would emerge over time. Supporting his own project, Koumaris felt that a real Ethnological Museum did exist under the auspices of the Anthropological Museum, especially after the latter had moved to its new premises in Goudi (1931). And he saw his claim justified on the grounds of the rights anthropology generally conveyed to ethnology in terms of both epistemic scope and representative institutions. He stated that:

> "ethnography" or "ethnology" is this very part of the overall anthropological project that copes with the collective soul of modern man. It has constituted the most original trends among the anthropological sciences from their very beginnings. Ethnographical or ethnological museums presented the first anthropological collections.[37]

Indeed, the integration of ethnology into anthropology followed a tradition originating in the mid-19th century that regarded ethnology as one of anthropology's subdivisions. At that time anthropology was defined as "the science of the whole nature of man" (Stocking 1971: 376) by anthropologists such as Paul Broca (1824–80) in France, and later by James Hunt in Great Britain and others elsewhere. In claiming rights to the use of the term 'ethnology' by the director of the Anthropological Museum, the concept of anthropology was broadened to encompass colonial orders (or desires) and, not least, colonial concepts of race.

Ioannis Koumaris was an admirer of empires—colonial empires in particular. In 1931 he visited the International Colonial Exhibition [*Exposition Coloniale Internationale*] in Paris (cf. Leininger-Miller 2001), and delivered a passionate report on the French colonial regime in the Greek press (Koumaris 1931e), describing Paris during the exhibition as a multicoloured city and providing details of the presence of the 'savages'. Omit-

[37] Report recorded in *PGAS* (3 December 1929): 67–70, 67.

ting any protests and criticism by opponents of colonialism,[38] he praised the French colonial order, boasting of its national, commercial and scientific achievements as well as its "civilizing mission over millions of savage people" (Koumaris 1931e).

> Missionaries, scientists, even women [*sic*!], together, with superlative disinterestedness strove for sanitation, progress, study, moral education of these literally savage members of the human family. The hygienic and educational methods they applied deserve admiration (Koumaris 1931e).

The Greek anthropologist claimed to have seen in Paris how colonized peoples expressed their "gratitude and love" towards their benefactors for the civilizing mission. While extolling the "magnificent civilizing operation of the colonial state", allegedly based on "care for the savages", he did not fail to allude to a "negative effect", whereby the assimilation of the latter caused a "physical intermixture" that, among other consequences, might eventually also cause the extermination of the natives, the "uncivilized savages" (Koumaris 1931e).

Given his enthusiasm about French colonialism, it is not surprising that his model for an ethnological museum in Athens was the contested Permanent Colonial Museum (today the Musée des Arts Africains et Océans) that was founded after the 1931 colonial exhibition in Paris. The future ethnological museum in Athens was thought to represent human races and bear the name 'Museum of Races from the Entire World' (Koumaris 26 March 1949c). In the eyes of the Greek anthropologist, the founding of an ethnological museum in Athens was a matter of national importance. Although Greece was not a colonial power (and therefore such a museum was not intended to stimulate national memory), it was, nevertheless, authorized as a part of the European world to have a share in the colonial legacy. The national importance he attached to this project is demonstrated in a letter addressed to the commission that was preparing the celebrations for the 100th anniversary of the establishment of the Greek state which was to take place on 25 March 1930. In this letter he argued that "a century old capital such as Athens must not lack the liveliest of adornment, a cornerstone for studying the life of peoples".[39] That Greece did not possess colonies was no obstacle to finding artefacts for such a museum, since appropriate exhibits could be acquired. Koumaris

[38] For instance, the Communist Party of France arranged a counter-exhibition pointing at the forced labour in the colonies. Cf. Zanella (2004).

[39] *PGAS* (3 December 1929): 71.

even had a detailed plan for the exhibits and compiled a catalogue of objects to be acquired.[40] Consciously or not, he emulated the German anthropologists of the 19th century who were anxious to acquire colonial artefacts as a means of sharing in the colonial order. According to the historian Andrew Zimmermann, the purchase or exchange of such artefacts was of vital importance in the German colonial system:

> The collections of anthropological objects depended on, and was a constituent part of a broader system of colonial exchanges. The objects clearly had an important function in the establishment of German colonial rule, both defining and representing the political relations between colonizers and colonized (Zimmermann 2001: 152).

Since Greece had no colonies, traders, patriots, diplomats and others who were resident outside of Europe were asked to contribute to the collection of objects for the Ethnological Museum. Through publications in the press[41] and correspondence with state and diplomatic institutions abroad, the director of the Anthropological Museum strove to attract the attention of the public and the authorities,[42] and to sensitize individuals of Greek origins who lived abroad to the need to collect, purchase and donate appropriate objects to the Anthropological Museum. As a public institution, this museum was essential not only for scientific purposes but also for the general education of the population. The visit to the Parisian colonial exhibition in 1931 was also part of this campaign. In order to acquire some objects, the Greek anthropologist personally contacted the organizers of the exhibition, as well as the Greek Ministry for Foreign Affairs, the Greek embassy in Paris and individual Greek diplomats.[43] However, despite his best endeavours, the undertaking was rather unsuccessful,[44] although not entirely so; in 1933, for example, the collections could be enriched by several objects originating from the native population of French New

---

[40] *PGAS* (3 December 1929): 67.

[41] See for example his "Popular University" column in the daily paper *Evening* [*Βραδυνή*], where the importance of an ethnological museum open to the whole population is underlined (1931d).

[42] In this campaign state institutions were also factored in, and the Ministry of Foreign Affairs was asked in 1929 to inform Greek authorities abroad about the project and to call for donations. *PGAS* (3 December 1929): 67–70.

[43] The *PGAS* published two letters (translated into Greek) addressed to the Greek embassy in Paris (dated 20 August 1931 and 24 August 1931), the first to the organizers of the International Colonial Exposition and the second to the dean of the University of Athens. See *PGAS* (30 December 1931): 59, 60.

[44] See *PGAS* (30 December 1931): 59f.

Guinea, donated by a Greek trader in Africa.[45] Some years later, in 1949, a 'seldom object' was obtained; the preserved head of a South American native treated by a technique common among natives in the past. This object was a gift from a commercial attaché at the French embassy in Venezuela—a 'distinguished philhellene'.[46] Negative responses to the appeals and the long-lasting campaign were not lacking either. In 1949 the Greek ambassador in Washington, for example, pointed out in a letter that the Greek migrants had different priorities for supporting their country, and were more interested in promoting their home regions rather than world culture in Greece. World culture was the lowest priority, especially at a time when the country was facing such a difficult situation[47] as the very last phase of the Civil War.

Regardless of any assessments concerning either the success or failure of the campaign, the fact remains that Koumaris's project of the Anthropological Museum gives clear evidence for a conceptual watershed in the understanding of anthropology. The most important aspects in this shift included first the intention for the museum (although not realized) to act as a vehicle for anthropology to reach the public and to popularize anthropological knowledge. A further aspect in this watershed corresponded with anthropology's claim to rights to ethnology and thus to the universal history of the human species embracing colonial racial anthropology. This was an innovative moment in Greek anthropology insofar as its realm had hitherto been confined to narrow national settings. The museum ceased to act as exclusively national heterotopia, although in the novel concept its national outlook was not undervalued. Rather, the connotations of the 'national' were transformed, or better, enriched with elements that earlier were not considered: the national was extended to share universal culture beyond the restricted terrain of antiquity. Although anthropology had never ceased to be a science in the service of the Greek nation, as Koumaris understood it, during his era its scope and epistemic content became more multi-layered, and the university chair was intended to operate in a way that was complementary to that.

---

45 Reported in *PGAS* (28 December 1933): 22.

46 Koumaris (26 March 1949c).
In the margins of his report he remarked that a Greek from Argentina had in the past attempted to purchase a similar item in Buenos Aires but that it had been too expensive.

47 Koumaris (26 March 1949c).

CHAPTER FOUR

ANTHROPOLOGY AT THE UNIVERSITY CHAIR

> ... capital breeds capital, and holding positions conferring social influence determines and justifies holding new positions, themselves invested with all the weight of their combined holders.
>
> (Bourdieu 1988: 85)

The establishment of the chair for physical anthropology at the University of Athens implied a step towards the accumulation of university capital by the director of the Anthropological Museum. The innovative conception of anthropology that was inaugurated with the museum was enriched at the university chair, leading to an understanding of the tenets of anthropology that went far beyond finding assistant sciences or drawing the lines to divide the fields of authority between anthropology and relative disciplines. The programmatic intention of the chair-holder was to establish a scientific discipline with a holistic scope to include all sciences of human and human life and, moreover, to successfully link with society. This chapter explores the initial agenda of the chair-holder and its implementation and outlines an essential contradiction between the programmatic target to teach anthropology in a broad sense, on the one hand, and the teaching of a restricted scope of physical anthropology in reality on the other.

4.1 *The Agenda: Teaching Anthropology in the Broad Sense*

The inaugural lecture given by the chair-holder for physical anthropology on 22 May 1925 was programmatic for his understanding of anthropology generally and as a university subject in particular.[1] On this occasion, the chair-holder disclosed to the Greek academic audience the universal epistemological realm of anthropology by reflecting on the state of affairs in European academia and specifying particular aspects of its intended implementation in the Greek scientific scene. The realm of anthropology

[1] The text of Koumaris's inaugural lecture appeared in two issues of *Medicine—Greek Medical Review*, the Greek Medical Association's journal. See Koumaris (1925a and 1925b).

was delineated in terms of a discipline situated in the "broader field of biology", thereby constituting the sub-discipline "biology of the human species". This statement shows the influence of the continental anthropological schools, especially of the German, French and Russian traditions in which anthropology was used in a physicalist, natural historical sense (Proctor 1988: 141). The definition of anthropology as 'biology of the human species' was clearly close to that of Ernst Haeckel (1834–1919) who conceived of anthropology as a 'third biology', next to botany and zoology. However, the Greek chair-holder did not embrace Haeckel's concept as a whole but, going beyond the limits of biological anthropology, delineated this discipline as "humanity's natural history in every time and every place". Anthropology in this sense embraced all disciplines concerned with "human groups (*Hominids*), their past and future as well as with collective human action" (Koumaris 1925: 75–6). This specification, again, betrays the influence of the German anthropologist Rudolf Martin (1864–1924) who, in the introduction of his influential textbook of anthropology, depicted this discipline as "the natural history of the hominids in their spatial and temporal expansion" (Martin 1914: 1). However, the Greek anthropologist embraced Martin's approach only partially. He kept the distinction between 'physical' and 'psychical anthropology', but challenged Martin's suggestion of limiting the scope of anthropology to the very first branch, arguing instead that anthropology should also consider "the psyche, that is, the intellect" of humans instead of being limited to the "nature or morphology of man" (Koumaris 1925a: 76). This statement is indicative of a post-Martin approach, elaborated by Martin's critics who extended the realm of anthropology beyond a concept that focused exclusively on the physical dimensions of humans (Proctor 1988). Instead, Martin's critics advocated anthropological sciences that encompassed culture and all expressions of human life, considering humans in their entirety. This criticism emerged among German scholars in the aftermath of World War I, when Martin's anthropology had begun to appear obsolete and out of step with the contemporary situation in the sciences, especially after the rapid progress in genetics. According to the historian of science Robert Proctor, by the 1930s, "many of Martin's younger colleagues were no longer satisfied with a science that confined itself to the measurement and description of human physical forms" (Proctor 1988: 142). Anthropological research tended now to shift "to how physical *and* cultural qualities together might be explained within the rubric of human genetics" (Proctor 1988: 155). This shift in anthropological thinking coincided with the end of a "sterile period" of anthropology, as Wilhelm Mühlmann

(1904–88) characterized the era prior to the collaboration of anthropology and genetics (Mühlmann 1986/1948: 110). In Germany this development found its most demonstrably institutional form with the founding of the Kaiser-Wilhelm-Institute for Anthropology, Human Heredity and Eugenics (Kaiser-Wilhelm-Institut für Anthropologie, menschliche Erblehre und Eugenik) in 1927 (Schmuhl 2005).

Despite reservations and criticism, Martin's approach continued to be influential insofar as it highlighted the mental (or 'psychic') dimensions of human ontology. Many of his critics embraced the distinction between 'psychic' and 'physical anthropology', but regarded them as two sub-fields of one and the same science. So did Ioannis Koumaris when he specified 'psychic anthropology' to include 'ethnology', 'psycho-sociology', 'folklore' ("or to be more exact, ethology"), 'religious studies', 'linguistics' and 'criminal anthropology'. Physical anthropology, again, was depicted as encompassing such disciplines as "zoological anthropology, anatomical anthropology and others" (Koumaris 1925a: 76).

Such a broad definition of anthropology was obviously a deviation from the proper subject of his chair, which was *physical* anthropology. His intention to use his professorship for advancing an extensive realm of anthropology became quite clear as soon as he outlined the relationship between medicine, anthropology and archaeology. In so doing, he essentially breaks programmatically with an era in which Greek physical anthropology originated in medicine and was primarily allied with archaeology. The chair-holder argued that the contribution of medicine to the emergence of anthropology was without doubt of great importance, especially with respect to offering significant cognitive and practical tools. However, anthropology overcame the narrow settings of medicine and "managed to prevail against it" while investigating man not as an individual but as a collective entity (Koumaris 1925b: 99).

The innovation in the subject does not mean, however, that the legacy of the first professional anthropologist, Clon Stéphanos, was disregarded. On the contrary, Koumaris acknowledged Stéphanos's bequest as a cornerstone for building further research on the Greek race/*fili*. He discerned a real challenge for the new generation of scholars to be trained at his chair in the task of filling the gap in the research on the Greece race, since, he argued, with the exception of the works of Stéphanos, "no other studies on this so important matter existed" (Koumaris 1925b: 103). After having acquired the necessary skills, future anthropologists would be able "to select somatic information on contemporary Greeks, but also on mores and traditions that have disappeared in the process of civilization"

(Koumaris 1925b: 104). Comparative investigations should be conducted into both Greek nationals and neighbouring peoples, and especially those residing in the border area who were of particular importance. The new scientists would be trained in conducting research not only on the 'dead', but also on 'living compatriots' who were:

> either members of isolated groups or populations living in border regions with other peoples, our current friends or enemies, or those who might have been once relative or strange to our *fili*. The future Greek anthropologists will look for elements of ancient civilization and preserve the remains of our remote ancestors (Koumaris 1925b: 104).

But, for all assertions to carry on Stéphanos's tradition by and large, a break up with Stéphanos's legacy had already started with Koumaris's directorship at the museum and this now deepened following the announcement of the programmatic goals of the university chair. The anthropological concept promulgated by the chair-holder at his inaugural address went beyond exploring collective entities, Greece in particular, using exclusively anthropological tools. Rather, it reflected a mainstream trend in the anthropological discourse of the inter-war period wherein (national) society was approached by means of disciplines such as demography (*population science*, in the sense of the German *Bevölkerungswissenschaft*), racial sciences and eugenics. The research interest was then shifted from tracing common origins and human history to interventions into current communities and the human species at large with the aim of enhancing their quality, staring into the future. This is the particular task of eugenics that was illustrated in Koumaris's inaugural address as a matter of practical implementation, targeting "the correction of human species by means of selection and liberating the world from superstition and the consequences of unreasonable traditions" (Koumaris 1925b: 100). It is clear that anthropology, defined in such terms, obtained new qualities that supplied it with topicality and a profound societal and national significance. Yet the chair-holder prioritized the practical implementation of anthropology as a modern (and 'disinterested') science in the service of society and nation. The social and national relevance that was ascribed to anthropology postulated its closeness to society and the diffusion of its tenets into the population. Thus anthropologists to be trained at the newly established chair were intended also to work towards popularizing anthropological knowledge and disseminating it beyond the limited sphere of academia. The Greek population should become familiar with anthropological tenets, and anthropology would elaborate concepts for society and provide its own models for dealing with social problems and developments.[2]

The inaugural lecture of the chair-holder for physical anthropology not only marked a watershed in the anthropological discourse in Greece, but also revealed a grandiose plan for the newly established discipline at the Greek university. The splendour of the programme lay in its conception of the anthropological sciences as an epistemic terrain to interlink theoretical knowledge, empirical research and educational training, and also to produce knowledge that would be useful for coping with topical societal issues. But its success was reliant upon its implementation in the specific academic environment of the University of Athens. The next chapter explores the trajectory of anthropology at the university chair, and argues that the anthropological studies there differed largely from the programmatic goals.

4.2 *The Implementation: Teaching the Restricted Scope of Physical Anthropology*

When the chair of anthropology was established, the rank of the anthropological discipline in the Faculty of Medicine was peripheral to other subjects since at the very outset it was an elective subject, and remained so for most of the period. It was also faced with the challenge of attracting a quite new student body, since, as the course calendars suggest, physical anthropology began to be taught at the University of Athens only after the chair for physical anthropology had been established (i.e. in the academic year 1925/6). Yet in contrast to the proclaimed broad scope of anthropology, which gives reasons to anticipate an appropriately multifaceted organization of the anthropological studies, the classes were limited to prehistory and human palaeontology throughout the entire 25-year period of the existence of the chair. In exact terms, under the rubric 'physical anthropology' there were twice-weekly lectures in "Elements of Human Palaeontology, Prehistory". But palaeontology was also taught at the Faculty of Natural

[2] Koumaris himself was striving to meet these goals, publishing a good deal in the popular press and in encyclopedias. He even wrote a regular column titled "Popular University" ["Λαϊκόν Πανεπιστήμιο"] for the conservative Athenian daily *Evening* [*Βραδυνή*] and published regularly in the liberal journal for popular science *Helios* [*Ἥλιος*]. See Koumaris's articles in the Athenian press on "Anthropology and Religion" (1932); and "Mendelism and Humanity" (1937). After 1940, he published numerous articles in *Helios* [*Ἥλιος*].

Sciences [φυσιογνωστικό][3] at which zoological and botanic palaeontology were taught by a professor for geology and another for palaeontology. These courses were addressed to students in the third and fourth year of studies.[4] (Unfortunately, the curricula, which could have provided significant insights into the contents of these classes, appear not to have been archived.) Information on the contents of the lessons are provided in Koumaris's autobiography, where he describes his classes as being designed from the very beginning to cover the "study of man in his entirety", that is, the consideration of both human 'nature' and the 'soul'[5] and even from the perspective of 'palaeoethnology' or 'prehistoric anthropology' and 'human palaeontology'; the former is mentioned as referring to "cultural traces left by prehistoric man" and the latter to different types of hominids such as Pithecanthropus, Telanthropus, and so on. As Koumaris himself remarks in his autobiography, at an "advanced stage" lectures addressed the "history of the earth", specifically "biogeny", the genesis of humankind and a "general overview of the zoological scale".

According to the course calendars, the curriculum included, in addition to the lectures, also practical courses, referred to initially as "Technical Part", later as "Tutorial" and from the academic year 1934/5 onwards as "Exercises".[6] In Koumaris's autobiography, the practical course was specified as practical exercises on some theoretical issues, such as how to verify paternity.[7] Furthermore, the tutorials included a recapitulation of the theoretical section and a demonstration of theoretical knowledge on models stored in the Anthropological Museum, while the lectures

---

[3] See the *Course Calendars of the University of Athens* for the academic years 1935/6 (Athens 1935: 70); 1936/7 (Athens 1936: 110); 1937/8 (Athens 1937: 129–30).

[4] The professor of geology, Theodoros Skoufos, taught palaeontology to fourth-year students from 1935 to 1938, while the professor for palaeontology Maximos Mitsopoulos taught third-year students in 1939/40 and again from 1950/1 onwards. Mitsopoulos's classes included theory four times a week and laboratory exercises three times weekly. Maximos Mitsopoulos taught geology; in the following academic year he again taught palaeontology.

See *Course Calendar of the University of Athens*, 1939/40 (Athens 1939: 77); *Course Calendar of the University of Athens*, 1948/9 (Athens 1948: 70–1). In 1949/50 *Course Calendar of the University of Athens*, 1950/51 (Athens 1950: 113).

[5] In the first volume of his autobiography Koumaris outlined the curriculum in approximately three pages with the title "Teaching of 'Anthropology' ". Koumaris (1951): 31–3.

[6] See the *Course Calendars of the University of Athens* from 1925/6 (Athens 1925: 56); 1926/7 (Athens 1926: 57) and 1934/5 (Athens 1934: 52–3).

[7] This is a matter addressed by Koumaris in several of his publications, for example, a series of articles entitled "The Search for Paternity" in the medical press in 1935–6; see Koumaris (1935).

addressed methodological issues and anthropological methods, especially body and craniometrical measurements, and 'general racial issues'.

The practical course is recorded in the course calendars as initially taking place on a daily basis. In the second year of the chair (1926/7), the syllabus mentioned only that the "hours would be specified", and in the subsequent period the tutorial hours were gradually reduced.[8] The reduction in teaching hours was probably dictated by a ministerial directive concerning compulsory and elective subjects. Indeed, a ministerial directive of 1933 with the title "Compulsory classes for the students of the Medicine Faculty"[9] prescribed three teaching hours a week for the elective subject of physical anthropology. The fact that the rank of anthropology within the Faculty of Medicine was not enhanced in the years following the establishment of the chair soon led to consequences which, in reverse, negatively influenced the rank of the subject and the chair at the faculty. The above-mentioned ministerial directive not only prescribed the reduction of the teaching hours of the elective subjects but also dictated that elective subjects should be addressed exclusively to first-year students. But, in practice, the ministerial regulation came to institutionalize a situation that was already a reality at the chair. In fact, physical anthropology had been addressed to students in their first academic year since 1929/30. Only in the short period from 1925 to 1929 and again in 1939/40 was anthropology offered for second-year students. In 1939/40 it appeared in the curriculum as a compulsory subject. This was the last university calendar published before Greece entered World War II. The next course calendar printed was for the academic year 1948/9. In the meantime the chair-holder had been appointed regular professor (in August 1944)[10] and, at least from 1948 up to 1950, he taught physical anthropology to students in the second academic year, but without changing the total teaching

[8] For the academic years 1928/9 to 1932/3 the tutorial was programmed for three hours a week, and from 1933/4 onwards was reduced to one hour weekly.

[9] See the *Official Governmental Gazette* [ΦΕΚ], no. 161, 24 June 1933: "On compulsory classes for students of the Medical Faculty", ["Περί υποχρεωτικών μαθημάτων φοιτητών Ιατρικής Σχολής", ΦΕΚ αρ. 161, 24 Ιουνίου 1933].

Cf. also the *Course Calendar of the University of Athens* from 1933/4 (Athens 1933: 94–5).

[10] In the first volume of his autobiography, Koumaris notes that his salary was raised to that of an ordinary professor on 6 July 1942. Koumaris (1951): 13.

In the course calendar for the academic year 1948/9 he was listed as having been an ordinary professor since 1944 (by virtue of the Governmental Decree from 12 August 1944, published in the *Official Governmental Gazette* on 29 August 1944).

hours. The tutorial was dropped in 1948/9 as well.[11] From that academic year on, the subject of anthropology was no longer specified in the course calendars. The chair-holder appears to have taught physical anthropology in the faculties of sciences, physics and mathematics as well, with his teaching hours amounting to a total of three hours weekly over three days.[12] The anthropology classes included 'laboratory exercises', addressed to fourth-year students.[13] These changes occurred in the very last period before the chair was abolished in 1950.

Regarding this decaying trajectory of the chair for physical anthropology it is not surprising that no anthropologist was trained to succeed the chair-holder. No doctoral or postdoctoral theses in physical anthropology were submitted during the long period of 25 years.[14] Thus, once the chair-holder for physical anthropology had retired, the medical faculty decided to abolish the chair "due to the lack of a successor".[15] Reflecting retrospectively on this disappointing outcome, Ioannis Koumaris explained the failure of the university chair on the grounds that the subject of physical anthropology had not been applicable in practice, which made it difficult to win the "souls of the students",[16] although the lectures were "positively received by the students and were well taught";[17] and "the scientific study of man in his entirety met with overwhelming interest".[18]

Because of the concentration of all positions for anthropology with one sole individual, not only was the teaching of physical anthropology now endangered, but the Anthropological Museum was also leaderless. A short-term solution was found, and following a one-year interruption (1950/1)[19]

---

[11] *Course Calendar of the University of Athens*, 1948/9 (Athens 1948: 48).

[12] It is not always possible to recreate the status of anthropology as elective or compulsory subject in the calendars or the curricula of the respective academic year because the system of registration for classes often changed. Thus they are sometimes listed as "elective" or "compulsory" subjects and sometimes on the basis of who taught them and the status of these teachers (as regular or irregular professors).

[13] *Course Calendars of the University of Athens*: 1948/9 (Athens 1948: 48 and 74); 1949/50 (Athens 1949: 50 and 87.

[14] This fact is corroborated in the Historical Archive of the Athenian University, *Proceedings of the Meetings of the Medicine Faculty*, in which doctoral and postdoctoral theses were discussed.

[15] Meeting of the Medical Faculty on 9 July 1950: Historical Archive of the Athenian University, *Proceedings of the Meetings of the Medical Faculty*, vol. 32: 221–6, 226.

[16] Koumaris (1951): 33.

[17] Koumaris (1951): 33.

[18] Koumaris(1951): 33.

[19] The subject of anthropology is included in the *Course Calendar* for 1950/1 but without reference to who would teach it. *Course Calendar of the University of Athens*, 1950/1 (Athens 1950: 67).

physical anthropology was again taught twice weekly at the medical faculty, by a histologist-embryologist, namely Themistoklis Sklavounos (1896–?), who was the son of the anatomist Georgios Sklavounos. Themistoklis Sklavounos also took charge of the Anthropological Museum,[20] a position he held until 1966/7.[21] However, neither his teaching of physical anthropology nor his position at the museum lasted very long. Themistoklis Sklavounos taught anthropology to second-year students at the medical faculty for two academic years only (1952/3 and 1953/4).[22] After that he appears to have taught only his own subject, namely histology-embryology. From 1952/3 onwards, physical anthropology was again a subject at the Faculty of Natural Sciences and Geography, and was taught twice weekly to the fourth-year students by a biologist.[23] It is ironic enough that genetics and biology, two disciplines which abetted all innovation in anthropology from the very first decades of the 20th century in terms of epistemic contents and closeness to social issues, were associated with the subject of anthropology only after the chair for physical anthropology was abolished. This outcome was intriguingly similar to the starting position of 1915, when there was no anthropologist by training either for occupying the chair, or the directorship of the Anthropological Museum.

The university chair for physical anthropology was founded in a period in which the appointed chair-holder was accumulating scientific capital as director of the Anthropological Museum and president of the Greek Anthropological Society, which he himself had founded one year prior to the chair's establishment. The latter anthropological agency, for all its particular character as a non-university institution, pursued a similar trajectory, starting auspiciously but decaying gradually; and, as the next chapter argues, this outcome was not independent from the failure of the university chair, although it was by no means directly determined by this.

20 *Course Calendar of the University of Athens*, 1950/1 (Athens 1950: 69).

21 *Course Calendar of the University of Athens*, 1967/8 (Athens 1967: 81); here the position was not occupied.

During the same period he became more actively involved in the Greek Anthropological Society, e.g. by chairing its meetings.

22 *Course Calendar of the University of Athens*, 1952/3 (Athens 1952: 67).

23 The subject was taught by Georgios Pantazis twice weekly in addition to once-weekly exercises: *Course Calendar of the University of Athens*, 1952/3 (Athens 1952: 109); *Course Calendar of the University of Athens*, 1954/5 (Athens 1954: 96).

CHAPTER FIVE

THE GREEK ANTHROPOLOGICAL SOCIETY

The founding of the Greek Anthropological Society was one of the most important steps in opening up the anthropological sciences to broader scientific and intellectual circles and in bringing the natural sciences and humanities together under one institutional roof. The GAS was founded at a time in which anthropological science in Greece was entering an innovative and promising phase. The profile and agency of the GAS should thus be considered in parallel to the trajectory and conception of anthropology at the university institutions (i.e. the Anthropological Museum and the university chair). The aim of this chapter is therefore to expound the anthropological discourse within the GAS in light of the developments at the museum and the university chair. The argument presented here is that the scope of anthropology was expanding at exactly the same time as interest in the discipline was diminishing at the university, an outcome that was due to the absence of a group of anthropologists in Greece who could demarcate the epistemic terrain of the subject and specify their professional claims in academia. In reconsidering the profile of the anthropological institutions with the retrospective knowledge of their failure, the research questions of this study can be reformulated so as to ask how we can speak of an 'anthropological scene' in Greece given the absence of professional anthropologists in the country. This question along with the profile and trajectory of the GAS will be examined in the following on the grounds of the membership of the GAS and its internal debates.

5.1 *Contours of a Trajectory: Promising Outset, Inglorious End*

Compared with the rest of Europe, the establishment of an anthropological association in Greece occurred relatively late,[1] and was consequently heralded as an event that would serve to bridge a gap in Greek academia. The start was promising and the GAS seemed to exert a pull on intellectuals

[1] For an account of the founding of anthropological associations in different countries, see Mühlmann (1986): 96.

and scholars. Indeed, right at its first meeting (held on 1 June 1924), the GAS counted 68 founding members, some of whom were reasonably prominent or even affiliated to state and educational institutions and at times exerted a strong influence upon politics. Quite illustrative of its initial appeal is the fact that the prime minister and sociologist Alexandros Papanastasiou (1876–1936) presided over its founding session.

The establishment of this anthropological association was undoubtedly due to the efforts of the then director of the Anthropological Museum (and one year later the holder of the university chair for physical anthropology) to create a scientific network, inside and outside Greece, in order to advance the anthropological discipline in Greek academia and to enhance his own profile within the discipline. Indicative of these efforts, from the very beginning of its founding the Greek Anthropological Society was associated with the International Institute of Anthropology in Paris, yet constituting the 'Office Héllènique de l'Institut International d'Anthropologie'.[2] In intramural terms, following the long tradition in academia whereby scientific associations and museums cooperated with each other, the GAS was intended to operate as a supplementary body to the Anthropological Museum. The complementarity between these two institutions was literally stipulated in the statute of the Greek Anthropological Society, which specified one of its goals as being to promote the Anthropological Museum and its collections.[3] On a symbolic level, this was expressed in the appointment of the GAS's 'Benefactors', which included the first director of the Anthropological Museum, Clon Stéphanos, and the Austrian speleologist Adalbert Markovits[4] who donated all of his collections to Greek anthropological institutions. There were also two 'Great Benefactors' who donated significant amounts of money to the GAS,[5] including the president of the organization who had provided a fund for the award of a prize, the so-called "Koumaris Award", for five studies

---

[2] *PGAS*, Founding Session and Statute (1 June 1924): 19.

[3] *PGAS*, Founding Session and Statute (1 June 1924): 21.
According to the statute, these two institutions would also share a library whose directorate was to be temporarily based in the building of the Anthropological Museum. *PGAS*, Founding Session and Statute (1 June 1924): 29.

[4] Markovits delivered several lectures for the GAS that were translated into Greek. See Markovits, *PGAS* (24 April 1928: 45–61); (3 December 1929: 114–34); (29 November 1930: 44–50); (30 December 1931: 115–24).

[5] From the session of 12 March 1965 Giorgios Foteinos was also added to the Great Benefactors after confirming he had donated an amount to the anthropological society in his will. See *PGAS* (12 March 1965): 9.

on the Greek race in 1949.[6] A more customary practice was to grant the status of honorary membership of the GAS to institutions or individual scholars who supported the Anthropological Museum and anthropological science in Greece. The list of honorary members mostly included representatives of foreign scientific institutions who had donated objects to the Anthropological Museum.[7] Others were appointed honorary members for their contribution to the promotion of the anthropological sciences in Greece. These included the French Institute in Athens (Institut Français d'Athènes, which was granted honorary member status in 1947), Prince Peter of Greece and Denmark (1908–80) and the Soviet anthropologist and director of the anthropological department of the Academy of Sciences in Moscow, Georgy Frantsevich Debetz (or Georgii Frantsevich Debets, 1905–69), who was appointed in 1965.[8] The prominent foreign members of the GAS included the contentious German anthropologist Ilse Schwidetzky (1907–97), who joined the association in 1937 on Koumaris's recommendation while she was visiting the Anthropological Museum in Athens.[9]

Yet the complementarity of the three anthropological institutions did not occur solely on the basis of the exchange of material resources, as the three organizations also exerted a mutual influence on their respective symbolic capital. At least at the beginning, the university chair for physical anthropology, after its establishment in 1925, vested the anthropological discipline and its representative(s) in Greece with scientific capital and prestige. But despite its auspicious beginnings, the appeal of

---

6 *PGAS* (1 June 1949): 12–14.

7 One honorary member, for example, was the Greek Archaeological Society; but the list also included August Rutot, the director of the Museum for Natural History in Brussels (who dispensed a collection of prehistoric objects to the Anthropological Museum), George Papillaut (Paris), André Forster (Strasbourg), Georg Kyrle (Vienna), Carl M. Fuerst (Lund, Sweden), Salomon Reinach (Museum St Germain en Laye, France).

The honorary members generally had no further involvement in the works of the Greek Anthropological Society.

8 Debetz (or Debets) is mentioned in the proceedings of the anthropological society as a "distinctive scientist and friend of Greece who in 1936 studied the cranium of Asini and researched the Aegean civilization, proving its autochthonous character by means of the anthropological examination of skulls". *PGAS* (12 March 1965): 10.

Asini is a village in the Argolida prefecture and carries the name of a Homeric town. The ruins of the acropolis of the ancient town are located on a steep coastal hill outside the town of Tolo.

On Debetz, see Kozintsev (1997).

9 *PGAS* (7 April 1937): 2.

On Schwidezky, see Lüddecke (2000).

the GAS increasingly declined and ultimately resulted in its collapse. This downward trend became noticeable in the aftermath of World War II and especially after the abolition of the university chair in 1950. Although the reasons for this decay were manifold, the decline in the university capital of anthropology and its representatives served to contribute greatly to its downfall. The descending course of the GAS is manifested in developments in its internal organization, the trends in new membership and the frequency of the meetings.

Illustrative of the auspicious outset of the Greek Anthropological Society is the fact that its founding session took place in the rooms of the Association of Political and Social Studies at which Alexandros Papanastasiou held a leading position.[10] In the subsequent period the GAS shared the same residential problem as the Anthropological Museum—as it had never had its own premises, its meetings were held in the buildings of other scientific institutions.[11] The majority of these sessions were held either in the Archaeological Society's building or in the medical faculty at the University of Athens. Occasionally they took place in the Academy of Athens and at least once (on 27 January 1932) at the Philological Club 'Parnassus'. Generally speaking, the GAS followed the Anthropological Museum in moving from one building to another, and, after the medical faculty took up its abode in the new university building in the 1930s, the GAS for the most part met there. However, at the founding session, Papanastasiou offered the rooms of the Association of Political and Social Studies to host the GAS permanently, which was intended to provide the basis for future cooperation between the two agencies. But in practice the sociological association never hosted the GAS, nor has any other collaboration occurred between the two scientific institutions. It would also appear that Papanastasiou himself never cooperated with the anthropological agency despite his encouraging speech on the subject.

Seen from a retrospective point of view, this tendency to maintain a degree of distance between the two institutions appears to be congruent with a visible trend among prominent members of the GAS. At the outset, the membership of the GAS was quite varied in terms of scientific background, political attitudes and involvement in state politics,

---

[10] In 1907 Alexandros Papanastasiou had founded the Sociological Association [Κοινωνιολογική Εταιρεία], based on the model of the German Kathedersozialisten. In 1916 it was renamed the Association of Political and Social Studies [Εταιρεία των Κοινωνικών και Πολιτικών Σπουδών].

[11] The proceedings always recorded the location of each meeting.

although its composition remained predominantly masculine throughout its existence.[12] A consideration of the profile of some its founding members is illustrative of the opening of anthropology to a broad scholarly community and also of the prestige carried by the young association. Founding members included prestigious personalities such as the diplomat, politician and intellectual Filippos Dragoumis (1890–1980); the first professor of hygiene and microbiology at the University of Athens and for a period the doctor of the royal family, Konstantinos Savvas (1861–1929), who also played an instrumental role in shaping state health policy in Greece; the well-known anatomist Georgios Sklavounos;[13] the politician, professor of medicine and temporary dean of the University of Athens, Konstantinos Louros (1864–1957); the Byzantinist Konstantinos Amantos (1870–1960);[14] Faidon Koukoules (1881–1956);[15] one of the most important Greek linguists, Nikolaos P. Andriotis (1906–76);[16] the jurist and professor of criminology at the University of Athens, Konstantinos Gardikas (1896–1984); two of the most prominent Greek folklorists, namely Simos Menardos (1872–1933) and Stilpon Kiriakidis (1887–1964), and many others besides. Among the members who joined the Greek Anthropological Society in the late 1960s were the two most important anthropologists by training in Greece after 1970, namely Aris Poulanos (*1924) and Nikolaos Xirotiris. The latter is currently professor of physical anthropology at the Department of History and Ethnology in the Democritus University of Thrace. Aris Poulianos[17] founded a new anthropological institution, the

[12] The GAS was a predominantly masculine association from the beginning and this remained the case throughout the entire period considered in the current study. From 1924 to 1962, only five women joined the association. In 1965, the year with the most new members, only five women were included among the 59 new members; a year later, in 1966, another woman joined the society. The absence of female members certainly coincides with the low representation of women at that time in the scientific professions in Greece, in general, and in the natural sciences and medicine in particular (Lada 2007).

[13] Sklavounos was also among the most active members and appeared to have chaired most of the meetings of the society until his death.

[14] Konstantinos Amantos delivered a talk at the session held on 22 February 1926 on "Slavs and Slavophones in the Hellenic Lands": see *PGAS* (22 February 1926): 10–31.

[15] Koukoules was mentioned above as the secretary of the Historical and Ethnological Society. He was in effect an active member and delivered a talk at the anthropological society: Koukoules, *PGAS* (12 April 1927): 59–68.

[16] See a talk delivered by Andriotis in *PGAS* (12 March 1929): 54–63.

[17] After studying biology in the United States (1942–8), Aris Poulianos completed his postgraduate studies in anthropology in the Soviet Union where he earned his PhD in anthropology. In 1964 Poulianos left the Soviet Union and returned to Greece.

Anthropological Association of Greece [Ανθρωπολογική Εταιρεία Ελλάδος],[18] which still exists today, immediately after Koumaris's death in 1970. However, Poulianos's activities did not exceed the frames of his association and he has never obtained an academic position.

Within the GAS, there was a general tendency among prominent members to take up nominal membership, rather than actively participating in its activities. At the same time, many nominal members who were distinguished scholars in science or politics[19] were not vocal at the GAS but active in other contexts in dealing with issues that overlapped with those of the anthropological discourse. This was very much the case with scholars who shaped the discourses of population politics and eugenics outside of the anthropological association. This category includes leading persons in the Greek eugenic movement whose activities will be explored in more detail below, such as Moisis M. Moiseidis (1880–?); Nikolaos Drakoulidis (1900–85); and the gynaecologist Nikolaos Louros (1898–1986) who founded the Greek Eugenics Society in the aftermath of World War II. Others, such as the economist Kiriakos Varvaresos (1884–1957), contributed to hindering the establishment of eugenic measures. Also uninvolved in the works of the GAS, despite being a member of the organization, was the leftist pedagogue and intellectual Dimitris Glinos (1882–1943), a declared opponent of racism who ultimately paid the price of exile for his leftist beliefs (in 1935 and again in 1936–40). This is indicative of the initial appeal of the GAS to scholars and their subsequent dissociation from its activities once its profile had begun to take shape.

In quantitative terms, the course of the GAS can be discerned in the trend of the new membership annually, with a parallel consideration of the frequency in which its meetings took place.[20] The highest figures for new members were recorded in the first years after it was founded, but from 1932 onwards the numbers began to decline (see Appendix). Inter-

---

[18] See the official website of this association: http://www.aee.gr/english/1contents/contents.html (last accessed October 2012).

In 1974 the association started publishing the journal *Anthropus* [*Άνθρωπος*] and later the series "Library of the Anthropological Association of Greece" [*Βιβλιοθήκη της Ανθρωπολογικής Εταιρείας Ελλάδος*].

[19] The extent to which passive members supported the anthropological society in other ways, such as through financial or other means, cannot be discerned from the proceedings (*PGAS*). According to the society's Statute, all members paid a membership fee or were excluded for not doing so. However, the *PGAS* do not allude to any resignations or expulsions from the society for this or any other reason.

[20] The *PGAS* also contain detailed lists of the members, recorded by name, profession, the date they joined the society and their status as either honorary or ordinary members.

estingly, the upward trend was repeated (though to a lesser degree) in 1941–4—the period in which the country was occupied by the Axis forces. The lowest figures were recorded in the aftermath of World War II. The downward trend continued and even worsened sharply during the 1950s and early 1960s. But an upswing can be observed in 1965, when 59 members joined the GAS. Although this figure was unique, in subsequent years the numbers of new members remained unusually high, at least in comparison with the stagnation of the 1950s. This trend was especially remarkable as the GAS did not meet regularly, if at all, during that period. Thus the quantitative upswing of the 1960s did not coincide with a recovery in the work of the GAS; nor did these high numbers recur.

Indeed, according to the *PGAS*, until the end of the 1930s the Greek Anthropological Society used to meet twice annually, and in some cases as often as three times in a single year. Yet from the end of the 1940s onwards the meetings became increasingly irregular, and after 1950 they took place at intervals of several years. After a break of almost three years (starting from 1949) the GAS met again on 18 April 1951, but it now had a new board: the embryologist Themistoklis Sklavounos (who had temporarily succeeded the retired Ioannis Koumaris at the directorship of the Anthropological Museum and who also taught his subject at the Faculty of Medicine), had filled the position of general secretary of the GAS, although Koumaris remained the emblematic lifelong president. The next time the anthropological society met was five years later, on 29 November 1956.

The minutes of the society continued to be published even in the two decades when the GAS either met irregularly or was no longer convening meetings. The 1964 edition of the *PGAS* was dedicated to the 40th anniversary of the founding of GAS, but its tone was far from celebratory. It included a brief retrospective account of the circumstances under which the society had been founded and how it worked, along with a brief presentation of the most important statements and theses submitted by its president.[21] The following editions reproduced papers written on other occasions; lectures that Prince Peter delivered in the 1960s at various faculties at the University of Athens, for instance, or papers from the natural sciences or the humanities collected from scholars. This occurred in 1965–7, the very period during which the number of new members reached its highest point and was effectively the 'swan song' of the Greek

21 *PGAS* (1964): 14.

Anthropological Society—the next volume of the *PGAS* was also the final one: it was published in 1970 and consisted of nothing more than a farewell by the 91-year-old president[22] as well as a statement he had made on the occasion of an international congress held in France in 1969[23] as to whether the Neanderthal specimen was indeed an ancestor of *Homo sapiens*.

The internal debates of the GAS provide evidence of the society's profile, the scope of anthropology as designed in this anthropological agency and allusions to the possible reasons for its decaying trajectory.

5.2 *The Internal Debates of the Greek Anthropological Society*

What did scholars with diverse scientific backgrounds understand by anthropology, other than that it was a science that claimed, and was expected to offer, scientifically grounded answers to topical questions on humans and human life? Is this possible perception, and expectation, a sufficient condition for affiliating scholars to an *anthropological* network? What constituted the anthropological element in this network in the long term, especially when no anthropologists by training had emerged in Greece? That a shared notion of anthropology was not the common denominator and the starting point of the GAS is clear from the celebratory address by the chairman, Alexandros Papanastasiou, and the subsequent discussion. Papanasrtasiou voiced his conviction that the founding of the Greek Anthropological Society was important not only in scientific but also in national terms. One of its main tasks was expected to be "ascertaining the various racial elements that constitute the modern Greek nation".[24] Papanastasiou drew his understanding of anthropology from Felix von Luschan's lectures, which he claimed to have attended "enthusiastically" while studying in Berlin, and "always wishing the issue of the Greek *fili*/race to be approached applying the same anthropological theories".[25] He anticipated that the newly founded anthropological society would conduct systematic anthropological studies so as to provide scientific evidence for the racial [*filetic*] continuity of the Greeks, "who,

[22] Koumaris, *PGAS* (7 January 1970a): 14–15.
[23] Koumaris, *PGAS* (7 January 1970b): 9–13.
[24] See the opening lecture by Alexandros Papanastasiou in the minutes of the founding session: *PGAS*, Founding Session and Statute (1 June 1924): 5–6.
[25] *PGAS*, Founding Session and Statute (1 June 1924): 5–6.

though they had always mixed with other peoples had never lost their consistency".[26] This perception is entirely comprehensible if it is borne in mind that there was no other anthropological tradition than that derived from the work of Clon Stéphanos, which was quite close to Luschan's, in Greece at that time. The founding session of the GAS was the first occasion at which the museum's director, and from then on the president of the GAS, sketched out his own understanding of anthropology before the scholarly community;[27] and this was the same all-inclusive concept that he outlined one year later in his inaugural lecture, when he acceded to the university chair. Papanastasiou, perturbed by such a broad definition of anthropology, suggested that "the Greek Anthropological Society had better to concentrate on its own field".[28] Yet Papanastasiou's expectations may well have been answered in the subsequent long-lasting period of the GAS's existence: racial issues were indeed becoming more and more relevant to its internal debates and the organization eventually became a quite important scholarly forum in Greece for presenting, and actually 'normalizing', racial (anthropological) ideas, especially with respect to the Greek race; these racial concepts differed significantly, however, from Luschan's approach.

Given that, on the one hand, all of the intellectuals and scholars who joined the anthropological society had different scientific backgrounds and, on the other, that no anthropologists were trained at the university chair, the question again recurs as to what could serve as the common denominator for the GAS's members and to how the scope of anthropology could be defined at this association over the longer term.

To begin with, the Greek Anthropological Society was clearly not a scientific institution for *producing* knowledge, which made it distinct from other European anthropological associations (cf. Zimmermann 2001). Instead, it emerged to become a forum for *presenting* expertise derived from diverse research contexts that often moved around the interface of medicine, the natural sciences and humanities. The gathering of such a plethora of scientific approaches under an anthropological roof was facilitated—and legitimized—by the then widespread understanding of anthropology as a holistic science. Still, the common anthropological denominator for their intersection should be created.

---

26 *PGAS*, Founding Session and Statute (1 June 1924): 6.

27 *PGAS*, Founding Session and Statute (1 June 1924): 7–15. Cf. Mühlmann (1986/1948: 162); Lüddecke (2000): 25–49.

28 *PGAS*, Founding Session and Statute (1 June 1924): 15.

In the following, I will trace the concepts of anthropology developed by the GAS, as outlined in the internal proceedings of this society over the course of half a century (from its founding year in 1924 up to 1970). Such an undertaking reveals, first, a number of shifts in the internal debates with respect to both the topics discussed and the way in which they were linked with anthropological tenets. While in the early years medical topics prevailed, even if these were presented in a rather descriptive form, from the 1930s onwards questions of heredity and race were increasingly coming to play a predominant role, particularly from the 1940s onwards. Over the course of time, race and racial heredity grew to become the subjects *par excellence* of the anthropological discourse, and hence will be addressed in detail in a separate section of this book. Heredity was interpreted and treated against different backgrounds. Clearly, the attraction of heredity issues reflected international trends related to both contemporary advances in science (in particular biology and genetics) and the ideological uses of biology as a mode for interpreting social phenomena. The latter was instrumental to considerations of human behaviour as being determined by inherited biological and genetic factors and went hand in hand with approaches based on the quantification of humans. On the grounds of the internal debates of the GAS, I will argue that scholars with different scientific backgrounds were connected under a common anthropological denominator by means of diverse concepts of human heredity correlating either to social deviance or to (congenital) bodily divergence from the ideal of human normality; all of these concepts were in turn based on the quantification of human characteristics. Yet the accommodation of diverse topics and approaches within the anthropological frames was facilitated by the fact that, over the course of time, the scope of anthropology elaborated in the GAS was broadening as the failure of the anthropological institutions was becoming evident. Finally, anthropology came to be treated not only in a broad sense, nor even as an all-inclusive science, but as a fetish. I will argue that this levelling worked ambivalently; on the one hand it became a nexus for scientists from different disciplines, but on the other it caused the alienation of others, and only in exceptional cases did it become subject to criticism. This disappointing outcome prompted the retired Ioannis Koumaris to embrace social anthropology a fortiori when this discipline was epitomized in the person of a noble anthropologist such as Prince Peter. In the final section, I will suggest a close connection between the declining influence of the GAS over the scholarly community and its political profile, the latter of which came to light for all the proclaimed neutrality of the GAS's president. The focus on the 1940s (i.e. during Greece's involvement in World War II and the Civil War), provides a number of especially revealing insights into the question of the 'neutrality of science' and the political profile of the GAS.

5.3 *Quantification of Humans: Criminals, Sound Children and Intelligence Tests*

The integration of statistics into sciences that study humans and their social, natural and biological conditions of life was a corollary of the 'fascination for figures' that has corroborated the myth of the objectivity of scientific knowledge.[29] Scientists sought to codify biological and physical markers of difference into statistical schemata by measuring mental and physical abilities or by detecting regularities in human behaviour and correlating them (either directly or by allusion) to social conditions of life. Measurements in this sense mean assessments of human ability and behaviour according to pre-assigned norms of order and disorder. Such quantified records were intended to be applied in the organization of society; they were basically judgements on the value of humans, wherein the value system corresponded to the rational organization of society.

The meetings of the Greek Anthropological Society became a forum for the presentation of several projects in the applied sciences that observed, measured and statistically recorded human constitution, behaviour and abilities in order to explore their subjects from a criminological, medical or psychological perspective. The contributors were usually scholars involved in Greek state and educational institutions in which some of them even held leading positions, as for instance the jurist and professor for criminology Konstantinos G. Gardikas (1896–1984). From 1933 to 1936 Gardikas delivered four lectures at the Greek Anthropological Society on topics relating to fingerprints, profession and criminality, the frequency of crimes per Greek region and the influence of time, season and climate upon criminality in Greece.[30] Two of these presentations were based on works published in 1934 in a prominent mainstream German periodical for criminology (Gardikas 1934a and 1934b). In addition, one of his female

29 On the development of statistics from the 19th century onwards, see Porter (1986) and (2003). Especially for quantification of humans, see Gould (1981).

30 Gardikas, *PGAS* (28 December 1933a): 28–33; *PGAS* (28 December 1933b): 33–5; *PGAS* (7 December 1934): 36–47; *PGAS* (23 May 1936): 21–41.

students presented her research on crimes of filicide in Greece.[31] All these papers were descriptive in character and demonstrated statistical correlations of criminality to environmental, biological and social factors. The approaches pertained to models of measurement and the classification of human behaviour, following a mainstream trend in criminology during that period. However, these papers did not establish explanatory patterns for criminality; nor do they suggest measures for the treatment of criminals. Classificatory patterns and the conclusions from such measurements were usable (and indeed were applied in varying degrees) for introducing, or reinforcing, norms of assessing human behaviour. This usage is consistent with what Ricardo Salvatore (2005) has termed "positivist criminology" in reference to the mathematization of human behaviour by means of observation and statistical records. Conclusions reached by this means were utilized as the basis for designing and reforming disciplinary institutions, and particularly the correctional system. Criminologists either placed their conclusions at the disposal of state institutions, or immediately became involved with such institutions themselves. Apart from his academic position as professor at the University of Athens (from 1930 to 1969), Konstantinos G. Gardikas was also involved intermittently in state administration under liberal governments, particularly in the Ministry of the Interior, and was in charge of organizing the police and the correctional system.[32]

Paradigms of paedology and psychology as applied sciences were discussed at the GAS with a pointed emphasis on their implications for society. Greek children and their somatic and mental constitution were of particular interest in the presented projects, as research on the physical and mental constitution of Greek children accompanied efforts to modernize the educational and care system for children (or to accommodate demands for increasing such care) from the first decades of the 20th century. In this period, institutions throughout the country were faced with the precarious and worsening situation of large parts of the population due to the proceeding processes of industrialization and urbanization. The contributors before the GAS included[33] none other than the paedia-

---

[31] Mazaraki, *PGAS* (1 June 1940): 8–23.

[32] The autobiography of his cousin, the doctor Konstantinos D. Gardikas, includes biographical information about him. See Konstantinos D. Gardikas (2002/1985: 409–19).

[33] Apart from these scholars, there were other, less prominent members of the Greek Anthropological Society who brought up the subject of the physical constitution and health of children and discussed their observations on the physical growth and the school

trician and intermittent head of the School Medical Service of the Ministry of Education, Emmanouil Lampadarios (1885–1942); the professor for experimental pedagogy at the University of Athens Nikolaos Exarchopoulos (1874–1960); and the psychologist and director of the Laboratory for Experimental Psychology, Georgios Sakellariou (1888–1964).[34]

Paedology was established in Greece at the start of the 20th century and is primarily associated with the work of Emmanouil Lampadarios. Generally, paedology emerged at the end of the 19th century, first in the United States and very soon after in Western Europe and Russia/the Soviet Union (Depaepe 1992), as a branch of medicine that deals with the study of children's behaviour and development. The realization of the first international congress on this subject, held in Brussels in 1911, indicates the rapidly increasing interest in this branch of science and the expanding network of its experts. Paedology was empirical in terms of collecting data; it was an applied science since its conclusions were predetermined for immediate application in practice. Experts in paedology explored the physical and mental constitution of children under laboratory conditions, and shaped categories in which to classify children according to their evaluated behaviour and abilities. Lampadarios was the first to present before the Greek Anthropological Society the conclusions from his anthropometric measurements, conducted with school children in order to ascertain their physical growth.[35] He was anxious to promote paedology and auxology (or auxanology) studies (i.e. the study of all aspects of human growth). In the period 1925–6 he directed the Paedological Institute of Athens, which had been established in 1911[36] as a department integrated into the School Hygiene Service in the Ministry of Education (Kazolea-Tavoulari 2002: 113). The establishment of this institute was one of the first attempts made by Greek state institutions to end the spread of childhood diseases and other generally transmissible diseases, which in

achievements of Greek children. These included the dentist Alexandros Krikos and the paediatrician Dimitrios Mpezos.

Krikos, *PGAS* (29 December 1940): 47–53.

Mpezos, *PGAS* (8 October 1946): 39–41.

Mpezos, *PGAS* (12 November 1967): 13.

[34] The research activities of Exarchopoulos and Lampadarios pertained to the realm of anthropometrics and 'racial-anatomy', according to Koumaris (1937). Likewise, Konstantinos Gardikas was presented by Koumaris as conducting 'criminal anthropological research'.

[35] Lampadarios, *PGAS* (21 February 1928): 19–37.

[36] The Paedological Institute was founded by a Royal Decree from 26 September 1911: *Governmental Gazette* [ΦΕΚ] No. 279, 5 October 1911.

some parts of Greece had claimed numerous victims among schoolchildren (Karakatsani and Theodorou 2011). The most important step taken towards the advancement of paedology in Greece was the measuring of pupils' bodies, carried out by school doctors using special tools such as cephalometers, thoracometers, stadiometers and ergographs. Lampadarios could even be described as a pioneer in exploring the psycho-somatic development of children—he advocated psychometrical methods in paedology, underlined the necessity of promoting paedological laboratories and asserted the need to introduce the conclusions of such research into teacher education and training (Lampadarios 1940).

Methodological questions and empirical methods of conducting research were a central concern among the leading scholars in paedodolgy. In one of his last papers (in 1941, after his retirement from the civil service) Lampadarios discussed the choice of credible anthropometric and biometrical indexes in order to distinguish individuals (particularly schoolchildren) of different "races or nations".[37] On this occasion he criticized some "dubious indexes" in anthropometrics and in the biometrical measurements that had been deployed in the past by health insurance agencies, the military and other institutions in order to collect information on the sound constitution of individuals and the normal function of their systems. Nikolaos Exarchopoulos, on the other hand, developed standardized tests to be applied by schools, and conducted measurements of children[38] with his research team at the Laboratory of Experimental Pedagogy and the Experimental School at the University of Athens (founded in 1929)[39] where he held a leading position. Exarchopoulos was concerned with examining the somatic development of Greek children compared with those of Northern Europe. He and a female member of his team (Gedeon 1930) published their conclusions in Greece and in Germany.[40] Access to their sample was easy due to Exarchopoulos's involvement in

[37] Lampadarios, *PGAS* (27 December 1941): 40–2.
In this paper he outlined the results of measurements he had conducted in Athenian schools. His sample included 800 pupils whom he classified into three categories according to their constitution: those of thin, of middling or mixed and of corpulent constitution.

[38] At the GAS he presented some conclusions from the research he had conducted between 1924 and 1938: Exarchopoulos, *PGAS* (22 April 1941): 5–10.

[39] Law No. 4376, *Governmental Gazette* [ΦΕΚ] No. 300, 21 April 1929. See more on Exarchopoulos in Kazolea-Tavoulari (2002): 130 and *passim*.

[40] See, for instance, Exarchopoulos (1928); Gedeon (1935); Striftou (1935). See also a large catalogue with publications of the Paedagogic Institute of the Athenian University as well as publications by other scientists in similar topics in *Zeitschrift für Jugendkunde* 5 (1935): 132–4.

educational institutions and his headship of the School of Secondary Education [Διδασκαλείο Μέσης Εκπαιδεύσεως] that had been founded in 1910 to educate teachers. They drew their sample from schools and nursery schools, as well as from private and state maternity clinics. They sampled 7,000 individuals, including newborn babies, infants, teenagers and young people up to the age of 21. Children with 'physical defects' were excluded since, like Lampadarios, Exarchopoulos preferred to take his sample from among 'physically sound' individuals.

In contrast to Lampadarios and Exarchopoulos, the psychologist Georgios Sakellariou additionally considered children with 'defects', notably delayed progress at school, in exploring the mental abilities of children. Sakellariou was a leading figure in establishing experimental psychology in Greece. He became professor of pedagogics at the University of Thessaloniki in 1926, one year after the university had been founded, and in 1935 he established and directed the psychological and pedagogical laboratory which included a department for paedology. From 1940 onwards, he was professor and director of the experimental laboratory at the university in Athens.[41] In 1950 he taught for a year at the Western Reserve University in Cleveland, Ohio.[42]

Sakellariou believed that heredity was responsible for the 'intelligence deficiencies' he had observed, and that the vehicle for the transmission of these deficiencies was of a social nature. He claims to have observed that families of children with alleged 'mental deficiencies' featured pathological social behaviours, and specifically, that 30 per cent of the parents were alcoholics, drug abusers, suffered from syphilis or were "people of the lowest level".[43] His line of argument kept pace with the environmentalist interpretations of intelligence that had increasingly penetrated the humanities from the 1920s onwards (cf. Carson 2003). In one of the papers he delivered at the Greek Anthropological Society, Sakellariou outlined the conclusions of research conducted in 1939–40 by his team in cooperation with school inspectors.[44] The sample was taken from marginal urban districts and from "the lowest social classes". He concluded that 92 per cent of the children tested had low intellectual capability and the rest

41 Results of his research were published in the journal *Psychological Abstracts*, the monthly periodical of The American Psychological Association. See, for instance, Sakellariou (1939); (1940a); (1940b).

42 According to Kazolea-Tavoulari (2002): 108–10. See also Tzavaras (2003).

43 Sakellariou, *PGAS* (1 May 1942): 2–6.

44 The conclusions of his research were published in a Greek series titled "Minutes of the Psychological Laboratory" starting in 1937: Sakellariou (1941) and (1946).

were merely of average intelligence. Sakellariou's research was conceptualized in a way that would allow it to be applied in practice. The objective of his research was to formulate propositions for "therapy for problematic children" with delayed school progress.[45] This, however, was only one part of a large-scale research project that also considered gifted children with exceptional and high intelligence quotients (IQ) (Kazolea-Tavoulari 2002: 111). The data he collected and analysed was intended to form the basis for reforming the educational system in that children would start school according to their 'intellectual age' rather than to their physical age. He suggested that intelligence tests should be established at all levels of education, and that they should also be used for enrolling students at university and even when choosing professions (Kazolea-Tavoulari 2002: 112).

Education and choice of profession were central concerns in Sakellariou's work, and his psychological laboratory appropriately included a Department of Occupational Guidance. He was convinced that vocational guidance should be based on measurements of intelligence from very early infancy, and in order to realize his project he conducted tests with children in schools and orphanages. However, his activities were not exhausted in tests with children, since he also carried out intelligence tests among the Greek military services. He himself claimed to have maintained a temporary office as consultant at the General Military Headquarters [Γενικό Επιτελείο Στρατού] from 1941 to 1946 (Sakellariou 1946). This assertion in his writings is particularly remarkable as during that period Greece was occupied by the Axis powers and, after liberation, the Civil War was unfolding throughout the country.

The assessment of women's capabilities also became a subject for Sakellariou's research. A large project on "Woman and Research" that was integrated into his Department of Occupational Guidance conducted research that was designed to assess the intelligence and intellectual capabilities of women in order to develop a guide for choosing the optimal profession.[46] On the basis of psychological experiments, Sakellariou claimed to have ascertained that men could analyse and delve into a subject better than women, although women were more precise than men in expressing their thoughts. Using the environmentalist approach, he explained this conclusion on the grounds that there were disparities in both the nervous

---

[45] Sakellariou, *PGAS* (1 May 1942): 2–6.

[46] Some results of this project were presented at the Greek Anthropological Society in a lecture entitled "Woman and Scientific Research".

Sakellariou, *PGAS* (12 December 1943): 80–5.

systems and the social roles of men and women. And, availing himself of rather popular language and drawing his explanation patterns on common sense, he claimed that men faced various challenges in life and had therefore been trained to manage them successfully, with the aid of an appropriately trained nervous system. Women, on the other hand, were distinguished by their exactness because of their role in caring for the cleanliness and aesthetics of the household and of clothing. In concluding, Sakellariou stated that men were allegedly superior to women in terms of scientific ability. In practical terms, women's 'deficiencies' had to be considered with regard to choice of profession. How paradoxical and out of touch with reality his assertions were become apparent when considering the fact that several women submitted their doctoral theses to Sakellariou's university department, or even collaborated with him. A significant number of members of the research board of the psychology laboratory were female,[47] and Sakellariou's close associate, Sofia Paraskeva-Sakka, following postgraduate work in the United States, went on to develop an outstanding career in Greece (Kazolea-Tavoulari 2002: 116–17).

The validity of intelligence tests as a means to "differentiate safely between what is due to innate capacity and what is the result of environmental influences, training and education" was challenged by the UNESCO's *First Statement on Race* (1950).[48] However, despite all the criticism that these tests have faced, they have yet to be entirely discredited. Studies of IQ and its implementation experienced a revival in the late 1960s when the American psychologist Arthur R. Jensen (1923–2012) stirred up the scientific scene with his contested article "How Much Can We Boost I.Q. and Scholastic Achievement?" (Jensen 1969). One of his critics was the geologist, palaeontologist and evolution researcher Stephen Jay Gould (1941–2002) who criticized intelligence tests generally, and Jensen's theory in particular as "a 'more direct' path [than the craniometric arguments] to the same invalid goal of ranking groups by mental worth" (Gould 1981: 108; cf. Jensen 1982).

With this criticism Gould highlighted the fact that the quantification of humans (which implies a totalitarian conception of human beings based

---

47 See for instance the monographs by Gkarmati-Theodoropoulou (1952) and Paraskeva (1954).
Kazolea-Tavoulari's monograph on the history of psychology in Greece provides information on the female collaborators in the Laboratory of Psychology: Kazolea-Tavoulari (2002): 107f.

48 Paragraph 9 of UNESCO's *First Statement on Race* (1950); also cited and commended in Montagu (1972): 81f.

on statistical and classification systems) in order to describe human behaviour by scientifically objective means actually results in the evaluation of human abilities according to established social norms. The measured traits (either somatic or mental) condition the positions that individuals (should) arrive at in their lives, the course of which are perceived as determined by hereditary factors imprinted in familial characteristics and which become apparent in social behaviours.

5.4 *Physical Anomaly and Heredity As Nexus to Anthropological Discourse*

> An anomaly is a fact of individual variation which prevents two beings from being able to take the place of each other completely. It illustrates the Leibnizean principle of indiscernibles in the biological order. But diversity is not disease; the *anomalous* is not the pathological. Pathological implies *pathos*, the direct and concrete feeling of suffering and impotence, the feeling of life gone wrong.
>
> (Canguilhem 2007: 137)

Other than ascertaining hereditary preconditions in manifest social behaviours and in somatic or psychic traits, several medical projects presented at the GAS addressed congenital somatic deviance under the categories of anomaly or pathology. As a rule, such projects drew on their contemporary stage of genetics and, for the purposes of the presentation, they sought to find a link with anthropological tenets by using the notion of heredity as a vehicle, either only nominally or in relation to the overall concepts of congenital anomaly frequently associated with social conditions of life. This link was justified on the grounds that once the appeal of genetics for anthropology had begun to grow, from the initial decades of the 20th century on, increasing numbers of anthropologists and anthropological schools worldwide were inclined to adjust their methods by recognizing blood and genetic codes as the main ways in which heredity was transmitted.[49] Anthropology gradually expanded its domain to the biosciences, embracing biology as the basis for the anthropological sciences. But when appropriate anthropological tools were absent, as occurred in

[49] On the attraction of genetics to anthropology see Leaf (1974) and Spencer (1997), 1st volume: 1073–5 and 2nd volume: 323–30.

the Greek case, the issue of heredity, with all its different connotations, became the medium that brought several researchers together under the roof of the anthropological discourse. Especially in the post-war period, studies on heredity or issues related to congenital traits became a nexus for diverse approaches by way of a common anthropological platform, although heredity was often interpreted and treated by the discussants in different ways. As a contributor asserted while introducing his paper on genetics, anthropology's "broad horizon allows the presentation of biological observations".[50] In diverse medical presentations, the link between medicine and anthropology was thought to be merely the congenital character of some bodily or physiological defects that were presented as cases of 'congenital anomaly'. In such cases the question of heredity either stood in the background or arose to the fore aporetically as to whether a certain 'congenital anomaly' was hereditary. For instance, a professor of forensic medicine and his assistant presented a study on *situs viscerum inversus totalis*,[51] that is, a case of 'heterotaxy', arguing that the topic was interesting for anthropology as a phenomenon of 'teratogenicity/monstrosity'.[52] The contributors clearly equalized heterotaxy and monstrosity by using medical terminology, although this was filled with popular connotations. Heterotaxies are indeed:

> complex anomalies... but they impede no function and are not apparent on the outside;... *Monstrosities* are very complex anomalies, very serious, making the performance of one or more functions impossible or difficult, or producing in the individuals so affected a defect in structure very different from that ordinarily found in the species (Canguilhem 2007: 134).

But the specific representation of a case of *situs viscerum inversus totalis* as an example of *monstrosity* (teratogenicity in original contribution) was helpful in approaching the field of anthropology, even if it was anachronistic, since teratology was an anthropological discourse of the 19th century, and even then, was charged with judgements of value.

Some physicians who discussed cases of congenital anomalies before the GAS were more hesitant in devising a connection between their case study and anthropology. In a presentation involving a case of supernumerary neck vertebra (also known as Klippel-Feil syndrome),[53] for example,

---

[50] Panagiotou, *PGAS* (12 November 1967): 8–12.

[51] This is a congenital defect in which the major visceral organs are distributed abnormally within the chest and abdomen.

[52] Katsas, *PGAS* (8 October 1943): 68–70 (discussion pp. 70–1).

[53] Krontiris, *PGAS* (30 December 1946): 54–61.

the contributor stated that the topic is "of a rather general interest" but that it might be interesting for an anthropological approach as well. Likewise, a rare case of cyanosis[54] in newborns was discussed because of its congenital character, which might render the case study interesting for anthropologists.[55]

Heredity gradually became the path for reaching the terrain of anthropology either associated with inborn diversity from the ideal human normal, or in correlation of any disease, for example the question of 'heredity and psoriasis'[56] or whether 'deaf-muteness' was a hereditary defect.[57] The latter case is worth a closer look because deaf individuals (in the Greek original: 'deaf-mute') were at that time still considered damaging for society. Earlier, the inventor Alexander Graham Bell (1847–1922) had even suggested that deaf individuals should be prevented from procreation.[58] The contributor before the GAS, namely Panagiotis Panagiotou, asserted that when 'deaf-muteness' is inborn, it is accompanied by 'feeble-mindedness'. In his presentation he transferred his contemporary discourse using the most topical bibliography at the time, which included works by the National Socialist Arthur Gütt (1891–1949), the so-called "father of the Law for the Prevention of Genetically Diseased Offspring" (Klee 2003: 210). However, Panagiotou himself did not suggest practical measures to be applied generally, or in the specific case of Greece; at least not at this presentation, since he also voiced his ideas and suggestions on other occasions,[59] and also in the eugenic movement in which he was actively involved in the post-war period.

A prominent physician who also gave a presentation at the GAS, and similarly believed that scientific competence should not be limited to a plain description of hereditary congenital defects, was Georgios Kosmetatos (1876–1974), who made suggestions for treating affected individuals in practice. At an earlier date (1943) he had also provided the audience of the GAS with his expertise when he addressed the contested theory of biological evolution by Friedrich Leopold August Weismann (1834–1914)—best known for his opposition to the doctrine of the inheritance of acquired

[54] Cyanosis is a bluish or purplish tinge of the skin and mucous membranes.
[55] Kodounis, Loukatos and Loutsidis, *PGAS* (8 October 1946): 30–9.
[56] Mpelezos, *PGAS* (18 December 1943): 85–9.
[57] Panagiotou, *PGAS* (18 December 1943): 89–90.
[58] On Alexander Graham Bell's involvement in the eugenic movement, see Bruinius (2006): *passim*.
[59] Panagiotou, *PGAS* (30 December 1946): 69–73.

traits and for his 'germplasm' theory.[60] Three years later he discussed "The frequency with which Daltonism[61] appears in the Greek *fill*",[62] and suggested introducing a genealogical health card in infancy in order to consider possible 'health defects' associated with the choice of specific professions. The concern of Kosmetatos was not how affected individuals would manage to organize their lives taking the effects of Daltonism as given; rather, using the specific case of Daltonism, he suggested the bureaucratization and surveillance of humans by recording their health condition over long periods.

A remarkable case of linking a medical topic with the anthropological frames was a presentation on the functions of the hypothalamus[63] by a professor of neurology and psychiatry. This case is remarkable because the lecturer enunciated the connection by a single reference to *b*-thalassaemia as "the Aravantinos-Cooley disease, which was first investigated anthropologically by Ioannis Koumaris".[64] Although this statement might be factually inaccurate, it clearly indicates the speaker's admiration for Koumaris. In medicine, *b*-thalassaemia is also known as 'Cooley's anemia', after the Detroit paediatrician Thomas Benton Cooley (1871–1945) who first described the form of anaemia known as *b*-thalassaemia in children of Italian origin in 1925. The Greek physician (and GAS member) Anastasios Aravantinos (1879–1948) had indeed dealt with thalassaemia; however, his international reputation in this field was never sufficient for thalassaemia to carry his name next to that of Cooley. In referring to the "first anthropological investigations by Koumaris", the contributor meant some of the observations on anaemia presented by the president of the GAS in elaborating the concept of "pseudo-Mongoloid faces".[65] Here Koumaris hypothesized a potential correlation between anaemia and the Mongolian racial ancestry of the affected individuals who, he thought, possessed "Mongoloid-like facial characteristics". The intention of the speaker to keep himself in the dictated frameworks of the GAS reveals

60 Kosmetatos, *PGAS* (18 December 1943): 72–80.

61 Daltonism is red-green colour blindness, also known as *deuteranopia* or *deuteranomaly*. The term 'Daltonism' is derived from the name of the chemist and physicist John Dalton (1766–1844) who first described this phenomenon.

62 Kosmetatos, *PGAS* (30 December 1946): 49–54, 53.

63 *Hypothalamus*: the part of the brain containing a number of small nuclei that serves, among other things, to link the nervous system and the endocrine system via the hypophysis.

64 Kouretas, *PGAS* (9 January 1944): 12–28 (incl. discussion).

65 Koumaris, *PGAS* (23 May 1936): 3–19; and *PGAS* (14 May 1938): 2–12.

his preference to refer to Aravantinos and Koumaris, but not to that Greek physician who did in fact lead the way in studying thalassaemia and enjoyed an international reputation—Ioannis Kaminopetros/Caminopetros (1898–1963). The historian of medicine, Ruth Schwartz-Cowan, provides an account of his pioneering contribution with the following words:

> Caminopetros may be the first to provide evidence that *b*-thalassaemia was a recessive condition in two papers he published in French in 1936 and 1938. By 1945 Caminopetros's hypothesis had been corroborated by pedigree and blood studies done by several Italian and American researchers ... (Schwartz-Cowan 2008: 192; cf. Caminopetros 1938a and 1938b).

The discussion following the presentation on *b*-thalassaemia suggests that Caminopetros remained aloof from the GAS. The president of the Greek Anthropological Society complained that Caminopetros had come to "anthropological conclusions" without making any reference to his own pioneering anthropological research, and that Caminopetros was simply uninterested in collaboration or discussion with the anthropologists, despite all the invitations he had received. (Caminopetros was a member of the GAS, having joined the organization on 25 November 1935.)

In the following I will argue that several of the members who gave presentations at the GAS often sought to find a link to anthropological frames by referring to Koumaris's theories. If heredity in genetic and biological terms was indeed close to anthropology's scope, and yet its association with totalitarian interpretations of human life was a noticeable trend in the anthropological sciences at least in the inter-war period and during World War II, the expansion of anthropology's scope to pose an integrative model for interpreting any form of human relations at any time, in any space and in any context, was an idiosyncrasy of the Greek case. Although there was hardly any opposition to this trend expressed in the internal debates, the silent disassociation of the association's members can be viewed as a tendency that ultimately proved to be of greater importance for the future of the organization than internal dissent.

5.5 *Anthropology As Fetish, Professional Conformity and Opposition*

The internal debates of the GAS disclose that, in a time-span of 46 years, the scope of anthropology was broadening, eventually becoming so wide that it included *any* form of human activity, while its conceptual limits were gradually obliterated. After the 1950s, the president of the anthropological society expressed his conviction that everyone could be, or even *was*, an anthropologist. Yet in the 1960s he distinguished between "scientists who were aware" and those "who were unaware" of being anthropologists.[66] Any issue could be interpreted as pertaining to the anthropological sciences. All sciences, arts, even theatre,[67] were thought of not simply as auxiliary or allied disciplines, but as belonging conceptually to the scope of anthropology—as *being* anthropology. More than this, in the conception suggested by the president of the GAS, anthropology aspired to become a cosmological doctrine. It was not the approach that decided the classification of an issue as anthropological but the reverse; every single issue related to human and human life, be it natural, biological or social, was deemed to dictate the scope of anthropology.

How did the members of the GAS cope with this idiosyncratic definition of anthropology? For members who were less prominent, affiliation to GAS facilitated access to a scholarly circle. Affiliation to a scientific association transmits a kind of social capital that Pierre Bourdieu has described as:

> the entirety of existing and potential resources connected with the possession of a sustainable network of more or less institutionalized *relations* of a reciprocative acquaintance and acknowledgment;... it is a matter of resources based on *belonging* to a group (Bourdieu 1992: 63).

Indeed, several members proved to be remarkably flexible in adjusting the topics they presented to the setting of the GAS, and in embracing the established terminology. This kind of conformity has long been a conventional strategy in the academic community, where scholars cite and even flatter each other for purposes of networking and in order to promote their own positions. After all, the president of the anthropological society, until 1950, was still an authority in Greek academia and drew his symbolic capital from his position as a university professor and director of the Anthropological Museum, even though his symbolic capital was obviously in decline after World War II.

Subsuming diverse topics into the scope of anthropology did not demand strong efforts, especially when the declining trajectory of the anthropological institutions left room for action for everybody willing,

66 *PGAS* (1964): 10.

67 In the presentation of a book by Katerina Kakouri (1912–90) on theatre (Kakouri 1946), Koumaris assumed that theatre also pertains to the scope of anthropology: *PGAS* (30 December 1946): 47–9.

as a sufficient condition for being positioned in the anthropological discourse was a simple reference to established terminology, foremost to race and the Greek race in particular, or/and to Koumaris's theories. Illustrative of the role personal connections played in disseminating theories and embracing views is a statement made in 1966 by the archaeologist Spiridon Marinatos[68] in a paper entitled "Could There Be a 'Mental Anthropology'?" In introducing his topic (in which 'mental anthropology' was not actually the main issue), Marinatos described the Greek race as an amalgam of several ancient races. In the next edition of the *PGAS* a note was added which clarified that "racial amalgamation in the Greek lands does not mean miscegenation with members of strange nations, as professor Koumaris has repeatedly ruled out".[69] Marinatos surely did not need to flatter the president of the GAS; he himself was sufficiently prestigious and networked. Friendship might have been his motivation in this case, provided, of course, that he had initiated the correction in the first place.

Several further examples indicate professional conformity: a speaker who gave a talk on "Volcanic Eruptions and Earthquake in the Aegean and Man in the Southeastern Mediterranean Sea"[70] found a link to the anthropological settings through an introductory remark on the broad character of the Greek Anthropological Society as "a Society for physical anthropology, prehistory, ethnology, ethology, psycho-sociology, criminological anthropology, eugenics, *etc.*" And, when a historical detail referring to Armenians in the writings of the ancient geographer Strabo (63 BC–AD 23) was discussed—the title of the paper being "Are the Armenians of Greek origins?"[71]—the link was elaborated out of a reference to a single word in Koumaris's racial theory.

In all of the presentations, any consensus cases of opposition were extremely rare. A striking case (the only one recorded in the *PGAS*) is worth outlining not only because it was unique, but because it questioned the anthropological concept as it had been developed at the Greek Anthropological Society and the Anthropological Museum. The criticism came from the physician (surgeon-urologist) Ioannis Rivouar (1886–?) and acquired serious weight because of the occasion on which it was voiced—

68 Marinatos, *PGAS* (12 March 1965): 59–63.
69 Editorial Note in *PGAS* (12 November 1966): 10.
70 Kritikos, *PGAS* (28 October 1941): 22–5 (discussion pp. 26–7).
71 Anagnostou, *PGAS* (21 February 1945): 22–3.

the festivities for the 20th anniversary of the Greek Anthropological Society in June 1944.[72]

The paper Rivouar presented at that anniversary session carried the ironical title "Redeeming Craniology" and addressed phrenology[73] and its founder Franz Joseph Gall (1758–1828). It was an allegoric and bitterly sarcastic text that satirized not only phrenology, which by then was generally discredited, but also the Greek Anthropological Society and those who had shaped the profile of institutionalized anthropology in Greece. His criticism addressed the fetishistic character that was ascribed both to the cranium as the reified epitome of the anthropological sciences as well as to the ubiquity of anthropology. In his own words:

> On the occasion of the twentieth-anniversary of this Society whose activity has to a great extent been based and will continue to be based on the real or alleged importance of the cranium, either those to be examined or those that do the examinations, I recall the crania of those unfortunate creatures that have become subjects to studies in quite recent times but also the cranium of the eccentric researcher who studied them.[74]

Using Gall and his phrenology as a pretext, Rivouar warned scholars that they ran the risk of crossing the fine line from science to quackery. This, he argued, was because research on brain diseases could easily unleash the uncontrollable fantasies of researchers. The lack of knowledge about many defects of the nervous system, as well as possible mingling with afflicted individuals, could influence the researchers themselves in such a way that they would reveal 'odd behaviour' likely to take the form of 'mental wobbles'. In his sarcastic way, Rivouar provided a great deal of information about Gall's working methods and the reactions of the public to his theories. He even showed a few caricatures produced by Gall's contemporaries (which were reproduced in the *PGAS*) that satirized and pictured him teaching surrounded by the crania of monkeys, philosophers, poets, actors, heroes, murders and tramps. Rivouar concluded his ironic

---

72 Rivouar, *PGAS* (1 June 1944): 34–42.

73 Phrenology was a term first established by Franz Joseph Gall's student Johann Christian Kaspar Spurzheim (1776–1832) to describe Gall's theory according to which intellectual dispositions of each individual can be recognized in the form of cranium and the traits of the head.
On the British, French and American schools and the political use of phrenology, see Goldstein (2003): 149–53.
On how Rudolf Virchow adopted phrenological approaches in Germany, see Goschler (2003): 326.

74 Rivouar, *PGAS* (1 June 1944): 34.

tale by asserting that Gall's collections included the crania of Homer and biblical figures such as Adam, Eve, Noah and Goliath. In the caricatures Gall was portrayed as practising 'crania divination' and 'cranium worship'. Turning to more recent scholarship Rivouar referred dryly to modern-day attempts to enrich collections of crania. Although no names were mentioned, this remark might have been an implied criticism of the campaign to enrich the collections of the Anthropological Museum. His remarks were allegorical but unequivocal: the collections of Gall's craniological museum, he said, were valuable; they could eventually cause the envy of our contemporary anthropologists who—"striving to fill their surrounding vacuum"—grant an infinite freedom to their thoughts and allow them to wander in several directions beyond 'crania divination' or 'cranium worship'. In doing so, Rivouar went on, they evoke their conviction that everything derives from anthropology. In closing, the paper suggested how much irrationality had been and still was inherent to the claims of the sciences (and of scientists) to rationality:

> the bare and empty cranium has been used as a cult object not only by the savages; it still represents an important object and its nonexistent soul is adored as a resource of human and racial felicitousness.[75]

After this surprising performance, the response by the president of the Greek Anthropological Society only served to fortify Rivouar's assertions. Koumaris reminded the audience that the "plagued outsider", namely Gall, had contributed a great deal to brain research. For all the satire, Gall was the first to have raised the question of the multiple cortex of the brain, on the basis of which cell architecture was later developed. His prolonged and emotionally charged response essentially affirmed Rivouar's assertion that Greek anthropology had become a fetish:

> Mr. Rivouar, who is bookish and does not confine himself to the narrow scope of his field, i.e. the medical sciences, had the collegial audacity to "cook anthropology's goose" on the occasion of the twentieth anniversary of the Society, choosing from his large and valuable library—without any remorse—the bitterest shafts of the pictorial quiver. But anthropology, as generous as it is pertinent to the science, and the weird wise men who have stoically withstood the shafts of satire from their contemporaries … appreciate this pleasant act of analysis of the activities of the legendary wise man Franz Gall; his figure greatly interests us.[76]

---

[75] Rivouar, *PGAS* (1 June 1944): 39 and 42.
[76] See Koumaris's response to Rivouar in *PGAS* (1 June 1944): 42–4.

Rivouar's performance was unexpected. He had joined the GAS in 1935 and in 1942 the records show that he was a member of its executive committee. His sarcastic talk took place in 1944, a year in which the profile of the anthropological institutions was being consolidated and the ideological tendencies within the anthropological discourse made manifest. The decay of the anthropological agency began to accelerate in the subsequent period.

5.6 *Embracing Social Anthropology in the Face of a Noble Anthropologist: Prince Peter*

It would not be off the mark to assert that the first appearance of ethnological thinking (in the sense of colonial anthropology) in Greek academic settings coincided with the emergence of a new concept for the Anthropological Museum from 1915 onwards. Yet the GAS became a forum for very early presentations of social and cultural anthropology when, during the early 1940s, enlightening insights into social anthropology had also been provided by its member, the publicist Lazaros Piniatoglou (1909–45). Furthermore, a series of lectures delivered by the social anthropologist Prince Peter of Greece and Denmark (1908–80) in Athens in the early 1960s was announced to take place under the aegis of the GAS.

All these activities were pioneering in terms of introducing ethnology and social anthropology into Greek academia given that social anthropology was established as late as the 1980s in Greek universities (Zoïa 1990; Papataxiarchis 2003; Gefou-Madianou 2000; Panopoulos 2003). In the period considered by this study, ethnology was still interpreted and practised in Greece as a discipline very close to folklore studies, constituting a kind of "national anthropology", in Herzfeld's words (1982: 8), because of its ability to serve national aspirations.[77] The relationship between

[77] All issues related to folklore that were presented in the GAS pursued a line of argument which clearly revealed the national priority at the heart of this discipline. For instance, in two papers on popular art in Crete, Rodoula Stathaki-Koumari expressed her conviction that all data she presented were "purely anthropological". Moreover, she regarded her conclusions on Greek folklore as corroborating the thesis about the autochthonous character, and the uniformity and continuity of the Greek *fili* since prehistoric times. For her, folklore was a discipline that collaborated with other sciences, such as archaeology, in serving national concerns: "Archaeologists and folklorists every day find new data to connect the past [of the Greek *fili*] with the present."

Koumari-Stathaki, *PGAS* (12 November 1966): 15–21. See also Koumari-Stathaki, *PGAS* (12 November 1967): 30–40.

folklore studies, social anthropology and ethnology was addressed by Greek scholars only in the first decades after World War II. An early sign of the rising interest was a paper included in the *PGAS* of 1966 by the prominent folklorist Georgios Megas (1893–1976), which outlined the definition of ethnology in respect to its distinction from folklore studies.[78]

This was the very period in which Price Peter familiarized the Greek scholarly audience with social and cultural anthropology. In several lectures given during the 1960s at institutions in Athens (such as university institutes, or the National Theatre), he provided an insightful summary of the relationship between physical anthropology, social anthropology, ethnology and their auxiliary disciplines. These lectures also provided a wide overview of theoretical and methodological approaches, and outlined the central (yet in part disputable) issues anchored in the studies of social anthropology. He clarified that 'anthropology' was a general notion for a scientific field that integrated biological and cultural sciences in a single discipline.[79] He suggested dividing anthropology into two main subfields: 'physical' and 'ethnological anthropology'. The object of the former was humankind as part of the animal kingdom, and he specifically addressed the natural traits which serve to differentiate humans from other living creatures as well as the distinctions between the human races. 'Ethnological anthropology', on the other hand, dealt with the life of man as organized in societies and nations. Hence, the focal point of this subfield of anthropology was culture and social relations. Contrary to a holistic and levelling notion of anthropology, Prince Peter called attention to methodological and conceptual aspects essential to the anthropological sciences, arguing that ethnography referred to the method of collecting data, and ethnology to the elaboration of the collected data and the synthesis of conclusions. In his scheme, social anthropology was a subcategory of ethnology that referred to social organization, social institutions, kinship, family, political organization and so on.[80] In several subsequent lectures[81] Prince Peter outlined the conclusions of his own research in

[78] In this, Megas outlined the definition of ethnology with respect to its distinction from folklore studies. Megas, *PGAS* (12 November 1966): 22–7.

[79] The scope of anthropology was first depicted in an extensive introductory lecture by Prince Peter at the University of Athens in 1961, at a celebratory meeting chaired by the dean of the university. Prince Peter, *PGAS* (1 February 1961): 24–38.

[80] Prince Peter, *PGAS* (1 February 1961): 24–38.

[81] He also delivered an introductory lecture to ethnology at the University for Economic and Commercial Sciences [Ανωτάτη Σχολή Οικονομικών και Εμπορικών Επιστημών]. In the following two years Prince Peter talked for Greek audiences at the National Theatre,

the Himalayas and South India,[82] and also referred to the relevance of the anthropological sciences for Greece in particular.

Prince Peter was not unknown in the Greek academic scene. The son of Prince George of Greece (1869–1957) and the psychoanalyst Maria Bonaparte (1882–1962), Peter was appointed 'honorary doctor of medicine' by the medical faculty of the University of Athens in 1957,[83] two years before he earned his PhD in social anthropology in London. His family was famous in Greece not only due to its royal descent, but also because of the scientific activities of his mother. Maria Bonaparte co-founded a circle of intellectuals and psychologists who sought to advance the study of psychoanalysis in Athens in 1947. She herself held lectures at the Greek Medical Society as well as several meetings which took place in her own house in Greece (Kazolea-Tavoulari 2002: 148–51). The relationship between psychoanalysis and social anthropology was addressed in the Greek anthropological discourse, but not in the specific lectures by Prince Peter in which he outlined issues related to psychoanalysis and psychology only roughly, as when he referred marginally to questions such as sexual instinct and aggressiveness, among other things. The contested relationship between psychoanalysis and social anthropology had been addressed much earlier by the publicist Lazaros Piniatoglou; in two lectures given at the anthropological society at the beginning of the 1940s, for example, Piniatoglou discussed the contribution of psychoanalysis in approaching the phenomenon of artistic creativity from a social anthropological perspective and was rather critical, if not disapproving.[84] He called attention to the limited capability of psychoanalysis to explain anthropological phenomena, and claimed that psychoanalysis was unable to provide sufficient reasoning for the phenomenon of artistic creativity as a whole. In his view, an overarching explanation of art as an anthropological phenomenon needed to consider further factors, foremost of which were the racial belonging of humans and the collective forms of their social organization.

Athenian university institutes, the Academy of Athens and the Archaeological Society. His topics included "The Historical Development of Ethnology"; "The Progress of Ethnology"; "The Contemporary Ethnology" and "A Scientific Theory of Culture."

Prince Peter, *PGAS* (3 February 1961: 24–38); (6 February 1961: 42–57); (21 March 1962: 37–51); (14 March 1962: 12–22); *PGAS* (19 March 1962: 24–36); (2 May 1963: 12–23); (21 March 1962: 37–51).

[82] See his most significant work *Prince Peter of Greece and Denmark* (1963).

[83] Introductory words by the rector of the University of Athens: *PGAS* (1 February 1961): 9.

[84] Piniatoglou, *PGAS* (15 April 1943): 46–9.

The question of the use of psychoanalysis in anthropology was actually a dispute as old as the suggestion made by Sigmund Freud (1856–1939) in his work *Totem and Taboo* (Freud 1950/1913) that these two disciplines should be connected. Freud's model became subject to different readings, interpretations and perceptions by anthropologists and the relation between psychoanalysis and social anthropology had, generally speaking, been tense. George Stocking described it as a "sometimes fruitful and often contentious relationship of two twentieth-century discourses that seek, in somewhat different ways, rational explanations of the apparently irrational" (Stocking 1986: 13).[85] In Greece, psychoanalysis was a contested subject. While it had been embraced enthusiastically in alternative pedagogy and in intellectual circles even prior to World War II, it met with scepticism in medical and psychiatric circles (Kazolea-Tavoulari 2002: 141–82). Yet the association of psychoanalysis with anthropology may ultimately be a novelty specific to the Greek anthropological discourse.

Lazaros Piniatoglou's contribution to anthropological discourse in Greece was innovative in terms of making Greek scholars familiar with social anthropological topics much earlier than this discipline found entrance into Greek academia. In addition to his presentations, he also raised scholarly knowledge of these topics by means of a monograph with the title *The Archaic Form of Royalism—An Essay on Social Anthropology* (Piniatoglou 1938) (a book classified by Ioannis Koumaris as 'folklore' and "social anthropology or psycho-sociology—this very recently established branch of 'anthropology' in a broad sense").[86] The association of social anthropology with Greek national doctrine, as he attempted to do, was also remarkable insofar as it was atypical with respect to the dominant antiquity cult. Indeed, in a paper on the 'Pythagorean taboo' delivered to the GAS Piniatoglou used social anthropological approaches to address "some primitive characteristics of the ancient Hellenic civilization".[87] He embraced the term 'taboo', as conceptualized by the British anthropologist James George Frazer (1854–1941) in his monumental work *The Golden Bough* (Frazer 1890), and pointed provocatively to the correlation of a "barbaric term" (namely taboo) with the "sacred name of Pythagoras", a correlation that "moralists afflicted with antiquity mania might regard as

---

[85] On the relationship between anthropology and psychoanalysis see also Heald and Deluz (1994).

[86] *PGAS* (22 April 1941): 2–4.

[87] Piniatoglou, *PGAS* (25 September 1942): 36–9.

sacrilegious".[88] Although he did not share this view, his goal in elaborating on this correlation was the innovation of the scientific approach to Hellenic antiquity by means of a "comparative anthropological method that is today definitively established as the most exact instrument for studying all social phenomena and even guarantees for the scientific thoroughness that past methods lacked".[89] Viewed from this perspective, 'taboo' was not a 'barbaric term' but rather an appropriate methodological means for a modern scientific approach to Hellenic antiquity, all the more so as, in Frazer's concept, primitive religion and magic in so-called primitive societies were thought of as a kind of applied science (Barnard 2000: 37–8). Past and present were tied together in this approach by Piniatoglou, who seemed to follow the tradition of the unilinear evolutionists, and his verbal attack on the Greek 'antiquity cult' (which must not be understood as an attack on Greek nationalism generally since he was a declared nationalist), was particularly remarkable.

The representation of social anthropology and ethnology by two scientists and public figures who enjoyed the admiration of Koumaris[90] (Prince Peter's authority probably being the greater of the two) contributed to upgrading the rank of these disciplines in the anthropological concept of the society's president. Ethnology had clearly acquired a leading position by the time the 40th anniversary festivities of the Greek Anthropological Society took place, given that it was positioned first in the *Proceedings* of 1964, the anniversary edition being entitled "Forty Years: The 'Greek Anthropological Society' (1924) (Ethnological-Prehistoric)".[91] Considering that this discipline had been integrated into the anthropological project with respect to the organization of the Anthropological Museum early on, this opening to ethnology was by no means a late conversion. However, taking up ethnology in the name of the anthropological association had a programmatic character. This was obviously not independent from Prince Peter's scientific activities in Greece and could even be read as an attempt to enhance the image of Greek anthropology. Such a conclusion

---

88 Piniatoglou, *PGAS* (25 September 1942): 39.

89 Piniatoglou, *PGAS* (25 September 1942): 39.

90 Koumaris published texts on racial issues in his newspaper *Greek Blood*, Koumaris (1945a). When Lazaros Piniatoglou died in 1945, he lauded his scientific achievements in a commemorative address. This text, with the title "The Scientist", was printed in *Greek Blood*: Koumaris (1945h).

91 *PGAS* (1964): "40 years—The 'Greek Anthropological Society' (1924) (Ethnological—Prehistoric)" ["Τεσσαρακονταετηρίς. Μία Τεσσαρακονταετία—Η 'Ελληνική Ανθρωπολογική Εταιρεία' (1924)—(Εθνολογική—Προϊστορική)"].

would not be too far off the mark if one agrees that representatives and delegations of groups (of whatever kind) are a precondition for the accumulation of social capital and, in turn, a manifestation of the accumulated social capital (Bourdieu 1992: 69). This is all the more so when the possible representative is a noble person and the institution in question may benefit from association with a royal name. Similarly, the president of the anthropological society and now emeritus professor for physical anthropology had nothing against surrendering the predominance of physical/racial anthropology to ethnology (or social anthropology) given that his successor as the chair of anthropology would be a member of the Greek royal family. In 1958, as he disclosed,[92] he had addressed a request to the senate of the University of Athens, recommending the re-establishment of the chair of anthropology for Prince Peter.

92 *PGAS* (14 March 1962): 9–11, 10.

CHAPTER SIX

A 'DISINTERESTED SCIENCE' IN WARTIME

The alleged political and ideological neutrality of science, and of scientists as its servants, is a common assumption within academia. Science is often illustrated as both shaping and pertaining to a field of objective facts that are unimpeachable and independent of politics and ideology. This sort of neutrality was constantly reaffirmed within the internal debates of the Greek Anthropological Society as being integral to and highly important for its profile. Yet at the same time, the commitment to society and most of all to the nation and the Greek *fili* was programmatically proclaimed. How could ideological neutrality coexist with societal or national commitment? How was neutrality possible, during the war period, and to what extent did it comply with wartime patriotism? The central argument of this chapter is that the internal debates of the GAS during World War II and the Greek Civil War portray its profile as conforming to prevalent political relations. Criticism of the occupation forces was voiced only after the liberation and revealed the need for rehabilitation. How did their rehabilitation take place in a situation in which power relations were still unstable and would continue to be so for many years to come? Finally, in contrast to all proclaimed neutrality, involvement in GAS's work came to act as a political position, especially when the academic community was deeply divided into hostile fronts, reflecting the general divisions in the country.

6.1 *The Neutrality of Science and the War Narrative*

The tension between 'patriotism' and 'scientific neutrality' was accentuated as a result of the conflicting loyalties held by several members of the Greek Anthropological Society after Greece had entered World War II. Many members—and particularly those who had studied in Germany and maintained networks with German scientific institutions—admired German science; but they also felt bound to honour their Greek homeland. The imperative of neutrality faced another acid test when the political situation in Greece became further aggravated in the Civil War of the 1940s which flared up during the occupation before tapering off in the second half of the decade after German forces had left the country in October 1944.

Greece entered World War II in October 1940, and the country was occupied by the Axis powers and the German Wehrmacht from 27 April 1941 to 12 October 1944. Throughout this period the Greek Anthropological Society continued to meet regularly, even as often as three times a year, despite all the difficulties that the war caused.[1] The occupation arose only implicitly as an issue in its internal debates, and mostly to the extent that the internal, organizational work was affected. A notable difference from the period prior to the occupation was that topics related to 'Greek *fili*/race' now proliferated. But neither the world war in general, nor the occupation of the country in particular, were referred to in the internal discussions, much less the conditions under which the Greek population was now living. The relatively rare exceptions to this are therefore enlightening with regard to how the actors involved in the anthropological discourse imagined patriotism and scientific neutrality.

In December 1941, in the worst winter of the occupation period, during which famine had claimed a large number of victims among the Greek population (Hionidou 2006), a dentist gave a talk at the GAS that raised the problem of malnutrition and its degenerative consequences for the Greek *fili*.[2] This lecture was prompted by his alarming observation that the dental make-up of Greeks was tending towards 'degeneration' at such a rate that it showed signs of threatening the Greek *fili* at large. On the basis of comparative observations with children in Greek villages, towns

[1] According to the *PGAS*, the meetings were usually held in the Anthropological Museum.

The minutes of the meetings continued to be printed without interruption, despite the many practical and financial difficulties that resulted from inflation and rising paper prices. Donations and aid provided by bankers, ministry officials and the National Printing House [Εθνικό Τυπογραφείο] ensured that the minutes could be printed. This seems, however, to have been anything but straightforward, and adverse events were not infrequent. For instance, a contretemps occurred in 1942 when the Germans confiscated all materials and paper from the printing house, including the manuscripts of the *PGAS*. Only after negotiations with the German embassy in Athens was the material given back. Koumaris expressed his gratitude to several institutions and notable individuals in politics and finance, amongst whom were the Minister of Education Nikolaos Louvaris personally, the Ministry of Finance and the National Printing House. *PGAS* (18 December 1943: 91); *PGAS* (27 June 1945: 24–5) and *PGAS* (31 December 1942): 73–4.

[2] Krikos, *PGAS* (27 December 1941): 51–6.

The author, the dentist Alexandros Krikos, had already presented medical topics with national implications during the 1930s, namely caries in the Greek *fili* in antiquity and the present.

Krikos, *PGAS* (11 April 1935: 3–18; and 28 December 1936: 50–4).

and cities, the speaker warned that the side-effects of malnutrition could inflict serious damage on the Greek *fili* that would be carried on to future generations over the longer term. In order to counteract this threat and to "rescue our fili", he suggested immediately increasing production and distributing plots of land to the landless population. With this suggestion, the contributor was embracing a prevailing view according to which the food crisis and famine in occupied Greece had come about because of the low-level of agricultural productivity. This view has only recently been challenged by historians, first by Giorgos Margaritis (1993) and later by Violeta Hionidou (2006). On the basis of supplementary local case studies and oral testimonies the latter concluded that:

> at times and in some places production was substantially raised. The [low] figures quoted by the report writers usually referred to the corresponding amounts of produce that the Greek government had managed to tax, though this was never admitted. And these were very small indeed. In essence, the published figures for productivity during the occupation years were closely linked to the taxation in kind that was imposed at the time.[3]

This recent revision suggests that Greek governments and the grave political situation were responsible for the famine; it also uncovers a set of factors that combined to bring about the food crisis. Such factors included the administrative division of the country (and subsequently of both the food production sector and the markets) into a patchwork of segments that barely communicated with each other; rising inflation; the destruction caused by military operations; the inability of the government to impose laws that could alleviate the effects of the war, and other factors (Hionidou 2006: 3 and *passim*).

In view of this complexity, the proposal made by the speaker at the GAS appears, at first glance, to simplify the situation. But given that the more vulnerable groups of society were the first to be affected by the famine (Hionidou 2006), his overall proposition for coping with the food crisis appears somewhat bizarre: he suggested distributing aid to the population, giving the highest priority to children and to 'productive individuals'. These two categories would receive the nourishment necessary for ensuring their existence, whereas support for the rest, "the sleeping individuals whose absence would not be grave for society", would be secondary.[4] In order to solve the severe problem of malnutrition the Greeks would have

---

[3] Hionidou (2006): 5 and at length in pp. 68–108. See also Clogg (2008).

[4] Krikos, *PGAS* (27 December 1941): 55.

to act on their own without anticipating any help from the 'well-meaning' occupying power:

> As well-meaning as the forces under whose occupation we are, they themselves face a harsh struggle as well. Much as they would do to help us, due to the appreciation we have won by our fight (and also because they are surely not willing to leave "a rotten apple in the same basket among the sound ones") the whole burden of solving the problem with respect to the next generations is left to us alone. The scientists of the countries on whom the fate of our people depends are experienced and familiar with such problems; they could help us as well.[5]

This lecture was one of the very few exceptional cases where references were made to the war in the internal debates of the anthropological society. A further exceptional case was the question of the Jews, which arose twice within the GAS while the persecution of Jews under the National Socialists was in progress. The concern expressed in these discussions was the racial relations between Jews and Greeks, the question at issue being whether these relations justified the persecution of Jews in Greece, as was taking place in Germany. The Greek Anthropological Society discussed the Jews for the first time in 1939[6] when its president put this topic on the agenda; the second occasion was a historical lecture on the Jewish community of Tripoli (Peloponnese), delivered in October 1943 by the prominent historian Nikos Bees.[7] On each of these two occasions both the lecturers and the discussants jointly argued that Jews in Greece should in no way be treated like those in Germany. Yet though the GAS's president condemned the persecution that Jews faced under the National Socialists, he also adopted a rather ambiguous attitude towards the issue as a whole, well known from his previous statements on Jews.[8] While, on the one hand, he approved of any attempts to maintain the purity of a race, including also that of the German race, on the other, however, he expressed his sympathy for the destiny of the Jews who he portrayed as a generally undesired and "wandering people". The surprising event, in his eyes, was not so much the repression with which the Jews had to contend, which

---

[5] Krikos, *PGAS* (27 December 1941): 54.
According to the proceedings, Krikos's presentation was not followed by any discussion. It cannot be excluded that some papers were not presented at a session but were only submitted for printing in the proceedings.
[6] Koumaris, *PGAS* (14 May 1939): 24.
[7] Bees, *PGAS* (8 October 1943): 50–9.
[8] Cf. Koumaris's statement one year earlier in his serialized Greek article "Race and Health": Koumaris (1938).

allegedly was quite usual and an "understandable practice in any historical period".[9] Rather, it was the 'intensity' of the contemporary measures against the Jews which made their treatment in Germany comparable with 'medieval religious fanaticism' that should "under no circumstances be replaced by racial fanaticism".[10] Jewish persecution by National Socialists along with the arrogant racial theories elaborated in Germany claiming the alleged superiority of particular races had "tarnished the racial ideal".[11] One year earlier, in an article published in the Italian fascist journal *La Difesa della Razza*, he had asserted that, in Greece, "similar radical means" were not needed for the regulation of racial issues since the close religious ties that bound Greeks together did not allow marriages between Greeks and Jews.[12]

It is astonishing how similar this line of argument was with those pursued two years later by the German archaeologist and Philhellinist, Roland Hampe (1908–81) in the popular National Socialist journal *Volk und Rasse*. Hampe argued that Greeks were so tightly bound together by traditional and religious bonds that any mixing with non-Christians, in particular Jews, would be inconceivable (Hampe 1941). He also asserted that Jews in Greece should not necessarily be treated like those in other Balkan states, as for instance in Romania,[13] since, unlike these latter countries which had so far neglected to segregate Jews and were therefore now making up for lost time, Greece had passed racial legislation much earlier. A further reason for not persecuting Jews in Greece according to Hampe was that they were unpopular, disdained and poor, and hence posed no danger whatsoever to either the economy or society, and, since Greek Jews lived in separate districts, they posed no threat in racial terms either. Hampe made these arguments while discrimination against Greek Jews was growing; at least this was the case in Thessaloniki, where they were subjected to increasing levels of discrimination from April 1941 onwards (Dordanas 2004).

---

9 Koumaris, *PGAS* (14 May 1939): 24.

10 Koumaris, *PGAS* (14 May 1939): 24.
Statement to Bees's lecture: Bees, *PGAS* (8 October 1943) and discussion pp. 59–60.

11 Koumaris, *PGAS* (14 May 1939): 24.

12 Koumaris (1938a). This article appeared in the section for correspondence. Apparently the short text is based on the Greek article on "Race and Health": Koumaris (1938). Cf. Koumaris (1939b).

13 On Jews in Romania see Müller (2005). With respect to genocide committed against the Roma in Romania, see Mihok (2008).

The attitudes towards Jews and particular forms of anti-Semitism in Greece covered a wide and variegated landscape. Certainly, anti-Semitism has many faces and it should not automatically be interpreted as evidence for the support of the genocide of the Jews; the idiosyncratic character of phenomena like anti-Semitism is formed under particular historical and political circumstances in each country. In the Greek case, for instance, the fact that the dictatorial regime of Ioannis Metaxas (1871–1941), which ruled from 4 August 1936 to 29 January 1941, did not persecute Jews in the same way as other minorities is certainly not without importance for the attitudes of Greeks towards Jews in the country. The historian Katerina Lagos (2005) has analysed the relationship between the Metaxas regime and the Greek Jewish communities, concluding that it was rather ambiguous. On the one hand, Jews were not recognized as members of the nation; yet on the other they were treated leniently, at least more leniently than Jews were treated in other Balkan states. This hypothesis can be corroborated by the research of other contemporary scholars, who came to conclude that the regimes which succeeded the Metaxas dictatorship and collaborated with the occupation forces had accused the Metaxists of 'philosemitism' (Papanastasiou 2006). It is thus significant in this context that the newspaper *Evening* [*Βραδυνή*], which sympathized with Metaxas's dictatorship, often published reports of the "Jewish struggle in Palestine".[14] This, however, is no explanation for the attitudes of the members of the Greek Anthropological Society; rather, it is a note that might be useful for understanding the kaleidoscope of diverse opinions on the Jews in Greece. In the debates of the GAS the predominant argument was that there was no reason for persecuting Jews in Greece due to the lack of racial intermixture between Greeks and Jews.[15] Judaism prohibited marriage outside the Jewish community, and Jews themselves had always striven to preserve their purity by living in closed communities, organized according to strict rules of endogamy, so that they presented an archetype of racial purity and preserved particular racial characteristics.[16]

Only a single opposing voice brought forward the counterargument that such mixing of blood was possible and clearly not condemnable. This view was voiced by Georgios Sklavounos, the teacher of many members of

[14] See for instance the editions from 13 and 14 October 1938.

[15] See Koumaris's statement to Bees's lecture in: Bees, *PGAS* (8 October 1943), discussion pp. 59–60.

[16] This was also a widespread view among Jews. See an overview on the racial constitution and purity of Jews in Salaman (1911).

the Greek Anthropological Society; it was one of the remarkably few statements he ever made in the GAS, even though he appeared to be present in most of its meetings. Sklavounos even claimed that Greece benefited from such miscegenation with Jews. Typical Greek qualities, he said, such as the "inborn proneness to mercantilism" and "speculation", and the fact that Greeks had lived in diaspora, bore testimony to the Jewish influence on Greeks. Yet by means of such mixtures the Greek race was strengthened rather than dissipated. This occurred due to the claimed ability of the Greek race to discard weaker racial elements and to maintain stronger and more vital traits. Moreover, he embraced the Jewish consideration of genealogy in the mothers' line and announced that he had addressed a request to the Academy of Athens to register the names of the mothers of the academy's members in order to ascertain their ancestry and to reconstruct their pedigrees.[17] In contrast to the outspoken Sklavounos, the speaker himself, Bees, sought to avoid making any statement on the contested question, asserting that the relationship between the Greek and Jewish races was "a huge issue" and that he preferred to avoid any further comments on the matter.[18]

The debate on "the Jewish Question" as a theoretical racial issue took place while the extermination of the Jews in Europe and even in Greece was approaching its gruesome climax. This particular meeting of the Greek Anthropological Society was held in October 1943, only a few months *after* the deportation of Greek Jews from Thessaloniki to National Socialist extermination camps had begun, in March 1943. Further deportations from elsewhere in Greece followed, and by the end of World War II approximately 60,000 Greek Jews had been deported and murdered (Varon 1994; Margaritis 2005). The Greek Anthropological Society was silent about these events. Neither at that meeting nor on another occasion did the persecution of Greek Jews become an issue. In the newspaper *Greek Blood* [*Ελληνικόν Αίμα*], published by Lazaros Piniatoglou,[19] a brief notice in the margins of the first page of the 15 October 1943 edition referred to the deportations of the Jews from Athens, and appealed to 'Greek patriots' to help 'Jewish compatriots' in their hardship. Recent studies on the

17 See Sklavounos's statement in Bees, *PGAS* (8 October 1943): 63–5.

18 Bees, *PGAS* (8 October 1943): 50–9.

19 Piniatoglou was actively involved in the resistance during the occupation. He was the co-founder of the resistance group named 'Greek Blood' [Ελληνικόν Αίμα] and publisher of the newspaper with the same name. In terms of his political attitude, Piniatoglou was a royalist and a fervent anti-communist, as was the profile of his newspaper and his own publications within it.

'Greek Holocaust' emphasize the 'silence' of the Greek population generally and scholars in particular concerning the destiny of Greek Jews, both at that time and subsequently. The members of the Greek Anthropological Society who were involved in the discussions described above might not have been indifferent to the Jewish destiny. But simply addressing possible racial bonds between Greeks and Jews at a time when the latter were being subjected to extermination means discarding responsibility and allowing extermination, rather than serving a disinterested science.

6.2 *The Need for Rehabilitation and Ex Post Facto Patriotism*

World War II emerged as an issue in the internal debates of the GAS only after the end of the occupation and while the Civil War was ongoing, and even then it was not a large debate, and much less a dispute. Rather, references to the war appear to have reflected an unuttered need for rehabilitation or at least a vindication of the previous attitudes of members of the anthropological society or physicians collectively, as a professional group. It may have been a coincidence that this debate began on the day that Athens was liberated from the occupation forces, 12 October 1944, the very day that the third meeting of the GAS for that year is recorded to have taken place. The opening of the meeting celebrated the liberation: "The Anthropological Society joins the entire nation in celebrating our freedom.... Greece is liberated from the contingent, undeserved and harsh triple slavery!"[20]

On the occasion of that joyous event, the GAS president felt that it was his personal duty to give an account to the members of his attempts to defend Greek national concerns during the war years. The most impressive among them was the announcement of his correspondence to "eleven of the most distinguished German anthropologists"[21] decrying the Italian attack against Greece. This letter, dated December 1940, was addressed to Eugen Fischer (1874–1967), Fritz Lenz (1887–1976), K. [Hans Friedrich Karl] Günther (1891–1968),[22] H[ans][23] Reinerth (1900–90), Egon von Eickstedt (1892–1965), Otmar Freiherr von Verschuer (1896–1969), W. [Hans][24]

[20] Koumaris's opening address in *PGAS* (12 October 1944): 54.
[21] *PGAS* (12 October 1944): 55.
[22] Hans Friedrich Karl Günther was known as 'Rassen-Günther'; see Klee (2003): 208.
[23] On the prehistorian Hans Reinerth, see Klee (2003): 487.
[24] On the racial hygienist Hans Weinert, see Klee (2003): 662.

Weinert (1887–1967), Otto Reche (1879–1966), Theodor Mollison (1874–1952), Wilhelm Gieseler (1900–76) and O[swald] Menghin (1888–1973). None of these recipients responded to his letter. On the day of the liberation of Athens a further open letter was addressed to German anthropologists in order to report the "afflictions of the Greeks". In doing so, the president of the Greek Anthropological Society intended to act as a scientist in being "objective and free from prejudices" and motivated by "professional duty"; it was not his intention to "celebrate the current total defeat of Germany ... political and national concerns have no place in the temple of science in which only the truth is sought for".[25] This letter, printed in the *PGAS* in both German and Greek, reveals a bitter criticism of the occupation, although it does not waste any words on the real afflictions of the Greek population. It is only the Hellenic spirit that seems to have suffered; the Acropolis, but not the people. While all accounts of the past needed to be balanced on the day of the liberation, even these serious events failed to break the normality of the scientific association. After the celebratory address, routine matters continued with a lecture on the "order of the letters in the new Greek phonetic alphabet"[26] and another concerning an ophthalmologic study of 'epitarsus'.[27]

Even though it had taken almost three years, in 1947 it was announced that the second open letter addressed to German anthropologists had been more successful than the first. Of the initial eleven, the three who had responded[28] were none other than the leading National Socialist anthropologists Eugen Fischer, Fritz Lenz and Otmar Freiherr von Verschuer.[29] All three claimed that they had never received the first letter and that the second one had reached them only in 1947. The president of the Greek Anthropological Society claimed he could recognize in their writings the character of each one, since he knew "in person all three great scientists"[30] whose lectures he had attended during the summer semesters.[31]

---

25 Koumaris, *PGAS* (12 October 1944): 57.

26 Charitakis, *PGAS* (12 October 1944): 62–7.

27 Fronimopoulos, *PGAS* (12 October 1944): 67–71.
The Epitarsus is an apron-like fold of the conjunctiva, attached to the inner tarsal surface of the eyelid.

28 *PGAS* (24 December 1947): 11.

29 The response letters were printed in the *PGAS* in Greek translation only.

30 *PGAS* (24 December 1947): 11; the letters are printed on pp. 11–14.

31 *PGAS* (24 December 1947): 11.

> [Eugen Fischer], the most famous current anthropologist ... absolutely sincere, apologizes to Parthenon and to the members of the Greek Anthropological Society ... Otmar Verschuer[32] loses himself in Gallic diplomacy, expressing his regret for the happenings. ... Fritz Lenz, famous for his classic treatise on heredity, almost protests, with a Teutonic obstinacy typical of his character, against wronging German scientists.[33]

The audience was reminded of the responsibility these scholars bore: thus, Lenz had taught the superiority of the German race and "many of these anthropologists were Hitler's advisers in the brutal anti-Semitic campaign".[34] Fischer, Lenz and Verschuer appear to have presented themselves in the letter as nothing but 'powerless scholars' who neither wished, nor were able, to exert any influence on politics. Lenz emphasized that Germans also had to cope with the consequences of this 'criminal war', which had been followed by the separation of the eastern territories of Germany (1945) and thus the "loss of a great part of the population". Eugen Fischer received the least criticism, although this was not because he enjoyed a reputation in Greece prior to World War II (his scientific achievements had been acknowledged by the Greek government in 1935 when he was awarded a prize "in recognition of his scholarly achievements and the advancement of the cultural relationship between our two peoples").[35] The reason for reprieving him was rather that Ioannis Koumaris himself seemed to have had a closer relationship to Eugen Fischer, which even continued in the post-war period. The involvement of the Greek anthropologist in the commemorative publication dedicated to Fischer's 80th birthday in 1954 ([Johannes] Koumaris 1954), confirmed his status as an associate of the group of 'Fischer's students', the so-called 'Fischerianer' (Lösch 1997: 486). Neither Eugen Fischer nor the other German anthropologists have ever been held liable for their activities during the era of National Socialism; all of them continued their academic careers after the war.

---

32 On the contact and collaboration of Verschuer and the notorious Dr Mengele (1911–79), see Lösch, (1997): 405–17.

33 *PGAS* (24 December 1947): 11 and letters in pp. 11–14.

34 *PGAS* (24 December 1947): 11.

35 Letter (dated 15 June 1935) by Eugen Fischer addressed to Professor Bieberbach, dean of the philosophical faculty of the Friedrich-Wilhelm-Universität Berlin.

Archive Humboldt Universität Berlin, Bestand UK F 62, p. 6. See also correspondence on the same issue between the dean of the university and Fischer in the same portfolio pp. 8, 51, 53, 55, 62.

6.3 *Civil War: Victims, Patriots and Badly Educated Peoples*

With the intensification of the Greek Civil War, 'neutrality' became impossible as each of the hostile fronts called the others to account. The academic community, including the members of the Greek Anthropological Society, was not immune from this conflict. The turbulent political climate was reflected in the first meeting of the GAS after liberation, held on 21 February 1945, three months after the so-called 'Fights of December 1944' [*Δεκεμβριανά*][36] and only a few days after the signature of the Varkiza Agreement (12 February 1945),[37] which led to the intensification of the Civil War (Mazower 2000). The archaeologist Spiridon Marinatos[38] complained of having been attacked in his house by five armed men, members of the National People's Liberation Army [ΕΛΑΣ-Εθνικός Λαϊκός Απελευθερωτικός Στρατός] on the night of 22 December 1944. Although he managed to escape, he said, the men returned in greater numbers three days later,[39] surrounded his house and ransacked his manuscripts. They returned some of the manuscripts to him several days later (on 1 January 1945).

Apart from such skirmishes, the anthropological society also lamented the loss of members in the "terrible nightmare of that December night".[40] Mention was made of the founding member Konstantinos Melissinos, who committed suicide when his wife was fatally shot, and a professor of the Polytechnic University[41] who "died of the consequences of captivity" (it was not further specified by which front). It was noted that another member of the university staff[42] had been found dismembered, and that

36 'Fights of December' [*Δεκεμβριανά*] is an established term for the armed confrontations between the leftist front and a broader front that included various forces, ranging from the conservatives to ultra right-wingers. The British army was also involved. The armed conflict started during a demonstration in Athens on 3 December 1944 and continued until January 1945.

37 The Varkiza Agreement was signed between the Greek government (under Nikolaos Plastiras, 1883–1953) and the National Liberation Front [Εθνικό Απελευθερωτικό Μέτωπο: ΕΑΜ] which coordinated the greatest part of the resistance. One of the conditions of this treaty was the disarming of the National People's Liberation Army [Εθνικός Λαϊκός Απελευθερωτικός Στρατός: ΕΛΑΣ], although some of its leaders refused to give up their weapons.

38 Spiridon Marinatos later forged a political career during the Colonels' dictatorship, 1967–74.

39 Marinatos, *PGAS* (21 February 1945): 13–20.

40 *PGAS* (21 February 1945): 11.

41 He meant Georgios Sarropoulos.

42 Namely Nikolaos Efstathianos.

other professors had been kidnapped, including the coroner, Spiridon Ardavanis-Limperatos, a well-known admirer of Italian fascism.[43]

These references actually constitute brief accounts of the events which shook Greek society and which also caused polarization among physicians, echoing a public controversy concerning the collective role of physicians during the Axis occupation and the Civil War. Physicians had indeed been involved on all fronts of the political and armed conflicts. After the Varkiza Agreement, some had sided with the government forces and their collaborators, while others had joined the Democratic Armed Forces of Greece [Δημοκρατικός Στρατός Ελλάδας] which sympathized with the Communist Party of Greece, and which also had a Hygiene Service.[44] In this extremely polarized climate, patriotism became a key notion for defending the collective image of physicians as a professional group both retrospectively with respect to the occupation period, as well as actually during the Civil War. The then contemporary medical press bears witness to these efforts. Various articles in the *New Medical Newspaper* [*Νέα Ιατρική Εφημερίς*], for instance, either reported on the physicians' selfless patriotic commitment to saving the lives of Greek fighters and civilians,[45] or condemned the murder and torture of colleagues during the 'Fights of December' and the Civil War. The director of this newspaper (a former military physician and

---

[43] In his serialized publication on "The New Methods of State Ideology in Italy" in the *New Medical Newspaper* in 1938, Spiridon Ardavanis-Limperatos presented the fascist social system and the care for a sound population, both children and family, in Italy. He also emphasized the differences between socialism and fascism, advocating the latter: Ardavanis-Limperatos (1935c).

Ardavanis-Limperatos was also an enthusiastic proponent of eugenics and frequently published his views in the *New Medical Newspaper* [*Νέα Ιατρική Εφημερίς*], where he also reported on related events such as the founding, in 1934, of the Greek Society of Criminal Biology [Ελληνική Εταιρεία Εγκληματολογικής Βιολογίας]. In the medical press Limperatos presented Lombroso's theories as the basis on which the Greek Society of Criminal Biology [Ελληνική Εταιρεία Εγκληματολογικής Βιολογίας] had been founded by Professor I. Georgiadis at the Institute for Forensic Medicine [Ιατροδικαστικόν Ινστιτούτο], University of Athens, in 1934. The society aimed to contribute to the implementation of eugenic measures in Greece: in doing so, it would coordinate its activities with various international organizations. Since the founding of this society occurred following international conferences on the subject of criminal biology in Italy and France, it is probable that the founders' main intention was to establish a scientific agency that conformed to international standards, through which they could insert themselves into international networks. See the serialized article by Ardavanis-Limperatos (1934).

[44] On the Hygiene Service during the World War II see: Sakelariou ([no year]); Pentedekas (2007); cf. also the autobiographies by the doctors Stefanos Chouzouris (1988) and Nikitas Agapitidis (1996).

[45] See some representative articles in the *New Medical Newspaper*: [Anonymous] (18 April 1945); Mprougkas (15 May 1945).

retired officer in the State Hygiene Service) reported on 14 March 1945 of "physicians massacred... [by] unscrupulous criminals of the December rebellion against the nation" ([K.B.K. (K.B. Kalogeris)] 1945). He underlined the responsibility not only of those physicians who had themselves acted against their colleagues, but also of those who had either collaborated in these crimes or tacitly tolerated them, thereby contributing to the cultivation of a climate of mistrust within the medical guild.

Once the occupation was over, and the Civil War had intensified, a new kind of war also started which Hagen Fleischer has described as the "wars of memory" (Fleischer 2008). Individual cases of resistance during the occupation were retrospectively interpreted as the collective patriotic activity of the entire medical profession. The most famous case was probably Georgios Sklavounos's statement to the Academy of Athens in 1942,[46] during discussions over the founding of a 'Greek-German Research Institute for Biology' in Piraeus.[47] This project was headed by Max Hartmann (1876–1962), the director of the Institute of Biology at the Kaiser-Wilhelm-Society in Berlin. The Greek–German Research Institute for Biology came into existence but it was short-lived, existing only from 1942 to 1944. Although important members of the Greek Anthropological Society were involved in both the planning and the realization of this project, no aspect of the topic was ever discussed in the anthropological society. The silence that reigned within the anthropological discourse is especially significant when members who were faithful to and active in the GAS, as Georgios Sklavounos was,[48] were involved in developments.

---

46 [Anonymous] (1945): "The National Commitment of Doctors during the Occupation: G. Sklavounos, Speech at the Academy of Athens in 1942" ["Η εθνική δράσις των ιατρών επί Κατοχής. Γ Σκλαβούνος, Ομιλία εν τη Ακαδημία Αθηνών τω 1942"], *Νέα Ιατρική Εφημερίς* (18 April 1945): 2.

47 The first idea for founding this institute emerged in 1936. Initiators and negotiators on the German side were the biologist Maximilian Hartmann (1876–1962), the botanist Fritz von Wettstein (1895–1945) and the zoologist and geneticist Alfred Kühn (1885–1968), directors of the Institute for Biology at the Kaiser Wilhelm Society (Berlin Dahlem). Konstantinos Tzonis was a potential director from the Greek side. The tasks of the institute were to conduct biological research on vegetables, and ultimately to advance bilateral cultural relations between Greece and Germany. After the outbreak of the war the institute and its investigations were recognized by the German side as being "important for the war", but there is little data available on its activities during the brief period of its existence. For information covering its initial conception in 1936 to its abolition in 1944, see Zarifi (2002 and 2010) based on the archival records of the Max-Planck-Society.

48 Sklavounos appears to have chaired the greatest number of the meetings of the GAS.

During the discussion on the founding of the Greek–German Research Institute for Biology at the Academy of Athens, Georgios Sklavounos voiced his objection to setting up such an institute at that point in time and suggested *postponing* the project until after Greece's liberation. He largely based this argument on the grounds of the enormous fiscal shortages that plagued Greece's economy, Greek scholarship and the population in general as a consequence of the occupation,[49] noting that the academy could resume negotiations after the war, when Greece would have equal rights in co-determining the conditions of the project. Sklavounos's statement reflected the ambivalent attitude of a Greek scholar who, on the one hand, was an admirer of German science, but, on the other, was opposed to the occupation and recognized its grave consequences for both Greek academia and the Greek population. He argued that:

> the German powers in Greece understandably seek to abrogate serious scarcities in the lives of the people as well as in science, and all Greeks follow their recommendations with much emotion.[50]

But on the other hand he reminded his audience that both the simple people and scholarship were suffering from the war; and, with regard to the latter, that all laboratories and research instruments had been seized by the occupying forces. In view of the economic disaster and its devastating effects upon the population, Sklavounos thought that the founding of such an institute was both unreasonable and unethical: how was it possible "to discuss at this meeting issues such as the founding of an Institute when, en route to the Academy, we see on any street-corner starving people dying?" After all, he asked, whose needs would the institute serve in a country where the population was dying of starvation?

> Mr Hartmann should offer bread to starving Greek scientists and the people rather than this project:... the Greek people are not discouraged but hope for the goodwill of the powerful Third Reich that, once divided into small states like Ancient Hellas, has managed by means of Hellenic education to rise to greatness, also in the arena of international contest.[51]

---

[49] Sklavounos's argument was not unfounded: it was proved in the course of time that the Greek side was not able to meet its financial obligations; the German organization, the Kaiser Wilhelm Institute, therefore had to carry the Greek share as well. Cf. Zarifi (2002): 218–19.

[50] [Anonymous] (1945).

[51] [Anonymous] (1945).

Sklavounos did not hesitate to underline the academy's responsibility, and warned that, if it accepted this project, its collaboration would be tantamount to subjugation. If the academy agreed to collaborate with the National Socialists while ignoring the extremely urgent situation of the people, then it would simply be an "Academy of Serfs".

In subsequent years, this latter expression became subject to various interpretations and usages. Immediately after Skalvounos's statement, the gazette *Medical Fight* (a professional periodical for physicians and dentists, which was part of the medical press during the resistance) reacted enthusiastically and published a large part of his statement in a special edition headlined "Academy of Serfs".[52] In the post-war period Sklavounos's statement before the academy was expropriated as a collective act of resistance by the medical professionals and even by the academy. His address was reprinted in 1962 in the journal *Review of Art* [*Επιθεώρηση Τέχνης*] with the additional subtitle "The Academy Declined" ([Anonymous] 1962). Even today, his statement is considered as a collective act of national resistance by Greek scholars. It has been incorporated into the history of the University of Athens and is still included on the university's official website.[53]

However, in contrast to this post-war and undoubtedly selective memory, it is unlikely that the academy rejected the project outright. According to a study by the historian Maria Zarifi, the academy (in line with Sklavounos's suggestion) officially proposed to Max Hartmann that the realization of the project should be postponed rather than turned down. However, the then head of the academy had already signed Hartmann's invitation to collaborate and later even became director of this institute (Zarifi 2002: 225). Furthermore, members of the Greek Anthropological Society were involved in this project, often in leading roles, such as, for example, the professor of zoology, Georgios Pantazis, and the biologist Konstantinos Tzonis. The directorship had initially been intended for the latter, who was director of the Laboratory for Biological and Biochemical Research at the Institute for Cancer Research [Ελληνικό Αντικαρκινικό Ινστιτούτο, Εργαστήρια Βιολογικών και Βιοχημικών Ερευνών].[54] Yet after

---

52 "Academy of Serfs" ["Ακαδημία Ραγιάδων"]. *Ιατρικός Αγών—Εθνικόν και Επαγγελματικόν Όργανον των Ιατρών και Οδοντιάτρων*, Αθήναι, 1, 2 (May 1942), (special issue).

53 See http://kapodistriako.uoa.gr/stories/113_th_01/index.php?m=2 (last accessed 22 August 2012).

54 Konstantinos Tzonis joined the Greek Anthropological Society on 3 November 1938. He worked from August 1936 to November 1937 at the Kaiser Wilhelm Institute for Biology under the leadership of Maximilian Hartmann who intended him to become director

Greece's entry into the war, Tzonis's relationship with the German head of the project seemed to change, and even though he continued to declare his willingness to participate in the project, the break was initiated by the German side (Zarifi 2002: 225). The reasons for dropping him and promoting the academy's director to his position are not known. After the war, Konstantinos Tzonis was depicted as a leftist, and even as a member of the Communist Party of Greece.[55] In a treatise concerned with the University of Thessaloniki and the resistance, Tzonis is mentioned as having been among the professors who most conspicuously influenced the resistance in Thessaloniki (Anastasiadis 1998: 262–77).

Although all these actors were involved in the Greek Anthropological Society and some were even among its most active members (especially Georgios Sklavounos), these events were not discussed within the society, the directors of which were anxious to keep the anthropological institution politically neutral, as if it existed in a political vacuum. In the settings of the Greek Anthropological Society, war, collaboration and war crimes seemed to exist beyond individual interests and were projected on the sphere of 'collective education', calling forth the need for a 'nurturing of the peoples'.

In the aftermath of World War II and during the Civil War, the president of the anthropological society strove to find a satisfying explanation for the 'Fights of December' which, due to their gravity, "bear no relation to the usual spur of the moment popular uprisings and disasters".[56] Responsibility for this, he concluded, could be ascribed to the 'bad education' of the Greek *fili* and the failure of their 'nurturing'.[57] The Greeks should be educated in order to cultivate their inborn virtues and minimize their inborn vices, above all their egocentrism, so that they would

---

of the planned institute and who sent him the draft of the project on 1 December 1937. Tzonis and Hartmann corresponded with each other from that time onwards and Spyros Dontas supported Tzonis as the possible director of the institute. Hartmann submitted expert reports when requested by Tzonis, who counted on Hartmann's collaboration for the Institute.

Archiv Max-Planck-Gesellschaft, Hauptabteilung III, Rep. 47, Nr. 1490: "Gutachten für die Gründung eines Griechisch-Deutschen Forschungsinstituts für Biologie in Athen".

55 According to information provided by Tzonis's son to Maria Zarifi (2002): 227.

56 Opening address of meeting: *PGAS* (21 February 1945): 12.

57 He had addressed this question extensively in an article titled "The Education of the Race" submitted to the church journal *Ecclesia* (Koumaris 1945c), but anyone reading this article will search in vain for references to the Greek Civil War or to any other war. Only the difference between 'nature' and 'nurture' was outlined, first generally and then more specifically with respect to the Greek race.

commit themselves more readily to the common good. The task of providing such an education should be given to the church and the school. Yet 'bad education' was the reason not only for the Civil War but for all recent wars, including World War II. Thus all peoples and races should be educated in the ideas of peaceful co-existence, with patriotism being an imperative for everyone. In an equivocal performance, when outlining the "Problem of Wars",[58] he recited the well-known "Song of the Germans" (Germany's anthem since 1922) suggesting that in a new post-war humanity any scholar, any people in the world, should embrace these words after replacing the reference to Germany by the name of their own country:

> any citizen will, and must, be taught "*Gedenke stets, daß du ein (Deutscher) bist! Du darfst es mit Stolz tun, es gibt kein besser Volk im Erdenrund*" [Remember always that you are a (German); you must do so with pride; there are no better people in the world];[59]
>
> and moreover, the famous: "*(Deutschland), (Deutschland), über Alles in der Welt*" [(Germany), (Germany), above all].[60]

Yet he did not hesitate to recite the most popular songs in National Socialism, as a negative example, arguing that neither Germans nor any other people should be taught

> *und so werden wir marschieren bis die Welt in Trümmer fällt, heute gehört uns (Deutschland), morgen die ganze Welt…*
>
> [And in this way we will march until the world is ruined. Today (Germany) belongs to us, tomorrow the whole world…][61]

This astonishing performance took place only a few months after the German occupying forces had left Greece and only three months before Germany's capitulation, and it was printed in the *Proceedings of the Greek Anthropological Society*. A close consideration of Koumaris's statements

58 Koumaris, *PGAS* (21 February 1945): 40.

59 (Parentheses in original.) A Greek translation of the text was written and the German original was left in a footnote. *PGAS* (27 June 1945): 31–44, 40.
This is a passage from the essay "Farewell" ["Ein Abschied"] by Tim Kröger (1844–1918), which is included in the collection *From an Old Chest* ["Aus alter Truhe"], Hamburg: Janssen, 1914. Kröger was a German jurist, agriculturist and author.

60 Koumaris, *PGAS* (21 February 1945): 40.

61 Koumaris, *PGAS* (21 February 1945): 41.
It was written by the National Socialist lyricist and composer Hans Baumann (1914–88) in 1933, when he joined the Hitler Youth. However, there are various versions; see Hillesheim and Michael (1993).

and attitudes during and after World War II can only be appropriately pictured in the light of his racial theories and the double-sided way in which he sought to cope with his sympathy for Germany, and the changing power relations in Greece and in Europe by adjusting his racial theories accordingly.

EXCURSUS: ANTHROPOLOGICAL CONCEPTIONS AND UNIVERSITY CAPITAL

In the period considered in this study the evolution of anthropology as a science in Greece was clearly defined by the distinct eras associated with the two professional anthropologists, Clon Stéphanos and Ioannis Koumaris. Each era is distinguished in terms of the epistemic concepts of anthropology, but also with regard to how each anthropologist sought to situate his discipline in society. The starting position of the anthropological discipline in Greece is associated with Clon Stéphanos's era and the introduction of anthropology as a scientific discipline with the founding of the Anthropological Museum. In epistemic terms, this period was characterized by the alliance of physical anthropology with archaeology, the natural sciences, linguistics and, to a lesser degree, ethnography. Stéphanos's anthropology bore a national outlook; it was confined to the corroboration of the dominant national doctrine and sought to buttress the antiquity cult using anthropological evidence. The personal change at the directorship of the museum entailed a conceptual shift in understanding anthropology and simultaneously inaugurated a long period during which the field of anthropology was shaped by the agency of Ioannis Koumaris. His agency was important not only in view of how long this period lasted; he was also provided with highly beneficial conditions for forming the scientific capital of this discipline at the Greek university while he headed all three anthropological institutions: the museum, the university chair and the anthropological society.

The shift from a physical anthropology allied with auxiliary sciences, as applied by Stéphanos, to anthropology in a broad sense under the agency of his successor, entailed a radical change in the scientific paradigm. The subject of anthropology was not simply approached by means of different disciplines, but was now reformulated to become "poly-scientific or interdisciplinary", as Canguilhem (1979: 147) defines a scientific subject that arises from the collaboration of diverse disciplines and comes to form an autonomous subject in which all these disciplines appear to be interwoven. Clearly, in the era of Ioannis Koumaris, the scope of anthropology varied in each of the three institutions, ranging from a physical anthropology that was restricted to the university chair, to a more extensive one that encompassed colonial racial anthropology at the museum, up to anthropology in a broad sense at the GAS. In the latter organization, anthropology's subject was more than 'poly-scientific' and came to bear a fetishistic outlook, since its subject was imagined as being dictated by the conditions that produce them; and these conditions were everything that was related to human activity. The broadening of anthropology's scope in this fashion occurred at the GAS and was inversely proportional to the declining symbolic capital of anthropology and its representatives due to the decreasing significance of the anthropological institutions at the university. The 'day after' Koumaris's era, anthropology found itself in a situation similar to that of its starting position: it was faced with an incapability to reproduce the scientific field because of the lack of potential successors. Shortly before retiring, the chair-holder became aware that he was the "only official representative of physical anthropology in Greece".[62]

This was the outcome of a much larger process, the starting point of which was rather auspicious. Indeed, at the very beginning of his career, the holder of the university chair for physical anthropology was accumulating university capital, and the three institutions he headed worked in a complementary way in accrediting him with the symbolic capital which serves as the primary resource of recognition and acknowledgement in academia (Bourdieu 1998: 22). This kind of symbolic capital empowers academic actors to stipulate both the "rules and the regularities of the game" in academia, as for example in determining which subjects are important at a given faculty (Bourdieu 1998: 23). But the field is a "stake of struggle", and holders of positions have to assert and secure their positions by pursuing appropriate strategies, even adjusting their strategies to the dynamic structures of the academic field (Bourdieu 1998: 25–6). Yet it would appear that this is just what the chair-holder of physical anthropology failed to do, and in effect his institutionalized acknowledgement was weakening and his symbolic capital decreasing. The most indicative affirmation of this failure was that the academic field of anthropology was not reproduced, since no professional anthropologists were trained in Greece, a situation that ultimately led to the abolishment of the university chair. But also, if the GAS can be seen as a mirror of recognition and acknowledgement, then the reduction in the number of its members is an indication of this trajectory. On a symbolic level, the epitome of this lack of acknowledgement was the failure of the "Koumaris Award for Studies on

[62] *PGAS* (1 June 1949): 12.

the Greek Race" to attract the interest of scholars: while there were plans to award five studies, only three prizes were awarded in reality.[63] The initiator complained about the lack of interest, asserting that young physicians were not willing to abdicate the lucrative profession of medicine in order to dedicate themselves to the anthropological sciences.

> Unfortunately, unselfish commitment to graceless and dry research today has become more and more rare. Prizes and awards are of no avail and cannot wrest fine young people from lucrative occupations, facile pleasures and frivolous life. As a result, the award continues to wait for the pioneers of science ...[64]

Without finding a link to the topical concerns of society, the anthropological discipline failed to meet the challenge it faced in the academic field. First of all, the chair of physical anthropology was required to assert the discipline's authority within the medical faculty, whose relevance for society was taken for granted. Indeed, while the social necessity of medicine asserted and justified the relevance of the medical faculty and professors for society (Bourdieu 1988: 63, Foucault 2006), a discipline like physical anthropology had to battle to achieve recognition as socially relevant within the same institution. It proved anything but an easy task to attract medical students with a curriculum in which human prehistory predominated at a faculty with a temporal social competence (Bourdieu 1988: 62). Nevertheless, the whole history of anthropology internationally points to the battle for the right to design society, wherein anthropologists believe that they are equipped with a clear, prior advantage: the academic authority over the cognition of humans in their entirety and even in a global context. At this stage of my analysis I conclude that Greek anthropology

[63] The first award (on 18 April 1951) honoured a team of three physicians for their joint study on "Biometrical Studies on the Face and Ears of Adult Greeks". The prized researchers were the professor Georgios Apostolakis, director of the anatomical laboratory of the University in Athens, the associate professor Petros Petridis and the dentist Orestis Louridis. *PGAS* (18 April 1951): 14–32.

On the second occasion the prize was awarded (on 29 November 1956) for a study entitled "The Aegean, Cradle of the Aryans and Hellenism: Cultural Antecedence and Radiation—Racial Unity—Racially Negligible Intrusion of Slavs and Turks". The winning author of this 200-page monograph, published in 1956, was General Xerxis Livas (1899–1965). A summary of the study was printed in the proceedings of the anthropological society: Livas, *PGAS* (29 November 1956): 26–43. English summary "The Aegeis [*sic*]—Cradle of the Arians [*sic*] and of Hellenisme", by General X. Livas.

The entire treatise was published the same year and a second edition appeared in 1963. Livas (1956/1963). The third prize was awarded to Aris Poulianos, for his PhD on "The Origins of the Greeks. Ethno-Genetic Research", see Poulianos, *PGAS* (2 April 1962): 20–31.

[64] *PGAS* (1 June 1949): 13.

failed to link itself to social questions in a way that may ultimately have upgraded its topical relevance. This failure was not due to the lack of such an aspiration. Finally, for all proclaimed or pretended neutrality, anthropology's relevance was supposed to reside in defining Greek communality in racial anthropological terms. The public statements of the main protagonist of the anthropological discourse and moreover the statements recorded in the *PGAS* disclose a singularity in thinking; his racist views are also unambiguous. However, it was not necessarily his political and other statements, or even his obsession with the Greek race, which hampered the attractiveness of the university chair for the students and the GAS for the scholars. Rather, the reverse occurred: the decline in his symbolic capital severely limited the legitimacy of his assumptions; it was not the assumptions as such that were discredited, but the incapability of their speaker to "manipulate the scientific field"; or perhaps it was due, in the final analysis, to the fact that "the scientific field is not manipulable", using Bourdieu's idea:

> For all possible expertise and administration of networking, the chances of an individual actor to enact the forces of a field according to his own desires, depends on his power over the field, on his scientific credit, or even more concrete, on his position in the structure in which capital is distributed (Bourdieu 1998: 22).

The identification of anthropology with racial science, and in particular with the study of the Greek race, might well have accounted for the increasingly downward course of institutionalized anthropology during his era. This was not because Greek elites were uninterested in racial concepts for Greek communality, but rather because the specific racial concepts in question seemed to be extraneous to mainstream Greek elites. In the following section the racial anthropological conceptions will be explored in terms of their epistemic and ideological resources, their contents and also with respect to their perception by those involved in the anthropological discourse.

SECTION THREE

CONCEPTS OF THE GREEK *FILI*: COMMUNALITY IN RACIAL AND EUGENIC TERMS

INTRODUCTION TO SECTION THREE: TERMINOLOGICAL METAMORPHOSES OF *FILI*—DIVERSE CONCEPTS OF COMMUNALITY

The anthropological sciences claimed to be able to respond and provide answers to topical social developments and even to design society. In the period considered in this study, this aspiration was reflected in the conceptualization of large groups in racial and eugenic terms. In the Greek anthropological discourse *fili* arose to become the main notion of reference to Greek communality. Although *fili* is usually translated into race, its semantics has been subject to alterations that have led the term to be marked by polysemy, even in its anthropological usages. In this section of the book I will address the semantics of *fili* in the racial anthropological and eugenic discourses and seek to ascertain the regimes of truth that are inherent to this. The starting argument is that, before being embraced by the anthropological discourse, the notion of *fili* experienced a number of transformations over a long period, which can be traced in its anthropological, racial and eugenic usages in perceiving Greek communality. A conceptual history of this notion, as will be attempted in following, can provide evidence for the representations of Greek communality and the ways in which the members of the Greek community are thought to be tied together, as historical and social subjects. I will argue that *fili* was conceptualized differently in the anthropological and eugenic discourse. While the notion came close to being identified with race in the former, bio-political, but not necessarily racial, connotations prevailed in the latter. The consideration of racial and eugenic concepts over this period allows for conclusions to be made regarding projects that sought to respond to transformations in Greek society during the process of its modernization and, in turn, the ways these transformations irrupt in the notion of *fili* and shape its semantics. In particular, the question arises as to what extent the racial notion of *fili*, as conceptualized in the anthropological discourse, was used to answer topical questions in Greek society and the extent to which it interpreted social targets in terms of racial terminology. Given that eugenic thinking in Greece, at least in the first decades of the 20th century, was embedded in socio-political reformist projects that advanced the establishment of welfare institutions in Greece and sought to cope with the rise of internal social differentiation and stratification, as outlined in the second chapter, to what extent did the anthropological notion of *fili* keep up with the attempts to meet this challenge? And what answers did it ultimately provide?

Tracing the conceptual history of the notion *fili*, as with that of any other notion, neither presumes nor implies a lineal evolution of meanings in the sense of passing from one meaning to another. Rather, it reveals the phenomenon that Michel Foucault calls the "displacement and transformation of concepts" (Foucault 1972: 4). By way of analogy with other terms, I will show that the conceptual history of *fili* is that "of its various fields of constitution and validity, that of its successive rules of use, that of the many theoretical contexts in which it developed and matured" (Foucault 1972: 4). This in turn provides the possibility to recognize the components and socio-historical developments by and through which this notion has been enriched, or, in a metaphorical sense, how a notion may function as a mirror for its age and the ages that it has covered until reaching us.

CHAPTER SEVEN

RACE AND GREEK ANCESTRY

> What is black authenticity? Who is really black? First, blackness has no meaning outside of a system of race-conscious people and practice. After centuries of racist degradation, exploitation, and oppression in America, being black means being minimally subject to white supremacist abuse and being part of a rich culture and community that has struggled against such abuse. All people with black skin and African phenotype are subject to potential white supremacist abuse.
>
> (West 2001: 39)

This chapter addresses anthropological concepts which sought to specify the Greek *fili* as a collective subject and aims to identify their sources. The central argument raised is that *fili* emerged to become a major means of researching and discussing descent in the Greek anthropological discourse, particularly as it pertained to Greek (Hellenic) ancestry, in racial terms. The concepts in question were racial anthropological in character and national in outlook, and they drew arguments from two basic sources: the international arsenal of racial theories and Greek nationalism. The former came to substantiate national ideology, while the latter was transformed into racial anthropological ideas. Both the racial and national connotations of *fili* were, however, subject to alteration due to their dependence on the transformation of racial approaches and national ideology in the course of historical-political developments in Greece and in Europe.

Dealing with this question faces multiple challenges, one of the most difficult of which is the complexity involved in specifying the notion of race itself. Indeed, attempts to define race reveal that it might be more efficient to specify forms of racial representations and racisms (even more appropriate in the plural) than a single concept of 'race', not least because the semantics of this 'magic' word has been subject to multiple transformations. The 'magic potency' of race derives from its conceptual metamorphoses, combining hard scientific data with irrational arguments, and nevertheless—or even therefore—becoming a persistent and durable concept. Enriched with diverse universal and particular implications, its ambiguity has increased since the 19th century. It was after this period that the idea of race emerged to become a means for shaping the taxonomies of humankind, for approaching world history and for designing social relations. This multi-functionality is symptomatic of its appearance in various contexts, even becoming interlinked with ideas that are no less important for representing internal relations in individual societies as well as the world order; ideas of humanity, nation, gender or class.

There has been more dissent than consensus among scholars regarding the conceptual specification of race. Physical and racial anthropologists are no exception to this, although they raised a strong claim to be providing objective knowledge. The dissent concerning the concept of race included objections, doubts and disagreements with regard to multifarious issues, ranging from questioning the measurability of races, the difficulties of specifying racial taxonomies, up to challenging the existence of races per se. But in spite of these disagreements, until the mid-20th century race still represented an important scientific category for expounding the descent and heredity of humankind in general and of contemporary peoples and societies in particular. The racial idea remains durable and operative, despite all forms of dissent and ambiguity. And however paradoxical it might appear, its tenacity and its enormous possibilities for becoming topical in diverse contexts are derived, to a high degree, precisely from this lack of clarity.

Race is a relational rather than an ontological notion. Even though its reality is doubted, the existence of 'race relations', as evidenced in the consciousness of the social agents or in institutionalized discourses and practices, is an indisputable fact. The racial idea is a reality in that it evokes social action and facilitates the shape of a societal or universal order and justifies communal ties, even though racial research has failed to shape trustworthy racial taxonomies. Indeed, the racial idea has proved to be instrumental in evoking the feeling of a common belonging and in facilitating (or justifying) the constitution of collective subjects. This can occur on the grounds either of racist exclusion and discrimination or of the consciousness of belonging to a singled out and excluded 'community of fate', as described by Cornel West (2001) in the quotation above.[1]

Racial anthropological discourse examined large groups on the grounds of objective, diachronic and unchangeable genealogies by drawing on common racial descent. Racial concepts are based on positivistic arguments

[1] This consideration came rapidly to the fore during the post-colonial age, when race was implied by post-colonial intellectuals with respect to the emancipation of the colonized subjects, the 'subalterns' (Bhabha 1994).

corroborated by 'hard data' (i.e. data derived from measurements that allegedly provide scientific and efficient means of producing objective knowledge). Yet both rational and irrational moments are inherent in the racial idea, regardless of the role played in its evolution by positivism and 'objective' science. The irrational moment effectively operates in racially justified 'groupness' (Brubaker 2004), while communicating the feeling of togetherness, and works even when racial taxonomies are contested. Irrational reasoning seizes upon myths of descent (often embedded in common sense) that are particularly effective when scientific instruments are unable to provide assistance. Racial descent successfully acts as a keystone for reasoning on communality. Its conceptions differ remarkably in the scientific discourses within individual societies and the scientific traditions in various countries. In each specific context, the application of racial descent reveals the internal dynamics of the societies in question and corresponds to the specific processes of nation-building, state-building or the modernization of society.

In the following I will explore the anthropological notion *fili* as a concept of communality in which racial, national and religious connotations alter their meanings interchangeably. The concept of *fili* has been subject to transformation, and the vehicle for this was the changing perceptions of Greek descent. But rather than drawing lines to distinguish between different meanings, I intend to elevate the polysemy of *fili*, which bears testimony to continuities, alterations and interruptions in its semantics; in doing so, I aim to sketch the epistemic and socio-historical basis on which racial anthropological concepts have been developed. Starting from its predominantly religious usages in the Ottoman period, I will examine the ways in which it became enriched by national connotations through means of the idea of Hellenism, which transmitted an additional hegemonic idea with regard to the Greek relationship with other Balkan states. Furthermore, I will argue that the racial connotations of the *fili*, on the one hand, reflected the state of affairs in the racial sciences in Europe, and on the other, that they were strongly linked to the German racial sciences, those pertaining to the National Socialist era in particular, to the extent that they either embraced its racial theories or sought to reject them.

7.1 *From Religious Universalism to Hellenism: A Bridge to Pass from Empire to Nation State*

Although the word *fili* was borrowed from Hellenic antiquity, its semantics in the early period of the modern Greek state was drawn from the context of the late Ottoman period. At that time the term was used to emphasize religious bonds among Orthodox Christians and to substantiate affinities beyond narrow geographical or linguistic boundaries. Its content changed as soon as the meaning of religion shifted from being predominantly universal to being predominantly national. National ideology, in turn, was also subject to transformations that are discernible in the altering semantics of *fili*.[2]

In 18th-century church texts, *fili* expressed the ecumenical character of a large religious community that included all Orthodox Christians in the Ottoman Empire under the leadership of the Patriarchy of Constantinople. This ecumenical religious community aroused a sense of unity among peoples with different languages, but with Greek retaining its position as the proper sacred language. Linguistic groups were perceived as sub-*files* in the larger *fili* of Orthodox Christians. In this context, the term *fili* was used for describing assumed affinities between peoples beyond the narrow settings of a single nation. While nationalisms were gaining more and more ground, the semantics of this notion was enriched by national connotations and came to describe distinct religious and linguistic communities that now sought to become nations. The incorporation of *fili* into the national concept in the early 19th century did not lead to an abrogation of its inherent religious connotations—religion was to serve as a vehicle (but certainly not the only one) in the rise of nationalisms in the late Ottoman era. With respect to the Greek case, religion, and specifically Orthodox Christianity, was compatible with nationalism as it emerged in the aftermath of the so-called 'Greek Enlightenment' (1774–1821) and the later founding of the Greek nation state (Matalas 2002; cf. Sigalas 2000 and 2001). Typical of the transforming usages and semantics of *fili* is the Greek–Bulgarian conflict after the founding of the Bulgarian Exarchate in the 1870s (the so-called 'Bulgarian Schism' in the terminology of Greek historiography). In 1872 the Patriarchy of Constantinople condemned the founding of the Bulgarian Church, calling the schism 'Filetismos' [*Φυλετισμός*].[3] The univocal national semantics came to the surface

[2] To date the most substantial study on the correlation of religious, linguistic and national semantics in the concept of *fili* is provided by the historian Paraskevas Matalas in his work, *Nation and Orthodoxy; Adventures of a Relationship* (2002).

[3] See Matalas (2002): 22; cf. the *Lexikon of the Neo-Hellenic Language* by Mpampiniotis, according to whom the term 'Philetismos' was first recorded in 1835: Mpampiniotis (1998): 1933.

insofar as *filetismos* referred to the emergence of an Orthodox religious authority that established its own 'sacred language' that was different from Greek. Moreover, this authority pertained to what was ultimately a rival national entity.

The overall transformation of relations during the collapse of the Ottoman Empire, as well as the decisive role played by Russia and the Western powers in the Balkans, had significant effects on the semantics of the term *fili* in the century that was to follow. These factors have been examined by several historians and will not be described in detail here. The central point of my argument is instead that, in a long-term process of transformation, *fili* emerged as an ambiguous term associated with 'nation' prior to its association with 'race'. Despite its increasing polysemy, *fili* in this context still referred to communities or entities, perceived as universal in essence and existing beyond geographical boundaries. The perception of the bonds that tied together sub-*files* or the members of each individual *fili* also changed, shifting from togetherness substantiated on religious grounds towards genealogies based on primordial bonds and common descent. Here the changing meaning of descent and the insertion of genealogies is a key element for understanding the conceptual alterations of *fili*. Starting from the late 18th and extending into the 19th century, 'descent' increasingly became a holistic notion for interpreting history, politics and society. It was no accident that relations between peoples, nations or states were interpreted more and more in terms of common or different descent. Following the general trend, *fili* communicated the idea of common descent, naturalizing and thus objectifying supposed bonds and ancestry. Particularly in the 19th century, naturalized *filetic* bonds and 'common descent' gradually emerged to become a central argument for justifying the cohesion of a single *fili* as well as the affinity between several *files* (Matalas 2002: 134).

A watershed in the conceptual evolution of *fili* was its association with Hellenism in the 19th century, by means of which both the ecumenical and the hegemonic connotations of *fili* came to the fore; these same connotations were also inherent in the concept of Hellenism. The ecumenical aspect arose in that all sub-*files* were required to unite, although in this case (and here the hegemonic aspect becomes evident) it had to be under the headship of the leading Greek *fili*. In fact, after a certain point in time, Hellenism and *fili* were used synonymously and shared an alleged 'assimilatory power', namely the ability to absorb other peoples, nations, *files* or cultures. In the following, this idea of Hellenism's 'assimilatory power' will be addressed in greater detail, since it formed the basis for national thinking about a 'strong nucleus of the Greek *fili*' from which racial anthropological theory created the idea of a 'strong racial nucleus'.

The idea of the 'assimilatory power' of Hellenism originated in a post-imperial order that emerged while the Ottoman Empire was collapsing and independent nation states had started to emerge in the Balkans. Similar to the lengthy, drawn out disintegration of the Ottoman Empire, the process of shaping national ideologies for the burgeoning nation states took place over a long period of time, during which diverse national ideas coexisted either in parallel or competitively. Remnants of ideas influential in the empire could still be traced in the 19th-century nationalisms implicit in the issues that emerged over the positioning of these young states in the world of national states; territoriality in terms of fixing state borders or/and implementing projects of expansion; coping with new hegemonic orders in the region; and efforts to obtain a leading role in this order in the making. Originally, the modern concept of Hellenism was not elaborated with respect to the Greek national narrative but was thought to depict a chapter of universal—for which read West European—history. Its origins were localized by Konstantinos Paparrigopoulos (1815–91) who, with Spiridon Zampelios (1815–81), was the most important 19th-century Greek national historian,[4] as well as the writings of the German historian Johann Gustav Droysen (1808–84), and specifically the latter's work on the *History of Alexander the Great*, which later became the first volume of his *History of Hellenism* (Droysen 1833; 1836; 1843). According to Droysen, it was Alexander the Great and his Macedonians who Hellenized, and thus civilized, the Orient. In his approach Hellenism was a stage of universal history, which played a civilizing role that would perhaps resolve the contradiction of Occident versus Orient.[5]

In the Greek historiography of the 19th century, this idea was perceived as linking the universal and the national spirit, thereby integrating universal history and national historiography into a single national concept. This linkage formed what Michael Herzfeld has termed "ecumenical ethnocentrism", describing this as the claim to an "ecumenical vision [that] grew

[4] By Konstantinos Paparrigopoulos (1815–91) in a text from 1881 on "History of the Names Greeks/Hellenes, Greek Nation, Hellenism": Paparrigopoulos (1970): 90. Cf. Sigalas (2000) and (2001).

[5] In his article on the development of the modern concept of Hellenism, Nikos Sigalas (2000) and (2001) depicts the diverse perceptions of this idea by 19th-century Greek scholars.

naturally from the transplantation of philhellenism to the intellectual soil of Greece itself" (Herzfeld 1982: 51). The universal (or ecumenical) dimension was further accentuated in that the embedding of Hellenism in a state territory did not form a necessary condition for its definition. In 1881 Konstantinos Paparrigopoulos wrote that the term Hellenism was embraced by those "who refused to confine the Greek nation to the narrow boundaries of free Greece", meaning the Greek nation state of that time (Paparrigopoulos 1970: 90).[6]

The idea communicated by Hellenism also sought to situate the young Greek nation in the new order that was rising in South-Eastern Europe as the Ottoman Empire continued to decline. Hellenism incorporated the relationship of the Greek nation to the Occident and the Orient, to neighbouring Balkan states, to the peoples of the erstwhile Ottoman Empire and to other potential nations, first of all, the Slavs and the Albanians. Hellenism at that time served as the ideological vehicle for coping with the Ottoman legacy and regulating the relationship with other (Balkan) peoples and nations by ascribing to them a relationship that was not exclusive, but assimilatory and hegemonic. This concept stipulated the intention to incorporate foreign peoples and nations into a larger 'Greek family' under the auspices of the Greek spirit. Considerable importance was attached to religion in this narrative, which, also on a symbolic level, reflected collaboration with different states in a European landscape of power relations in transformation. For example, the Russophile Konstantinos Oikonomos[7] claimed that Greeks and Slavs (by whom he meant Russians rather than Bulgarians) were siblings on the grounds of their common Orthodox religion. Oikonomos was convinced that Russians should learn and adopt Greek as their church language.[8] The journal *Century* [*Αιών*], which advanced his ideas, described him as the "precursor of the national religion that joins under the spirit of Hellenism many alien nations such as the Albanians, Illyrians, Bulgarians, Serbians, Dracaenas and others, by this means increasing its power to the utmost".[9]

---

[6] Cf. the assertion by Sigalas that the semantics of Hellenism shifted to embed the geographical boundaries of the Greek national state after the Crimean War (March 1854–February 1856): Sigalas (2000) and (2001).

[7] On the Russian Party in Greece and the involvement of Oikonomos, see Hering (1992), first volume 93f. and *passim*, and 2nd volume, *passim*.

[8] On Oikonomos, his ideology and his conviction that Russians should learn Greek and adopt the Greek language as the language of the church, see Matalas (2002): 135 and *passim*.

[9] Cited in Matalas (2002): 147.

Once integrated into the national narrative, religion was clearly detectible in Greek irredentism, which was also advanced by those who were not Russophiles. Adherents of Greek irredentism, the so-called 'Great Idea', believed that the boundaries of Hellenism were identical with those of the Byzantine Empire. Zampelios, who current historians view as the instigator of the Great Idea, considered the "ethnological and geographical boundaries of the Greek people identical with those of the Gospel" (Zampelios 1852: 66), and believed that Orthodox Christians should be united in a single *fili*. With these assertions Zampelios proclaimed that the universal and hegemonic character of Hellenism transcended national limits; however, unlike the Russophile Oikonomos, he was opposed to the emerging Pan-Slavic movement.

A new impetus for defining the relationship between the Greeks and other Balkan peoples, in particular the Slavs, by drawing on the idea of Hellenism came from the attempts to refute the famous assertion of Jakob Filipp Fallmerayer (1790–1861) that due to mixtures with Slavs and other peoples, not a drop of Hellenic blood flowed in the veins of contemporary Greeks (Fallmerayer 1830–6). In her monograph on Fallmerayer, the historian Elli Skopetea explores these perceptions of Greek historiography, and shows that, in his hierarchical scheme, the Greeks were placed at the lowest rank (Skopetea 1999: 149–63). Due to his views Fallmerayer became a 'persona non grata',[10] and has remained unsolicited, even until recent times. What were the arguments raised by those who sought to disprove Fallmerayer and restore the idea of national continuity that he had challenged? Scholars who drew on the idea of Hellenism did not allege the purity of the contemporary Greek nation, but instead addressed their arguments to the affinities between the Greek world and Slavs or other *files*, and especially those whose representatives resided in the territory of the Greek state (Matalas 2002: 144). That there had been inter-mixing with the Slavs or other *files* was not contested, but it was described in such a way that it revealed Greek superiority over them.

Spiridon Zampelios and Konstantinos Paparrigopoulos also pursued this line of argument.[11] While accepting that such mixing of races had occurred, they claimed that this had resulted in the 'Hellenization' of the Slavs and other peoples rather than in the 'Slavization' of Greeks or in

[10] Todorova (1999): 182. On later perceptions of Fallmerayer in Greece, see Todorova (1999): 181–2.

[11] On Paparrigopoulos's line of argument against Fallmerayer see Veloudis (1970).

the alteration of the Greek *filetic* character more generally. This happened because of the superiority of Hellenism, namely the Greek spirit and culture, and the ability of the Greek *fili* to assimilate other *files*. Moreover, the Greek *fili* could benefit and become stronger by means of such mixing. Paparrigopoulos argued not only that racial mixing occurred, but that it was also necessary, since it advanced the intelligence of the peoples.[12] Zampelios additionally claimed that the first Slavs who arrived in Greek territory identified themselves with the Greeks due to their "common love for liberty and independency" (Zampelios 1852: 393–402).

Using similar arguments, in 1850 Paparrigopoulos wrote that the point was not to prove that the modern Greek nation could be conceptualized in terms of a direct, linear history from Pericles. Mixing between races indeed took place, but the point was to verify the idea that this had resulted in the creation of a nation that was now a great political entity. The continuity of the Greek *fili* was consequently evident not in its purity but in the superior spirit of Hellenism that had persisted and had become even stronger over the centuries (Paparrigopoulos 1850/1851a: 201). Neither Zampelios nor Paparrigopoulos contested the presence of Slavs in the Peloponnese, although each arrived at different estimates of the magnitude of the Slavic presence. Slavs did exist in Greece, but they had been acculturated and absorbed into the Greek element. Paparrigopoulos colourfully illustrated the process of this claimed assimilation in the following words:

> [Slav settlers in Greece] were finally tamed by the Greek *fili*; they accepted the language and the religion and blended like the water of the river, which maintains its colour and quality, and although still visible at the river's mouth, finally disappears in the infinite liquidness of the sea (Paparrigopoulos 1848/1986: II).

Apart from the Slavs, the Albanians were also among the most discussed peoples with regard to their relationship to the Greeks, as their assumed affinity supposedly went back to antiquity. As successors of the Illyrians, the Albanians were thought to be affiliated with the Greeks via primordial ties dating from very ancient times. The indigenousness of the Albanians settled in Greek state territory was asserted by Nikolaos G. Nikoklis in his

12 Paparrigopoulos made this argument in an article with the title "Introduction to History of the Renewal of the Greek Nation" which appeared in two issues of the journal *Pandora* [*Πανδώρα*]: Paparrigopoulos (1850/1851a) and (1850/1851b); here (1850/1851a): 201.

doctoral thesis submitted in 1855 to the University of Göttingen,[13] which was attacked by none other than Fallmerayer himself.[14] This thesis in fact expanded upon assertions that Paparrigopoulos had originally made some years earlier, in 1850, in a work entitled *Introduction to the Renaissance of the Greek Nation,* that two *files,* the Greeks and the Albanians, had originally inhabited Greece. He had also asked the question: "But does the Albanian fili constitute a separate nation? Has it shared feelings, shared concerns?" (Paparrigopoulos 1850/1851a: 201), to which he concluded: "It did have once, but has no longer." Paparrigopoulos considered Albanians to be a linguistic entity, but even this bond of language was about to disappear in the process of their assimilation in "the conquering march of Hellenism" (Paparrigopoulos 1850/1851a: 201). He thought that the process through which the Albanians had been assimilated was still in progress, in contrast to the Slavs whose absorption had been completed much earlier.

Acculturation of the foreign *files* rather than the transformation of the Greeks eventually became a focal point for describing affinities not only with Slavs but also with other neighbouring Balkan nations. All of these neighbouring nationalities were regarded as related to the Greeks by means of cultural historical ties, and thus were gathered under the umbrella of the Hellenic spirit. They were thought of as relative *files* that had been merged, constituting a sole *fili* in which the Greek element prevailed.[15] The idea of this consolidation was based upon the allegedly de facto existing cultural and spiritual superiority of Hellenism, and thus of contemporary Greeks as its authorized representatives.

In essence, the idea of Hellenism and the assimilatory power of the Greek *fili* was a specific method to cope with the changing order in the territory of the decaying Ottoman Empire that advanced Greek hegemony. But Hellenism and the Greek *fili* became subject to alteration in the course of the consolidation of the Greek state and Greek nationalism. Throughout all of these transformations, the hegemonic element was ultimately retained. Before the founding of the Turkish state, which signified the definite end of the Ottoman era, the idea of Hellenism was also projected over the

[13] The topicality of this issue can be seen in the fact that this treatise was reprinted in Greece as recently as 2000: Nikoklis (2000/1855).

[14] On Fallmerayer's polemic towards Nikoklis, see Skopetea (1999): 161–3, referring to Fallmerayer (1857–60).

[15] On the politics of Oikonomos and the journal *Century* [*Αιών*] which advanced his ideas, see Matalas (2002): 146.

relationship of the Greek state (and Greek elites) with the Ottoman territories. Thus although Hellenism mostly prioritized the population in the interior of Greece, it was not restricted to Greek national territory, nor was it confined to a visionary sphere of ideals. Rather it had political consequences, which emerged once the idea became employed in the service of the ambitions of elites, who, in part, aimed to maintain or gain influence in the region of the former Ottoman Empire. Indeed, at the beginning of the 20th century, Asia Minor, a region with a large Greek population prior to the forced population exchange imposed by the Treaty of Lausanne (1923), had increasingly been viewed as a favourite area for research associated with the aspirations to assert Greek influence.[16] The diversity of the local population in terms of language, religion, customs, social organization and so on was not always thought of as a disturbed homogeneity that needed to be repaired, but as a resource from which Greek hegemony could draw power and legitimacy. This intention seemed to have formed the basis for a project concerning the founding of the so-called Ionian University in Smyrna. The project was designed by Constantin Carathéodory (1873–1950), a mathematician of Greek origins who later made an outstanding career in Germany (Georgiadou 2004; Agelopoulos 2010), in the period before the Greek defeat in Asia Minor (1922). Carathéodory's proposal on the founding of the Ionian University was addressed to the Venizelos government; it included the establishment of a school of ethnology to serve social modernization in the region, with the broader aim of consolidating relations between Greeks and their neighbours as well as the provision of measures that would allow for the active involvement of disadvantaged 'ethnic groups' in local society (Agelopoulos 2010). (In practice the project was never realized, having been cancelled after the burning of Smyrna in August 1922.)

The goal of similar research projects occasionally involved producing 'ethnological maps' of the region. As disclosed by a recent study (Agelopoulos 2010), the research interests of Greek scholars addressed Muslim communities to the same extent as Christian communities, and considered their internal, ethnological differentiation. This occurred in a period

[16] See Kalfoglous (2002/1899), Karolides (1992/1886), (1913), (1922–9).
Investigations on the composition of the population in Asia Minor against a nationalistic background were initiated, for example, by Ntinos Malouchos (1895–1928) in 1919 and were further conducted by other scholars such as Konstantinos Karavidas (1909–81), P.M. Kontogiannis, Georgios Skalieris and others. Kontogiannis (1995/1921) Skalieris (1990/1922) and (1995/[no year]).

in which religion was redefined in national terms and had ceased to provide the link for connecting large communities, with so-called ethnological features, such as language and descent, assuming an instrumental role in distinguishing large communities of the same religion. For those who were motivated by Greek nationalist aspirations, the 'ethnic differentiation' among Muslims in the region could be helpful in disconnecting Muslims with different 'ethnic origins' from the Turkish influence, thereby facilitating Greek hegemony in the region. This intention is unambiguously revealed in a statement made by the social scientist Konstantinos Karavidas (1890–1973) in an unpublished study.

> The form of contemporary civilization, which is advancing even more rapidly towards the ideals of democratic humanism, must sooner or later be imposed by the peoples of Western Asia Minor; this is what makes the presence here of the Greeks so important and their responsibilities so much heavier towards themselves and towards the whole country, which deserves a better fate.[17]

The similarities between this statement and the hegemonic and universal idea of Hellenism are intriguing. Karavidas voiced this view in 1922, on the eve of the Greek defeat in Asia Minor which was followed by the forced population exchange imposed by the Treaty of Lausanne (1923), an event which can be viewed as marking the definite end of the long decay of the Ottoman age.

It was at this point that the meaning attached to *fili* began to lean towards a concept of nation in which cultural and historical boundaries were instrumental to its definition, but which in fact existed parallel to naturalized elements that drew upon concepts of '*genos*' ([γένος], heritage; German: *Geschlecht*).[18] The idea of *genos* arose in the same context as *fili* and came to mean the Greek communality exclusively, especially from the later period of the Ottoman Empire when the two notions tended to be used interchangeably. The naturalistic connotations of both terms, transmitted by the idea of descent, are clearly apparent in the ways both notions are used in this period. Culture and representations of the past (in terms of historical bounds) are undoubtedly also subject to transformation and appear in several ways, interlinked with biologistic conceptions of communality and history. In the same period, biological and racial

[17] Karavidas, in *Archive Karavidas*, cited in Agelopoulos (2010): 187.
[18] Zelepos (2002): 76, footnote 153.

meanings, along with culture and language, penetrated into the semantics of *fili*, increasing its ambivalence and ambiguity.

7.2 *The Arrival of Race via the National Vehicle*

The ambivalence in the semantics of *fili* increased as soon as the same word was applied to both race and nation. In political and public language, as in the sciences—including physical anthropology and biology—race was literally translated into *fili* and vice versa.[19] The rather popular term '*ratsa*' (*ράτσα*: obviously borrowed from the Italian '*razza*') was introduced in the 20th century and, from the 1930s onwards, was increasingly utilized mostly with reference to German National Socialism,[20] without, however, ever replacing the term *fili* in the scholarly language. Polysemy finally conditioned the semantics of *fili* while cultural, linguistic, religious and biological components appeared to be merged in both nation and race. Alongside the specific reasons pertaining to the Greek case with regard to the conceptual sources of *fili*, polysemy was also transferred to the idea via the notion of race. Starting in the 19th century, when racial thinking was gaining more dynamism internationally, relationships and affiliations between peoples and states were increasingly interpreted in racial terms. This occurred in racial and anthropological studies, as well as in the historical sciences and the humanities.

Influential models of racial taxonomies were elaborated not only by physical anthropologists, but also by scholars in the humanities, including sociologists such as William Zebina Ripley (1867–1941) (see Ripley 1899). The diplomat Arthur de Gobineau (1816–82), whose treatise *Essai sur l'inégalité des races humaines* contributed significantly to shaping racial thinking (Gobineau 1853–5), was certainly not a racial anthropologist

19 An example of such a translation is a book on malaria by Ioannis Kardamatis (1857–1942), a pioneer in the anti-malaria campaign in Greece. His book *Filetic/Racial Mine* was translated into "Entorpeumiento de razar": Kardamatis (1934). See a serialized presentation of the book in the Greek medical press, *Medical Newspaper* [*Ιατρικη Εφημερίς*], 12 August 1934: 1; 19 August 1934: 1 and 4; 26 August 1934: 1; 9 September 1934: 3; 16 September 1934: 3.

20 This is verified in publications in the contemporary press, such as, for example, a 1933 correspondence in the Athenian daily *Free Forum* [*Ελεύθερον Βήμα*] from Rome: [Correspondence], "The Congress at Nuremberg and Nazist Theories: Hitler on Art and Race" ["Το Συνέδριον της Νυρεμβέργης και οι Θεωρίαι του Χιτλερισμού. Ο Χίτλερ περί τέχνης και ράτσας"], *Ελεύθερον Βήμα*, 6 September 1933. Cf. the reference made by Dimitris Glinos to '*ratsa*' in his treatise *War Trilogy* written in 1938 while in exile: Glinos (1976): *passim*.

by training. In all these approaches biological and cultural aspects were merged in racial reasoning (Stocking 1982/1968). Similarly, in the Greek case, cultural and racial arguments were used in a complementary way in order to corroborate ideas on origins and descent (cf. Herzfeld 1982: 77). Gobineau advanced the idea that culture and civilization were conditions for race.[21] His theory on the decline of civilization and culture in consequence of miscegenation also had a pivotal impact upon the endorsement of the racial connotations of the Greek notion of *fili* insofar as the defenders of the Greek national idea were anxious to disprove his assertions. Gobineau's theory was perceived as important for Greek concerns since it disqualified the affiliation of contemporary Greeks with the ancient Hellenes. He argued that the once high Hellenic civilization had decayed due to miscegenation with inferior races (Gobineau 1853b: 36–57). Moreover, he assumed that the Greeks had been composed of Aryan racial elements from the very beginning (Gobineau 1855b).

Gobineau's influential theory appeared at the same time as Fallmerayer's treatises on the Slavic influence on the Greeks. This coincidence evoked a discourse on an alleged "conspiracy against Hellenism" (Assimacopoulou 1999). The refutation of these authors' views—a concern among adherents of, and sympathizers with, the Greek national idea for many decades—contributed not only to the strengthening of racial connotations in the concept of *fili*, but also to the merging of national and racial terminology in Greek discourse generally. Ultimately, those (whether Greek or otherwise) who strove to restore the challenged unity, continuity and superiority of contemporary Greece as an entity with roots in Hellenic antiquity did not question the idea of race *as such*; instead, they shifted the focus from racial purity to the constancy of a superior and diachronic Greek *fili*/race. When, for instance, the archaeologist, philologist and founder of German Byzantine studies Karl Krumbacher (1856–1909) criticized the attempts of his contemporaries to define the Greeks through the "mechanical methods of ethnography and anthropometry", he questioned the effectiveness of the available technical means rather than the ideas behind their implementation (Krumbacher 1886: 7–8). Krumbacher criticized the racial evaluation of contemporary Greeks by means of statistical accounts of percentages of blond individuals and dark skin, blue and brown eyes and detailed measurements of crania, and argued that these methods were not sufficient for drawing safe conclusions; only technically

[21] On how Gobineau's theory was received, see Seidler (2005): 705–26.

advanced methods could provide such evidence, and these were unlikely to be available in the foreseeable future.

The increasing association of racial (anthropological) thinking with the Greek national discourse that occurred from the late 19th century onwards had, as a side effect, the dissemination of racial ideas via the vehicle of nationalism. In effect, *fili* emerged as a term with multiple meanings, each of which might prevail on a case-by-case basis. In all its fused and interchanging semantics, this notion began to assume a range of diverse, accumulated meanings. Enriched by racial connotations, *fili* came to share a distinct quality with race in having "no other content than that of the historical accumulation of religious, linguistic, and genealogical identity references" (Balibar 2004: 5).

Indeed, in the Greek scholarly and public discourse, especially in the early 20th century, *fili* was frequently applied to describe collective, national genealogy, indicating or presuming the naturalistic bonds among the members of the nation as a given. Individuals were affiliated by these bonds to a community that was deemed diachronic and universal; diachronic insofar as it allegedly existed beyond historical time, and universal since internal inequalities and differences were veiled behind the mask of the supposed homogeneity of equal nationals. The influence of this perception of *fili* as a diachronic and naturalised horizontal community of equal nationals surfaces in an exemplary fashion in the writings of the cosmopolitan socialist thinker, sociologist and physician Georgios Skliros (1878–1919).[22] In his work he undertakes a juxtaposition of two perceptions of communality, one as an entity of equal nationals (named Greek *fili*) and another as a society with internal inequality. In his 1919 treatise on *Current Problems of Hellenism*, Skliros dealt with the *filetic*, historical and sociological conditions of 'modern Hellenism'. He defined the 'filetic component' in universal terms as:

> all psychical, moral and intellectual qualities of our fili as they surfaced in the early period, along with the general qualities shared by the entire fili, the

---

[22] Georgios Skliros (1878–1919) was born in Russia and spent his life in Germany and Egypt. In Russia he was involved in the socialist movement. After the suppression of the 1905 revolution he left for Germany and studied medicine and sociology. His Greek treatise, *Our Social Question* (Skliros 1907) is thought to be the first attempt to approach and analyse modern Greek society and history by sociological means. He moved to Alexandria in 1912 and lived there until his death on 24 December 1919.

See also his monograph on the contemporary problems of Hellenism, Skliros (2000/1919).

> entire people, all social classes, from the potentate, plutocrat and scientist to the uneducated petty bourgeois and poor worker (Skliros 2000/1919: 62).

What Skliros meant by a 'filetic component' was shared primordial qualities, such as common psychic qualities and a common mentality, on which the coherence of the Greek collectivity was deemed to be based, regardless of internal inequalities or differences (Skliros 2000/1919: 62f.). *Fili* was thought of as a large, homogeneous community (an imagined community) that was not identical with society, since the latter had interior borders drawn along social lines, as with those between social strata. The social strata accounted for the internal relations, including social inequality, and were examined separately in his treatise under the heading of "Sociological Aspect". Despite all of its naturalistic implications, *fili*, as interpreted by Skliros, did not encompass race; yet his understanding of the notion clearly provides an example of the merging of diverse semantics at the intersection of cultural and naturalistic elements that are intercommunicated by the idea of common descent. In the formation of large communities, the idea of common descent operates according to an inclusiveness–exclusiveness relationship. The idea of descent encompasses a wide range of myths, including the supposed power of blood to explain the evolution of individuals and societies. Discourses of descent provide the foundation for evoking and designing racial communality in which internal homogeneity and the relation of inclusiveness—exclusiveness are translated in terms of 'racial purity'. The latter has been a central concern, and a contested issue, in racial anthropological approaches since the way in which anthropologists dealt with the question of racial purity was determinative of how they perceived (racial) communality.

7.3 *Dilemmas of Racial Purity: From Assimilatory Hellenism to Racial Hybridization*

> ... the feelings connected with race are one of the powers shaping the world
>
> Jacques Barzun (1937/1965)[23]

Integrating the idea of racial purity into a racial anthropological conception of Greek *fili* would prove to be anything but an easy task given that

[23] Barzun (1965: 2). The first edition of this book appeared in 1937 entitled *Race: A Study in Modern Superstition*.

such an undertaking was faced with a dual challenge: first, to cope with the perplexities inherent in contemporary racial theories generally, including the dissension on the reality of racial purity, and second, to dispute the widespread views in Europe which questioned the racial purity of the Greeks in particular. This was the concern of the most important scholars involved in the design of (racial) anthropological concepts within Greek scientific institutions, namely Clon Stéphanos and Ioannis Koumaris. Both men clearly prioritized Greek *fili* in their work, which linked anthropological concepts with national aspirations. Koumaris was probably alone, however, in his efforts to theorize race and the Greek *fili*. Regardless of the differences in their methodological and even conceptual approaches, they shared a conditional rejection of racial purity, the implicit advocacy of the racial constancy of the Greek *fili* and the conviction that it had remained racially constant from antiquity up to the present time. The possibility of racial purity was not completely rejected, however, but was instead considered as conditional and negotiable, depending on the point of reference: race as a universal idea; the Greek race in particular and its relations to neighbouring races; and the significance of racial issues for domestic society.

This kind of flexibility was allowed, and transmitted, by the ambiguity with which racial purity and pure races were conceived in their contemporary racial anthropological discourse. Indeed, already in the 19th century, 'pure races' were considered an ideal condition and their substantiality was contested even in the anthropological sciences. An essential influence on the destabilization of the notion of pure races was conveyed by the theory of 'racial types' and 'racial varieties' elaborated by Paul Topinard (1830–1911),[24] a theory that was embraced and further developed by 20th-century anthropologists.[25] As Stocking has shown, "the reality of race was not so easy to abandon as this. If races were only abstract conceptions of 'continuity in discontinuity, of unity in diversity', they nevertheless

[24] The idea of 'type' and variety was not new to physical anthropology and biology. In his book *Anthropology*, Topinard himself refers to the origins of this idea and to previous works on the issue: Topinard (1878): 198f. According to Stocking (1982/1968: 322–3), Topinard's thinking on this subject is as important as it is influential because he laid bare some of the assumptions that underpin the notion.

[25] See an anthology of the most influential concepts of race, racial varieties and types from the 18th century up to the First World War, edited by Count (1950). For Eugen Fischer's contribution to the development of the concept of race, see Lösch (1997). For an excellent critical presentation of the concepts of racial varieties and types and their development, see Stocking (1982/1968).

existed, even for Topinard" (Stocking 1982/1968: 50); races thus existed as 'hereditary types'.[26] Topinard was convinced that distinct racial qualities could be detected, although he took it for granted that modern populations were heterogeneous. This was because 'racial types' were thought to represent the original 'pure races' from whose mixture modern populations were derived (Stocking 1982/1968: 56). Topinard argued that racial types exhibited a relative constancy because they had been "perpetuated without change throughout the upsets and the mix-ups of history and prehistory" (Topinard 1950: 176). Later on, anthropologists and racial scientists also became preoccupied with the delineation of race, racial type, variety and people and identifying which of these was an ideal category, as well as the question of how such a category was substantiated in reality. For all doubts and dissent (which increased after the rediscovery of the Mendelian Laws of heredity), racial purity did not lose its significance as a motif (*Leitidee*) for coping with human inequality and for describing the internal homogeneity of societies in racial terms. In the scientific discourse, such ambivalence became apparent in a shift in the research interest from 'pure races' to 'human hybrids', 'mongrels' or 'bastards'; in other words, in a shift from presumed racial homogeneity to borderlines and conditions of miscegenation.

The ambiguity of the racial concepts and their conditional validity are evident in the attempts of the two Greek anthropologists to specify the Greek *fili* in racial anthropological terms. Stéphanos and Koumaris sought evidence for the existence of a distinct Greek *racial type* that was sufficiently constant to have survived over the centuries, even if it was constituted by different racial elements. In doing so, racial 'constancy' and 'purity' were often implied metonymically. But even then, shifts in the line of argument by each of these anthropologists were readily observed. In his 1884 treatise, for instance, Stéphanos assumed that the Greek/Hellenic population had been constituted from diverse racial and anthropological elements in both ancient and contemporary times. But during the years of his directorship at the Anthropological Museum the major concern of his research activities turned increasingly towards the detection of 'racially pure' Greek regions whose populations possessed the anthropological characteristics of the ancient Hellenes. In the progress reports submitted to the dean of the university, he stated that he sought to detect

[26] Stocking (1982/1968: 58–60) gives a detailed account of the development of Topinard's thinking on type and race.

"ancient Hellenic elements, relatively unmixed" in several Greek regions.[27] Furthermore, he believed that he had localized the oldest European racial elements preserved in Greece and to have found evidence for the racial anthropological unity of the Greek population. He came to the conclusion that the presence of Albanians and Slavs among the Greek population was rare and that their influence upon the Greek racial anthropological stock was insignificant; hence it was unable to disrupt the unity of the Greek population. According to his investigations, a large number of the Albanian inhabitants of Attica, Corinth, Boeotia and Carystia were rather low-dolichocephalic, which led him to conclude that the Albanian population in these areas was of Greek origin but had been Albanianized over the course of time. The Slavic influence upon the Greek population was also deemed to be restricted. With these findings he believed that he had verified the coherence of the Greek population hidden behind differences in language, religion and consciousness.[28] Aside from these questions which he sought to answer by anthropometrical investigations, Stéphanos was not concerned with becoming involved in the advanced racial issues of his contemporary period or in theorizing the Greek race, and much less in embedding his conclusions in topical issues related to society. The antiquity cult was his major concern.

Ioannis Koumaris elaborated a more dynamic concept of (Greek) race than his predecessor, and he pursued an ambivalent argumentative strategy in doing so: on the one hand he proclaimed the unity of the Greek race in its diversity, and on the other he insisted on keeping it pure by means of applying racial politics. The dynamic element in his concept lies in his intention to provide suggestions for designing a racially homogenous large community, the Greek *fili*, and for improving society. While he rejected racial purity as a reality, he thought of pure race as an ideal target that prompted the imperative of purification by applying the racial politics which were presented in his racial hygienic agenda. The arguments for the rejection of racial purity were elaborated in his theory on the "strong racial nucleus of the Greek *fili*" which, I will argue, was worked out in the intersection of the theory on the assimilatory power of Hellenism and the concepts of racial hybridization that proliferated in his contemporary period.

---

27 Letter addressed to dean of the University of Athens (24 July 1887), *HAUA*, File 3/5–21: Anthropological Museum.

28 See letter addressed to dean of the University of Athens (24 July 1887), *HAUA*, File 3/5–21: Anthropological Museum.

In a paper presented before the GAS in 1939 (perhaps one of his first comprehensive conceptions of race) Ioannis Koumaris asserted that the term 'pure race' had little meaning in practice because all races had been subject to transformations over the course of time. This occurred due to transmutations that were caused, among other things, by miscegenation. Racial mixtures could also result in the creation of superior races.[29] Therefore, attempts to define a current race by relating it to its initial (primordial) state could hardly be successful. At the same time, most of his texts reveal a conviction that primordial elements always remained traceable in a race, no matter whether it was thought to be a pure race, a racial type or a racial variety. For that reason Koumaris also suggested that racial purity should still be viewed as a goal that was worthy of being pursued, even if it was an essentially utopian aim. In arguing in this way he embraced the same double-sided argument that many other adherents of the racial idea in Europe had followed. It is not surprising therefore that the influential ideas of 'racial types' and 'racial varieties' were readily identifiable in his writings. The first of these was the conviction that physical differences between human races were virtually primordial, and that the most important of these differences were recorded in the human skull and brain. The theory of racial types was applied as a way to reinforce the racial bonds among members of the national community. Race, he argued, was not identical with nation, although it was instrumental in determining its constancy. The primordial racial qualities were inherent to the nation and persisted, latently or manifestly, even if the external characteristics of the nationals were transformed by miscegenation.

> Race does not mean the whole nation but the basic type that is primordial in physical and psychic terms. This initial type exists latently and arises through endogamy, coming into effect according to the laws of heredity. There is always a biological basis which prevails despite mixtures. This is the "group variety" of the race, particular features of which distinguish it from other groups with divergent features. The deviance of such a group from its initial/primordial purity (derived from the isolation in which small groups live) cannot be doubted. However, even through changes, the relation between the physical and ethnographic traits persist (Lester and Millot).[30]

The theory of racial varieties provided racial anthropological reasoning that could be used to explain the phenotypic divergence of contemporary

[29] Koumaris, *PGAS* (14 May 1939): 10–26.
[30] Cf. Koumaris (1951): 36, apparently referring to Lester and Millot (1947).

Greeks from the ancient Hellenes who were thought to be the ideal type. This theory was therefore helpful in refuting assertions that contemporary Greeks belonged to an impure and inferior race which differed from the race of the ancient Hellenes because its continuity had been disrupted by mixing with inferior races.[31] The counterargument here was that if races were not identical with nations, then the latently persisting (primordial) racial elements in a nation could (or even must) be advanced. A large and superior race that had in the past been subject to mixtures was in a position to purify and regulate itself in such a way as to maintain its basic 'racial nucleus' unchanged.[32] Miscegenation carried a temporary character since, over the course of time, racial elements could be separated, revealing the prevailing primordial racial nucleus. Due to the temporary character of racial mixtures, the Greek anthropologist believed that human groups (understood as equivalent to races) were in a state of 'fluid constancy' (i.e. despite miscegenation, the proper nucleus of the predominant race always persisted in a stable form and was capable of surfacing).[33] In the light of this theory, racial purity is not rejected but relativized, being replaced by the constancy of the essence of a race over the course of centuries. With the shift from purity to constancy, the imperative of the homogeneity of an entity is not violated but, in contrast, is buttressed by racial anthropological arguments.

Although formulated as a general racial concept, the theory of 'fluid constancy' and the 'strong racial nucleus' was in the main addressed to the Greek race. The similarity with the 19th-century national idea of Hellenism is prima facie manifest. Koumaris's theory on the 'strong racial nucleus' was an elaborated racial anthropological version of the former assertion that Hellenism and the Greek *fili* possessed the ability to acculturate and absorb foreign peoples or nations without losing their proper Hellenic qualities. The common approach to the assimilatory ability of Hellenism and the strong nucleus of the Greek race was the assumption that the Greek *fili* was unified and distinct, despite the possible diversity of its single constituents. This coherent and unique character was supposedly verified in the persistence and perpetuation of primordial features (either cultural or racial). In the racial anthropological version, this

31 This assumption (the most prominent proponent of which was Arthur de Gobineau) evolved during the 19th century and was embraced by many adherents of racial ideas even in the 20th century. See an extensive account by Stocking (1968/1982): 50.

32 Koumaris, *PGAS* (14 May 1939): 10–26.

33 Koumaris, *PGAS* (14 May 1939): 10–26. Cf. Koumaris (1948).

axiomatic assumption formed the basis for the belief that contemporary Greeks were representatives of a race "with its own soul and especially with its own variety, dating from prehistoric times" (Koumaris 1948). The ideas of assimilatory Hellenism and of the strong nucleus of the Greek race shared the hegemonic character ascribed to the Greek *fili*; however, they differed on at least one essential point: while the national conception of Hellenism threw the ultimate result of the mixture into relief, the theory of a 'strong racial nucleus' went a step forward and sought to retrace and specify the initial (primordial) racial type—the unchangeable core at the heart of the mixed races or amalgamated racial types. In the latter version, collective cultural and intellectual qualities were thought to be transmittable via blood and to reside invisibly in genealogical codes.

Finally, beside the similarities with the idea of assimilatory Hellenism, the 'strong racial nucleus' drew a great deal from the then contemporary racial discourses on hybridization, especially those rooted in the United States and Germany. In the United States, starting from the late 19th century and culminating in the 1930s, 'racial hybrid' emerged to become a pivotal term for dealing with race relations (Trubeta 2006). A hybrid was the result of miscegenation wherein the single components could come to the fore at any time. In Germany, especially during the era of National Socialism, 'miscegenation' was considered to come under the term 'bastardization', which in turn constituted a special research paradigm.[34] In both the North American and the German cases, the aim was to cope with society's internal heterogeneity, perceived in racial terms, which in the former instance was thought to be caused through migrants and African-Americans, and in Germany through Roma, Jews and other minorities.

Ioannis Koumaris elaborated his concept of the 'strong nucleus of the Greek race' against the background of the shift in anthropological research from 'racial purity' to 'hybridization' and 'bastardization'. He did not seek to identify every component of the hybrid result of miscegenation, but looked only for the strongest one. And while his main concern was the Greek race in particular, accommodating the assimilatory power of Hellenism, the strongest component was the initial, Hellenic one. But how was it possible, among all this diversity, to detect the dominant racial nucleus by anthropological means when most racial anthropologists found

[34] The issue of 'bastards', especially with respect to the Roma, was the main topic of Robert Ritter's bio-criminological researches. For a close examination of Robert Ritter's work see Zimmermann (1996); Hohmann (1991). For an overview of Eugen Fischer's post-war writings on 'bastards', see Lösch (1997).

it difficult to define the distinct features of even one race?[35] Koumaris himself admitted that classical anthropometric methods did not always lead to safe conclusions about specific races, and this included the Greek race. Nor were modern taxonomies always able to clarify the innumerable and subtle physical differences, much less the psychic ones, between different races.[36] Nevertheless, the conviction prevailed that no intricacy could disprove the existence of diverse racial groups.[37] In this narrative, the 'psychic uniformity' of a racial entity rather than its morphology was considered to be the decisive factor for determining the typical character and consistency of a certain race.[38] This assertion was close to German conceptions of race which highlighted psychic instead of physiological traits. Indeed, the ethnologist Ernst Grosse (1862–1927) defined race as a large group of people bound together by shared and hereditary somatic and mental traits that distinguished them from other groups.[39] But it was the achievements made in genetic science[40] which ultimately led to the conclusion that heredity was imprinted in invisible genetic codes rather than in bones. As soon as racial heredity became a matter of invisible properties, common descent became a conceived reality that, even if it was not always visible to the naked eye, was nevertheless safely fixed in blood and in a collective psyche. During the 1930s, especially in German anthropology, racial theories oscillated between rational and irrational reasoning, elevating the status of 'racial character' and the psychic disposition of humankind to focal points of anthropological science. This development was associated with the turning aside from German physical anthropology towards a racial science that was connected with the person of Eugen Fischer. Among other innovations, Fischer managed to "broaden anthropological discussion to include questions of mental disposition, social behavior, and 'racial character' " (Proctor 1988: 156). Fischer made a significant contribution to establishing "an anthropology that would combine traditional physicalist concerns with the exploration of the psychic, racial constitution of mankind" (Proctor 1988: 156).

---

35 Cf. Lösch (1997): 152.

36 Koumaris (1951a). Cf. a similar problematic with respect to Jews by Salaman (1911).

37 Koumaris (1951a).

38 Koumaris (1948) and Koumaris (1951a: 37).

39 See Grosse's perception by Koumaris in: Koumaris, *PGAS* (14 May 1939): 10–26, 11. Here Koumaris referred to Grosse's notion of race. See also a later reference in Koumaris (1951a).

Cf. Race definition by Grosse (1900): 117. On Ernst Grosse, see Lösch (1997): 13, 143.

40 See, for example, Wundt (1900–21); Eickstedt (1936).

Regardless of the (precise or vague) conclusions drawn by anthropological measurements, Koumaris, who was an admirer and student of Fischer, considered the psychic uniformity of the Greek race to be so obvious that closer explanation or verification was not needed.[41] He believed that "among modern Greeks a certain similarity of the general type is perceptible even though not verifiable by current anthropological methods".[42] And despite the large scale of variation, the Greek race did produce a characteristic type, even if it was hardy identifiable by anthropological measurements.[43] Racial psychology would be deployed for ascertaining the distinctive character of each racial group.[44] Koumaris admitted that he was surprised that most racial scientists and anthropologists were ignorant of this crucial aspect (i.e. the distinct psychic components of a race). It was his strong conviction that the most secure method for detecting a race was 'racial intuition'; this being the safest method available when "anthropometric tools and statistics fail or cause confusion".[45] Every anthropologist possessed an intuition for his own race, he argued, so that the most appropriate anthropologist for studying a certain race would belong to the race in question—the Greek anthropologist for the Greek race, or a "Negro for the African races of his area, *etc*".[46] This assertion seems to be a projection of his own mixophobia to human nature at large, in which the rejection of racial mixing appears as "an instinct, a manifestation of natural laws, an instrument for the providence of differences that nature wanted to preserve" (Taguieff 2000: 293). The physically indiscernible 'racial intuition' operates analogically to the consciousness of a common belonging that allegedly keeps the members of a race bound together. Located at the interface of scientific reasoning and common sense, this assertion is easily adaptable and applicable for justifying communality. Consciousness, similar to 'racial intuition', is transmitted by the power of the blood, as a *deus ex machina*, to substantiate national continuity and coherence. Within the Greek Anthropological Society this view was not only voiced by its president. For instance, a member of the GAS stated that "despite manifold instances of miscegenation lasting more than three thousand years we are aware of our nationality to the same

[41] Koumaris (1951a).
[42] John Koumaris (1948): 127.
[43] John Koumaris (1948): 127.
[44] See the entry "Race" [φυλή] in the *Encyclopaedia Helios* [*Νεώτερον Εγκυκλοπαιδικόν Λεξικόν Ήλιος*]: vol. 13 [no year]: 284–7.
[45] Koumaris (1951a): 39.
[46] Koumaris (1951a): 38–9.

degree as our ancestors in any past period".[47] Even the feeling of obligation towards the community appeared to be transmitted by the power of the blood. As a testimony to this fact, the above-mentioned contributor referred to the high awareness of obligation towards the nation and the sense of national integrity that, as stated, had always characterized all Greeks, regardless of social classes or level of education.

In the following I will argue that the theory of the strong racial nucleus of the Greek *fili* was elaborated in the shadow of the myth of the Nordic race and that it sought to cope with the Orientalistic prejudices that were inherent to this. The correspondence with the myth of the Nordic race generated shifts in the idea of Hellenism. The erstwhile acculturative power of Hellenism tended to blur, whereas questions such as the autochthony of Greek race, the premise of its homogeneity and the redefinition of the relationship to neighbouring states were gaining priority.

7.4 *To Cope with the Ottoman Legacy in the Shadow of the Nordic Racial Myth*

The perception of the idea of Hellenism once it had been integrated into the theory of the strong racial nucleus of the Greek *fili*, and was then associated with theories of racial hybridization, generated shifts from the erstwhile supposed assimilatory power of Hellenism, causing it to become a doctrine of *filetic* continuity 'in a direct line' from antiquity up to the present. The point that was now at issue concerned the affinity of contemporary Greece to antiquity, rather than the unification of diverse peoples under Greek hegemony. This shift in argument, which was also apparent in Clon Stéphanos's approach, corresponded with developments in Greek nationalism, where, at the end of the 19th and especially in the 20th century, the antiquity cult was becoming a predominant national doctrine. The theory of the strong racial nucleus of the Greek *fili* was well situated in mainstream nationalism, but its arguments were actually formulated in the shadow of the National Socialist myth of the Nordic race, which explicitly addressed Greek ancestry. At the heart of the theory of the 'strong racial nucleus of the Greek *fili*' rose, first, the argument of the autochthony of the Greek race in the current territory of the Greek state and therefore in European soil, and secondly, the attempt to disprove

---

[47] Dendias, *PGAS* (1 May 1948): 10–27, 25.

the Orientalistic prejudices grounded on Greece's Ottoman past. Indeed, National Socialists questioned not only the racial purity of the contemporary Greeks but also their autochthony in European soil and furthermore burdened them with civilizationary deficits allegedly caused by their Oriental descent. The most famous theoretician of the Nordic race, Alfred Ernst Rosenberg (1893–1946), followed Fallmerayer in asserting the Levantine character of the contemporary Greeks.[48] Popular National Socialist periodicals, such as *Volk und Rasse* (published between 1926 and 1946) or *Volkstum in Südosten* (from 1939 to 1944), often hosted articles which asserted the racial decay of Greeks and their current debilitated situation, similar to that ascribed by the National Socialists to the other Balkan states. "Racial diversity" in the Balkans was interpreted as maintenance of the Ottoman legacy and thus as lack of Europeanness and hence of civilization (cf. Trubeta 2004).

The deconstruction of this argument was certainly a concern not only of anthropologists, but also of all those who were anxious to restore the continuity of Greece from antiquity to the present time. Counterarguments were provided by international scholarly research, such as the works by the professor for Indo-Germanic philology Albert Thumb (1865–1915) which had become particularly popular among Greek intellectuals. His treatise *The Modern Greek and his Ancestry* (Thumb 1967/1915), in which the author argued against the Slav influence upon Greeks and for their racial and historical continuity, was often cited by Greeks. For instance, in a lecture delivered in 1943 to the Greek Anthropological Society, his writings were characterized as "the most exhaustive studies that reject Fallmerayer's theory and explore the issue of the Greek *fili* from a historical, geographical, anthropological and political point of view".[49] The same arguments were also pursued by the American professor of physical anthropology Carleton Stevens Coon (1904–81), who claimed that "[i]t is inaccurate to say that the modern Greeks are different physically from the ancient Greeks; such a statement is based on an ignorance of the Greek ethnic character" (Coon 1939: 604).

In the Greek anthropological discourse, counterarguments were integrated into the theory on the strong nucleus of the Greek *fili* and sought to corroborate the basic assertion that racial mixing had indeed occurred,

---

[48] "The Slavization of Greeks and their Levantine Character", Koumaris (26 March 1949c). On the Levantinian influence upon Greeks, see Rübel (1942).

Cf. Almgren/Hecker-Stampehl/Piper (2008).

[49] Pezopoulos, *PGAS* (15 April 1943): 12–17.

but only between peoples racially related to Greeks,[50] thereby normalizing the idea of miscegenation among the Greeks and other races insofar as a hegemonic role was ascribed to the Greek race. The racial type arising from intermixture, then, was primordial in character, since it had all the features of the initial basic racial elements, all of which were Greek and indigenous (Koumaris 1948). Furthermore, the initiator of this theory, Ioannis Koumaris, asserted that the birthplace of the Greek race had always been located in the Aegean Sea: all "component elements of the Greek race are in a sense 'indigenous' throughout the Greek Archipelago" and all were related to each other by virtue of their belonging to the "Mediterranean cycle".[51] In this racial narrative the Aegean Sea represented a bridge connecting Europe and Asia, or, more specifically, the east-meridional part of Europe and thus the intermediate region of Eurasia, in whose natural environment a distinct and autochthonous civilization could emerge. Koumaris distinguished three principal racial types which had developed in South-East Europe and in Asia beyond the region of the Euphrates, all of which were detectable both in ancient times and in contemporary Greece: the long-headed Mediterranean type; the 'Anterior-Asiatic' type (from German: *Vorderasiatisch*); and the Nordic racial type (Koumaris 1948). According to this model, the Nordic race could also be observed among the Greeks, bearing testimony to their affinity to a superior European race. Blond individuals were considered as evidence of the existence of this racial type. The Nordic type, although not represented in all of ancient Greece, nevertheless existed among the pre-Hellenes, particularly in Crete, Macedonia and other Greek regions (Koumaris 1948). This did not, however, require embracing the Aryan myth or the hegemonic idea of the Nordic race. On the contrary, it advanced the Greek influence upon the Mediterranean space: Greek racial influence could explain the existence of blond individuals in several parts of the Mediterranean, particularly in North Africa.[52] Even so, this influence was not a

50 Koumaris (1948).
For a perceptive account of this view see Dendias, *PGAS* (1 May 1948): 10–27.

51 Koumaris (1948).
He presented much of his theory to the Greek Anthropological Society in 1942 in a lecture on the morphological variety of the Greeks. Koumaris, *PGAS* (1 May 1942): 7–28. An extended version of this model appeared a decade later, again in Greek, Koumaris (1959); and after some years in France Koumaris (1961). A full revision of the same theory was published in the following years in the *Journal of the British Anthropological Association*, "Man"; (John Koumaris 1948); see also the Greek version of this article in the journal *Helios* [*Ἥλιος*]: Koumaris (1949b).

52 Koumaris, *PGAS* (1 May 1942): 7–28.

recent phenomenon, nor was it the result of the impact of Nordic racial characteristics; it should instead be traced back to prehistoric Hellenic racial elements, essentially to the Pelasgi and their spatial dispersion into North Africa. In fact, the Nordic influence over the Mediterranean had been subject to earlier challenges by anthropologists such as the Viennese Victor Lebzelter (1889–1936) and the American Carleton Stevens Coon. The former had expressed his doubts about explaining the appearance of blond individuals in South-Eastern Europe by means of a possible 'Nordic impact'. Instead, he raised the question of whether this was a racial type *sui generis* (Lebzelter 1929: 63). Coon also believed that the Nordic element, although weak, could be witnessed in Greece at this time, as it had been in the past, and even with the same appearance

> as it probably has been since the days of Homer. The racial type to which Socrates belonged is today the most important, while the Atlanto-Mediterranean, prominent in Greece since the Bronze Age, is still a major factor. It is my personal reaction to the living Greeks that their continuity with their ancestors of the ancient world is remarkable, rather than the opposite (Coon 1939: 607).

While the detection of ancient Hellenic elements and the Nordic racial type in Greece solved the problem of the European and autochthonous character of contemporary Greeks, the other two racial types included in Koumaris's typology—the Mediterranean and the 'Anterior-Asiatic' type—were more troublesome since they were burdened with grave civilizational deficiencies that needed to be rectified. The prestige of the inferior 'Homo Mediterraneus'[53] could, however, be advanced by means of its association with the primordial Hellenic racial element. The pre-Hellenic Pelasgi were thought to be the primeval bearers of the fundamental nucleus of the Mediterranean racial type, detectable in Greek-settled countries. The argument about the long-headed Mediterranean type had also been used earlier, in the 19th century, by Clon Stéphanos,[54] and was used again in the 20th century by other, non-Greek, anthropologists who sought to buttress the thesis on the Hellenization of Albanians. The same correlation was elaborated by the above-mentioned anthropologist Victor Lebzelter, who argued that:

---

[53] Concerning the hierarchy between the long-headed and superior 'Homo Europeus' and the likewise long-headed but inferior 'Homo Mediterraneus', see Stocking (1982/1968): 61.

[54] While Clon Stéphanos dedicated a large share of his 1884 treatise to the Albanians, during his directorship of the Anthropological Museum he devoted a much larger part of his subsequent research to them.

> it is not true that the ancient Greek population disappeared. The ancient Greek population was dolichoid and lingers on in the current dolichoids. The Albanians who live in Greece are also in a high percentage dolichocephalic. This might be due to mixtures with the Mediterranean Greek population or because the Albanians themselves, who inundated mediaeval Greece, were to a great degree Mediterranean (Lebzelter 1929: 78).

In the Greek context, the 'Anterior-Asiatic' racial type as a component of the Greek race was defined in the racial model presented by Koumaris, and was an explicit response to the theory on the Anterior-Asiatic racial mélange, inverted by German anthropologists. In all its versions the latter theory sought to determine the relationship between Western Europe and the Orient in racial terms. It was first formulated by Felix Ritter von Luschan (1854–1924), but was later elaborated by National Socialist anthropologists. According to Luschan's idea, the origins of the 'white race' were located in the Caucasus (Petersen and Luschan 1889: 248). Semites and Aryans emanated from the same Caucasian race and formed diverse racial mixtures before they spread to inhabit other regions in the world, including Europe. Following its initial launch, Luschan's theory went through several transformations and its content changed remarkably as soon as the idea of the Nordic race began to gain ground (Fuchs 2003: 248–50). During the era of National Socialism, the idea of the Anterior-Asiatic race increasingly became embedded in the ideology of the Nordic race, which thus led to a bipolar racial typology (based on value judgements) that reflected a global dichotomy between a progressive Occident and a backward Orient. If the dual association of Greece with antiquity and the Ottoman Empire rendered the country and its population an attractive field for anthropological investigations in the 19th century, the question of its racial composition was definitely answered during the National Socialist era with the racial disassociation of contemporary Greeks from the Hellenes and, instead, their affiliation to the Ottoman legacy. This view did not, however, remain unchallenged. Critics included the Austrian racial anthropologist Friedrich Keiter (1906–67) whose views evoked an argument between him and the eugenicist and racial anthropologist Walter Scheid (1895–1976).[55] This dispute first arose when Keiter criticized Scheid's attempt to explain the whole of European cultural history as a phenomenon of diverging racial-biological behaviours, which allegedly derived from the psychic diversity of Occidental and Oriental peoples

[55] See Scheid (1939); cf. Keiter (1941).

(Scheid 1939: 201). This meant that there was a contradiction between a superior Occidental (and Nordic) race and the Oriental (or Levantine) and inferior Anterior-Asiatic race. Scheid classified Greeks diachronically among the latter. Keiter, on the other hand, raised the question as to whether Scheid's model actually simplified reality, at least in equalizing the ancient Hellenic culture with the Oriental one.

In coping with this discourse, the strategy that Ioannis Koumaris pursued was not to deny the affiliation of the Greek race with the Anterior-Asiatic race, but to liberate the latter from the inferiority with which it was burdened. Rather than proving the Oriental descent of the Greeks, Koumaris believed that the presence of the Anterior-Asiatic racial type in the Greek people demonstrated the affinity of certain Asian races with the primordial Hellenic element. Anterior-Asiatics encompassed the first currents of Hellenic peoples, such as the Lelegers and the Karers, who originated from Asia.[56] These 'Pre-Hellenes' were thought to constitute the original background of the Greek world, having, in prehistoric periods, inhabited the Aegean and the intermediate region of Eurasia, notably Asia Minor. The idea of the Anterior-Asiatics was utilized by the Greek anthropologist to act as a vehicle for harmoniously connecting the Asiatic with the Mediterranean and the European regions including the territory of the current Greek state. The same region was identical with those of the Great Idea.

Regardless of which position the arguments were formulated upon, in this explanatory scheme history was perceived as having been determined by race relations, and racial affinities were utilized to explain the position of contemporary peoples/races throughout the world, viewed as a universal kaleidoscope of peoples and races. The Anterior-Asiatics, as Koumaris argued, essentially featured a large variety of different racial types. Semites were thought to be distinct from those Anterior-Asiatics who had inhabited Greece in the past. Similarly, 'Armenoids' (whose Europeanness was controversial)[57] were only one racial type among several others

[56] The view that Lelegers and Karers originated from south-west Asia Minor and then migrated to Greece preoccupied 19th-century German classical philologists and Hellenists; see, for instance, Karl Wilhelm Deimling's doctoral thesis (1862). The 1861 Proceedings of the Academy in Berlin published the view that the Lelegers were an aboriginal people in Asia and associated them with the Illyrians and thus with contemporary Albanians: "Herr Kiepert las 'Über den Volksstamm der Leleger'", *Monatsberichte der Königlichen Preus. Akademie der Wissenschaften zu Berlin* (Erste Hälfte, January–June 1861): 114–32.

[57] Koumaris, *PGAS* (1 May 1942): 7–28. Koumaris (contradicting Luschan) later asserted in another of his writings that the Hittites, and the Anterior-Asiatics in general, were absolutely distinct from the Armenoids; Koumaris (1959). On cranial forms among Hittites

that constituted the Anterior-Asiatic race. The brachycephalic (and non-European) 'Anterior-Asiatic' type coincided to some extent with the Hittites.[58] But, according to Koumaris, the presence of Hittites in Greece was limited to Crete. In effect, the most solid racial affinity between current Greeks and Anterior-Asiatics was found to be with the pre-Hellenic racial types. This specification of the Anterior-Asiatic race and its correlation with the Greeks and Europeans was not merely a matter of 'general anthropological interest', but in fact obtained topicality insofar as the historical glance at the bonds between races was used to substantiate relationships among contemporary peoples. The views and theories outlined above actually corresponded to highly topical social and political issues and even tacitly referred to certain population groups and their presence in Greece: given that a great part of Koumaris's theories were elaborated at the end of the 1930s and during World War II, the assertions on the Semites are indicative of an effort to disassociate the Jews from the Greeks in a period in which Jewish extermination was justified by racial arguments. Considering that Koumaris himself opposed the persecution of Jews by the National Socialists, as has been shown above, with his theory on Aryans and their relation to the Anterior-Asiatic race he sought to buttress his opposition by racial arguments, or, at least, to disconnect the Jews from the racial Greeks.

For all of the objections raised regarding the idea of the Nordic race, the Greek anthropologist remained an adherent of both German racial theory and of racial politics. In the following chapter I will argue that his rejection of racial purity was a defensive reaction to the National Socialist attack against Greece, whereas he was a fervent advocate of racial purity when the matter was the persistence of the Greek race and its diachronic characteristics. In arguing for the purity of the Greek race to be cared for by state politics, his mixophobic conception came close to a model for a totalitarian racial state.

7.5 *Back to Racial Purity via Mixophobia: In Light of National Socialism*

The uncomfortable dilemma of racial purity or miscegenation was not resolved with the assumption that a 'pure race' was little more than an ideal category whereas current populations, as a rule, were composed

---

and Armenians (the so-called 'Armenoid race') and the disputes as to whether they belonged to the European races, see Heberer/Kurth/Schwidetzky-Roesing (1959): 250–1.

[58] Koumaris, *PGAS* (1 May 1942): 7–28.

of mixed racial types. In the racial discourse, the ideal of racial purity acquired tangible topical dimensions when it became embedded in fantasies of homogeneity, the target of which was to avoid miscegenation. In the Greek anthropological discourse this ambiguity was translated by the initiator of the anthropological theory of the Greek *fili*, Ioannis Koumaris. According to the latter, even if the strong racial nucleus of a superior race ultimately prevailed for all occasional racial mixtures, racial purity continued to pose an ideal that dictated the need to safeguard the current homogeneity of the Greek *fili* while keeping races apart. This belief is unequivocally imprinted in the following statement: "We favour the interruption of any miscegenation between Greeks and foreign races, regardless of whether they are 'blacks', 'yellows', or even 'whites' from another race."[59] In this statement, which was made in 1940, Koumaris appeared to be more decisively against miscegenation than even a year earlier when he had argued that the first task for racial state politics would be to examine whether racial mixtures were ultimately advantageous or harmful.[60] He fortified his refutation of racial mixing even after UNESCO's first "Statement on Race" in July 1950 (Koumaris, 1951a and 1951b).

Ioannis Koumaris voiced his advocacy for racial purity and policies of purification during the same period in which he was anxious to disprove the National Socialist theory on the superiority of the Nordic race and the assumed inferiority of the Greek race and its alien character in Europe, a theory revived during the National Socialist era. From 1938 onwards he made public his views on the purity and purification of the Greek race not only in Greece; he also published in Italian fascist periodicals such as *La difesa della razza* (1938 and 1939), in the German National Socialist journal *Ziel und Weg, Zeitschrift des Nationalsozialistischen Deutschen Ärzte-Bundes* (1939), as well as in the journal *Quadrivio* [*Κουαδρίβιο*], which was printed in Italian and Greek at the beginning of the Italian occupation and which was primarily addressed to Greek intellectuals.[61]

---

[59] Koumaris (1940a); this article is based on his lecture delivered in the same year for the Greek Anthropological Society: Koumaris, *PGAS* (1 June 1940): 3–7. See also Koumaris's German-language article "Rasse und Gesundheit" [Race and Health] in *Ziel und Weg* which includes the same expression; Original: "Wir sind gegen die Geschlechtsverbindung zwischen andersrassigen Menschen"; Koumaris (1939a): 386–7. Cf. Koumaris, *PGAS* (14 May 1939): 17.

[60] Koumaris, *PGAS* (14 May 1939).

[61] Koumaris (1941); Koumaris (21 March 1945a).

An account of alleged Greek racial policy, including a comparison with Italy and Germany, was provided by Koumaris in a second article in *La Difesa della Razza*, Koumaris (1939).

In terms of coping with the question of racial purity, Koumaris clearly pursued a double-sided line of argumentation. This is because the idea to which he was opposed was not racial hierarchy as such, but the inferiority ascribed to the Greek race as opposed to the belief in the superiority of the Nordic race. He explicitly disapproved of racial hierarchy among *European* races (which Greeks also belonged to), whereas he advocated a racial hierarchy in distinguishing Europeans from non-Europeans. For instance, he took for granted the racial inferiority of Pygmies and Africans.[62] His conviction that races were equal but divergent and thus ought to be kept apart should be regarded in the light of his own words: "No blood is superior to Greek blood. They all are, however, divergent, and consequently inimical."[63] The closeness of this assertion to Pierre-André Taguieff's thesis of the 'differentialist racism' of the post-war period is intriguing (Taguieff 2000, cf. Lentin 2000). This kind of racism does not deny difference but absolutizes it, raising the goal of keeping races separate to the status of a duty. However, a point of divergence between Koumaris's view and post-war differentialist racism was that, in the latter, the 'difference' was described in cultural terms as the 'incommensurability' of cultures, even though biological determinism provided the basis for such a notion of culture. In contrast, the Greek anthropologist espoused a racial anthropological concept of difference in which 'blood' was not hidden behind 'culture' but which was at the forefront of difference in race. For Koumaris, race became a mode for interpreting all phenomena related to human and human life, natural and social; it appeared as a kind of obsession derived from the importance the Greek *fili* possessed in his ideological edifice.

In his racial imagination, the biggest threat to the Greek *fili* came from the 'lost' and the 'contamination' of Greek blood. In order to counteract this threat he suggested intentional state policies to be applied and for the Greek population to become aware of the duty towards the *fili*. He argued that "the major concern of a modern social politics should be fanatical resistance against miscegenation in order to maintain the present status quo of the 'fluid constancy' of the race".[64] The most dangerous risk for contaminating Greek blood was the procreation of diseased individuals, but, more than this, the threat emanated from what he termed

62 In Koumaris, *PGAS* (14 May 1939): 10–26. On the 'problem of pygmies' in anthropology, cf. Heberer, Kurth and Schwidetzky-Roesing (1959): 198.

63 Koumaris (3 June 1945d).

64 Koumaris (3 June 1945d).

'exogamy'—marriages between (female) Greeks and foreigners (understood in racial terms). He equalized 'exogamy' with 'blood transfusion' which caused 'infection': "Greek blood needs no transfusion" and "strange blood [causes]... infection".[65] The vehicle for bringing 'exogamy' into being was migration. Therefore, American migration policy was his prototype of a successful racial policy because, he argued, it had been able to select and promote racially superior elements by means of a policy that prevented the entry of racially inferior individuals while allowing entry to superior ones, such as those coming from Great Britain (Koumaris 1939). With respect to Greece, he considered foreigners living within the country to pose a minor, if any, threat to the Greek race because their presence was irrelevant both in terms of size and their potential for mixing with the indigenous population. Racial mixing with non-European races (such as "Turks, Mongols, and coloureds") was allegedly rare in Greece and thus did not need to be considered in legislation.[66] The extent to which his elitism allowed him to arrive at a reliable estimate regarding the presence of foreign populations in Greece is questionable; this very elitism is illustrated in the scornful way in which he described a chance encounter with non-European foreigners in Athens:

> While sitting on the magnificent throne of a... shoeblack, I heard Arabic being spoken. It was a black-haired man with Anterior-Asiatic characteristics, a Syrian, and another one, even darker, a Persian. A third man, a friend of both, came inside; he was a pure Greek [*Ρωμιός*], a globetrotter and polyglot. I had the bright idea of starting a conversation. The Greek said: "I am whatever you like, I speak any language, but I am Greek and a Greek citizen, whatever difficulties this might entail". "Stuff it, Sir, the races make wars, no matter if Greek, Syrian or Persian, they are all same", said the Syrian in bad French. What a would-be fashionable and international guy! The Persian was reading his newspaper, smiling, but also he found an advocate: "Oh yes, see this gentlemen from Persia, now he is Greek, he took a Greek wife, no troubles any more", the little shoeblack said and the Persian smirked.
>
> Viewing these dark physiognomies and looking back to the darkness of the past, I admit that I felt great unease.[67]

65 Koumaris (3 June 1945d).
66 Koumaris (1939).
67 This anecdote was in the introductory passage of Koumaris's article on "The Emigrants and Our Race" which appeared in 1940 in his column "Popular University" in the Athenian press. Here he expressed his conviction that the greatest danger for the purity of Greek blood was 'exogamy', meaning miscegenation caused by marriages between Greeks living abroad and foreigners. Koumaris (14 October 1940): 3.

The 'threat' of miscegenation was deemed to emanate from mixing with 'inferior races'; yet he considered emigration, rather than immigration, to pose the greatest risk for the Greek *fili*. The problem of 'losing Greek blood' allegedly arose as soon as Greek emigrants forgot their origins, ceased to cultivate their Greek consciousness and married racial strangers. With emigrants in mind, he appealed: "Do not permit the loss of a drop of Greek blood."[68] He believed that the state, along with communal, religious and other Greek authorities and their representatives abroad, had an especially important role to play in informing Greek emigrants about how to protect the Greek race from "infusion with 'new blood' ".[69] It was his conviction that 'endogamy' (meaning marriages exclusively between members of the same race) was the proper form for maintaining hereditary racial traits, and an imperative for the persistence and perpetuation of the 'essence of any race'.

The interlinking of emigration and race was assumed in 1940, in a serialized publication entitled "The Emigrants and Our Race" (Koumaris 1940), which appeared in the Greek press. Following Greek liberation from the Axis powers, Koumaris localized the threat of miscegenation and claimed that the phenomenon of "losing Greek blood by exogamy" was increasing. He warned the public that the "increasing number of legal and illegal mixed marriages" between Greek women and foreign men in the aftermath of the occupation had emerged as a major problem that challenged Greek racial purity.[70] He wrote colourfully: "Aeroplanes carried beautiful Greek women, daughters of Prime Ministers, to other countries to meet blond spouses" and "bridal processions accompanied the prisoners of war" (Koumaris 1945g). However, more than all these 'irrefutable facts', the true motivation for addressing the public was to comment on the information "provided by a distinguished member of the Greek Anthropological Society that ... ten thousand Greek women intend to leave for Italy in order to legalize their relations with Italians" (Koumaris 1945g). This fact was allegedly "monstrous, no matter whether they [these women] will be expatriated or be allowed to come back" (Koumaris 1945g). The threat of losing and contaminating Greek blood was a matter of 'women's exogamy'.

68 Koumaris (16 October 1940): 3.

69 Koumaris (1948): 127.

70 This was the message he communicated in "Mixed Marriages" in the liberal periodical *Focus* [*Εστία*] and a letter addressed to the Athenian daily, *Nation* [*Έθνος*], both in 1945. See also Koumaris (1945b).

These remarks were in fact part of the broader stance the Greek anthropologist had adopted (from his own, racial, perspective) on an issue that arose during that period and concerned the gendered role ascribed to women during the occupation, and subsequently in the ongoing Civil War. As a matter of gendered political action, female sexuality was perceived as a symbolic vehicle for political collaboration and antagonism between the hostile fronts. From the perspective of the left-wing resistance, the question was whether women who consorted with soldiers from the occupation forces ought to be considered (and treated as) collaborators. Executions of women accused of such behaviour were by no means rare during and after the occupation. Those from the anti-communist front, meanwhile, labelled the women involved in the armed or civil resistance as inclined towards promiscuity. The accusation of promiscuity against politically active women increased during the Civil War; however, this ultimately originated from the inter-war period, at which time promiscuity was imputed to communist ideology at large, with the communists (both masculine and feminine) being collectively accused of striving to abolish morality and ethical values. Yet despite these proclaimed concerns regarding female sexuality, women involved in the left-wing resistance during the occupation were often murdered by means of sexual torture, by opponents or collaborators.[71]

Koumaris was rather apprehensive regarding the possible racial implications of the relations between Greek women and foreign soldiers, a concern arising from the pivotal role he ascribed to women in maintaining the Greek hereditary disposition and reproducing the *fili*. While imagining the nation in racial terms, 'mixed marriages' appeared as the central aspect of the gender connotations of race. Sexual reproduction was the means through which the hereditary make-up of future generations could either be soundly transmitted or modified, and women were thought of as the guardians of racial purity and as a vehicle for multiplying the Greek race. Hence, in order to counteract the imminent 'racial mixtures' in the aftermath of the war, Koumaris urgently advised fostering ethical values among Greek women and educating them for their mission as guarantors of the maintenance and perpetuation of the Greek race. This was allegedly necessary, since "the naïve woman is unaware of the fact that first of all the Greek mother has all along prevented the Greek race from

[71] See Kostopoulos (2005): 134–9.

vanishing".[72] Regardless of his obviously disparaging attitudes towards women (which were also expressed on other occasions, such as when he alluded to female mental inferiority),[73] in speaking about female sexuality through symbols of blood he made use of the typical conjunction of race and sexual technologies that emerged from the 19th century onwards (Foucault 1998; Young 1995; Stoler 1995). Females and the female body were thought to be vehicles for the transition from sanguinity to sexuality and vice versa, once human reproduction became a matter to be regulated by state policy.

State intervention by means of passing appropriate laws was, in Koumaris's eyes, of vital importance for safeguarding "the future of the Greek *fili*"[74] and its racial purity, while the basis for any legislation should be the 'science of heredity' [*κληρονομολογία*]. Interpreting any repressive legislation in racial terms, Koumaris celebrated a recent revision of the Civil Law (in 1940) as an example of racial politics,[75] especially the widened scope of an amendment paragraph that prohibited marriage between "Christians and individuals of other religions".[76] In his opinion this restriction was a measure of state racial policy that was allegedly aimed at protecting the Greek race. Anticipating possible objections on whether religion could be subsumed under racial issues, he drew on his racial theory on the psychic traits of a race and remarked:

> It is wrong to believe that the [diversity of] language, religion and national consciousness are characteristics of secondary relevance with respect to race, to blood. In fact, there exists a major connection between "psychical" and "physical" features.[77]

Here the Greek anthropologist was subsuming religion under the 'lateral racial characteristics' that were no less important for the distinction

72 Koumaris (1945g).

73 In a letter entitled "On the Intellectual Difference between Male and Female" published in 1929 in the medical periodical *Medical Annals* [*Ιατρικά Χρονικά*], Koumaris argued that women are different from but not inferior to men. However, "the lower, or more exactly, the different intellectual value eunuchs feature [apparently in comparison to men], should be considered along with all their other female characteristics to explain their divergence". Koumaris (1929b).

74 Koumaris, *PGAS* (14 May 1939): 10–26.

75 Koumaris (16 October 1940): 3.

76 Article 1353 of the 1940 Civil Law: [Αστικός Κώδιξ, Αθήναι: Εθνικόν Τυπογραφείον 1940: 403].

Cf. Faidon Vegleris's statement in the public debate of the Greek Eugenics Society against the ban on marriage among individuals of different religions.

77 Koumaris (16 October 1940): 3.

between races. He depicted the racial relevance of religion in a brief text on race policy in Greece that was published in 1938 in the Italian fascist journal *La Difesa della Razza*.[78] This publication attracted the attention of the correspondent of a German daily newspaper, Cologne's *Kölnische Volkszeitung*, who provided a very short summary in German with the title "Care for the Race in Greece" ["Rassenpflege in Griechenland"].[79] The German correspondent underlined the "particularly remarkable information" provided by Koumaris, according to which mixed marriages had not taken place in Greece due to the strong religious ties that held Greeks together and which excluded religious strangers. In several of his publications (whether in Greece or abroad), Koumaris sought to transmit the message that the Greek state was pioneering in terms of racial policy. Such allegedly racial measures included repressive legislation like the ban on marriage for persons suffering from Hansen's disease which was passed in 1920; restrictions on marriage for those who suffered from syphilis and trachoma, along with many other examples. He further claimed that a law passed at the end of the 1920s prohibited foreigners from settling in Greece, while Jews were only allowed to travel through the country.[80] All of these issues were actively being discussed during this period by those who were advocating the implementation of eugenic measures; Koumaris simply took a stand by using a racial terminology. He also sided for the establishment of a health certification process prior to marriage in several of his publications (e.g. Koumaris 1931 and 1938). In a serialized Greek article on "Race and Health" (Koumaris 1938), he explicitly advised embracing the models of pre-marriage medical certification that had been implemented in North America and in other European states, especially in Scandinavian countries, rather than the "extreme German model that aimed at racial cleansing".[81] A shorter version of the same article was published in the German journal *Ziel und Weg*, the official periodical of National Socialist physicians.[82] The German version differed significantly from the Greek publication in accommodating National Socialist sensibili-

[78] Koumaris (1938). It appears as anonymous correspondence; however, Koumaris mentions it as his own publication, Koumaris (1951): 34. The text is apparently based on Koumaris (1938).

[79] [Correspondence], *Kölnische Volkszeitung*, "Rassenpflege in Griechenland" (27 November 1938).

[80] Exactly the same argument was embraced by the German archaeologist and Philhellene Roland Hampe (1941).

[81] Koumaris (16 October 1938).

[82] Koumaris (1939).

ties. Thus any criticism towards the German model was missing, as was his preference for the Scandinavian model.[83] In his Greek publications he favoured establishing voluntary examinations for the time being, along with the setting up of consultation agencies that could provide advice on health and heredity to couples who were about to get married.[84] These measures had the option of being accompanied by additional arrangements, such as giving financial benefits to couples who agreed to premarital health control, or penalizing those who ignored it. For his part he was inclined to the former option, arguing that "this is the only means the populace is able to understand".[85] He was convinced that the success of premarital control in the service of "the betterment of the race was worth any material damages".[86]

Koumaris was never criticized for his suggestions on measures relating to population politics, and it is worth noting that he was far from alone in making such proposals; rather, he followed a mainstream tendency in his contemporary period, as the following chapter will discuss in greater detail. The criticism he earned referred to his racist attitudes, but even then this was remarkably subdued. In the domestic arena, he was criticized in 1945 in the columns of the journal *Communist Review* [*Κομμουνιστική Επιθεώρηση*], the official magazine of the Communist Party of Greece. Under the title "The Pseudo-Science of Racism and its Agents in Greece", the columnist accused those university professors, and in particular Koumaris by name, who had conformed to National Socialist and fascist ideas, and had defended them in 'scientific publications' or in statements in 'scientific societies' (Aggelidis 1945).[87] With respect to Koumaris, the criticism was explicitly addressed to his racial theories and, in particular, to his publications during the war period, such as the article "Greek Blood" and another on "Racial Politics in Greece and in the World" which had appeared in 1941 in the journal *Quadrivio* [*Κουαδρίβιο*]. The columnist of the *Communist Review* disproved of Koumaris's endeavours to qualify racial science and racism as "objective and uncommitted scientific

83 After the war he denied any responsibility for this publication, claiming that the summary was translated and published without his permission or his knowledge. Koumaris, *50-ετηρίς* [50th anniversary] (1951): 87.

84 Koumaris (16 October 1938): 1.

85 Koumaris (1931b).

86 Koumaris (1931b).

87 Koumaris's response and further statements by the editors of the *Communist Review* [*ΚΟΜΕΠ*] were published in this journal: see Aggelidis (1945) and *Communist Review* (1945).

knowledge, free from political intentions". Instead, he himself underlined the strong political character of racial theories and racism which "serve hegemonic aims" (Aggelidis 1945). His assumption that what the professor of anthropology represented was nothing more than 'racist nationalism' was not entirely off target. Indeed, Koumaris himself did not reject this categorization; however, he assessed it differently, from an opposite position, by calling himself a 'filetist'. By this he meant a scholar who advocates scientific racial theory, and simultaneously a patriot who places his racial science in the service of the nation. In his own words:

> Filetism...[means] the commitment to the protection, care, cleansing, and prevention of miscegenation of any race by its members.... I am sorry that in my attempts to defend our race even by polemic means against inadmissible foreign racial politics I have been misunderstood by extremely oppositional Greek politics. I have, however, the hope to persuade with my writings that even the poor Greek science includes a bit of patriotism. Even if this patriotism induces [me] towards hunting a chimera, nevertheless, this makes the intention no less honourable (*Communist Review* 1945: 42).

It was this supposed patriotic commitment that mobilized him to defend the Greek race and the racial ideal in his publications (including those in the Italian fascist journal *La difessa della Razza* 1939), "which even met with the approval of his Jewish friends".[88] However, the contribution to the journal *Quadrivio* was allegedly published without his permission; nor did he have any note of this at all.[89] In a further abuse of his work, as he explained in his autobiography, another contribution he had made to the official organ of National Socialist physicians, the German journal *Ziel und Weg*, was also printed without his knowledge.[90]

Despite the attempts to restore his image retrospectively according to the outcome of the war, Koumaris did not discard the racial idea as such, which he continued to advance in the aftermath of the war. Race preoccupied him as an obsession and formed a holistic concept for interpreting all human relations and the world as a whole, Greek society included. World

---

[88] *Communist Review* (1945): 42.

[89] Koumaris (1951): 87.

[90] Koumaris (1951): 87.
The matter concerned an article with the title "Race and Health" ["Rasse und Gesundheit"] (Koumaris 1939). He argued that the initiative for the publication had been taken by the Athens correspondent of the German journal, who summarized his Greek article with the same title and submitted a German version to the German journal.

freedom could allegedly be ensured by implementing racial politics, targeted at keeping races apart. In his opinion mutual hostility between races was the reason for wars, and these would vanish if state frontiers were to be drawn on the basis of anthropological knowledge (Koumaris 1948: 127). But most importantly, race was seen to be a means of situating Greece in the world around it. Therefore his rejection of UNESCO's statement on race (1950) came as no surprise. Even in the period immediately after World War II, when the international scholarly community was discussing abolishing the notion of race altogether, the Greek anthropologist continued to defend it as the cornerstone and superstructure of his whole theory and worldview, and thus as the only concept that was able to provide a proper understanding of the existing condition of his homeland.

> We insist on the efforts to preserve that [the Greek] race. Races exist and will continue to exist; and each one defends itself. Because every infusion of "new blood" is something different and because children of mixed parents belong to no race, the Greek race, as all others, has to preserve its own "fluid constancy" by avoiding mixture with foreign elements (John Koumaris 1948: 127).

This statement in the British anthropological journal *Man*[91] caused the social anthropologist M.D.W. Jeffreys (University of the Witwatersrand) to react provocatively by asking the question, "What is it that distinguishes the Greek race from the Jewish race, from the Catholic race and from the Mohammedan race?"[92] Jeffreys also posed the rhetorical question as to whether the Greek anthropologist was aware of recent studies such as that by Ashley Montagu (1947). The response Jeffreys received is expressive enough: "if the Indo-European race is a fallacy, the Greek race is undoubtedly not so".[93] In the following period Ioannis Koumaris voiced his opposition to the UNESCO statement on several occasions. But he was not alone in doing so—in fact, the UNESCO statement was revised several times for a variety of reasons including the reactions it received from the scholarly community (cf. Montagu 1972/1951). Disapproving responses to the statement from several prominent scientists, such as H.V. Vallois

91 Koumaris (1948).
92 Jeffreys (1949): 36.
93 Koumaris (1949): 140.

(1951),[94] offered an occasion for the Greek anthropologist to celebrate his own views.[95]

Race, and the Greek race in particular, became an obsession for Ioannis Koumaris. The essential point in his model of race, and especially that of the Greek *fili*, was the absolute rejection of any racial mixing, and this was manifested in racial, national and gendered ideology. According to Pierre-André Taguieff's typology of mixophobics, Koumaris falls within the category of 'absolute mixophobics' (i.e. those who 'sacralize' race) (Taguieff 2000: 289–305). According to Taguieff's six-point formulation, the race concept and the racial hygiene ideas of Koumaris comply with the doctrine of the opponents of mixtures:

> 1. Every race is a human type that is assumed to be stable. This is the postulate of the stability of types. 2. There exist superior and inferior human types. This is the postulate of the inequality of types. 3. Every type has a specific "blood" property. This is the postulate of racial "monothemism". 4. The value of a race is based on the purity of its blood, the racial value of a mixed population is based on the proportion of blood of the superior race. 5. The mixing or crossing of races is a mixture of blood. Reproduction is like a "blood transfusion" by means of which traits and deficiencies are passed on. Immigration is perceived both as a massive ethnic "blood transfusion" and as an "interracial transplantation". 6. The mixture permanently destroys the respective quality of the "blood", and thus the specific value of the mixed race. The mixture works inevitably in favor of the inferior race; it produces mediocrity (Taguieff 2000: 291).

For Koumaris, all population issues could be reduced to the question of the extent to which certain developments in society would cause racial mixtures, and how far such mixtures were a risk for the Greek blood. Certainly, this is prima facie the core of the racial hygienic idea generally, whose embracement he programmatically announced at his inaugural lecture as chair for physical anthropology at the University of Athens when he defined it as an "allied science to anthropology" (Koumaris 1925). The following chapter on eugenics and racial hygiene may illuminate the fact that his denominated 'racial hygiene' was exceptional in this context since his concept failed to find the link with topical questions in society related to reproduction technology and life management by racial

[94] In the first part of his autobiography he made a note celebrating the recent reactions that had even been expressed in the British journal *Man*, the same journal in which he himself had been attacked a few years earlier. Koumaris (1951): 88. See reference to Montagu in Koumaris (1951a).

[95] Cf. the debate on the UNESCO statement: *Man* 51, no. 25–8 (1951): 15–18.

hygiene. In his narrative, racial hygiene was defined in an ideal, if not metaphysical, *Topos*, beyond the real concerns of society and, no less importantly, beyond the preoccupations of other actors. Thus 'hygiene' indicated a metaphorical understanding of 'cleanness'. It was a matter of genuineness, an idea that was carried in the 'symbolics of blood',[96] and it encompassed both the religious and the racist connotations of the dualism of 'purity/cleanness–impurity/uncleanness'.

Koumaris's racist anxieties concerning the possible infection of Greek blood through miscegenation caused by marriages between Greeks and racial aliens in emigration did not have any considerable impact over the circle of those involved in the anthropological discourse. On the contrary, the shifts in the line of arguments from the assimilatory power of the Greek *fili* to its constancy, and finally homogeneity, were witnessed in the perceptions of racial anthropological ideas and the representation of the Greek population as being in principle racially consistent. In the following and final part of this chapter I will outline the argument that the idea of the racial homogeneity of the Greek population kept up with the gradual solidification of trends in official national doctrine of the 20th century, and the rising minority question that was dictated by the relations between Greece and its neighbouring states.

7.6 *Perceptions of Racial Theories: Absorbed, Absent and Phantom Minorities*

The assumed homogeneity of the Greek population was a precondition for the discussion of those defined as being aliens, both racially and nationally, on Greek soil, a debate which became enflamed after the ultimate decay of the Ottoman Empire and which was epitomized in the minority question. This effectively served as the background against which the major concern of the Greek (racial) anthropological discourse was to substantiate the alleged homogeneity of the current Greek population, rather than providing racial arguments for excluding aliens. Indeed, from the inter-war period onwards the presence of national aliens in the Greek population was officially denied in the national narrative, even though repressive measures against them were applied in practice (Carabott 2003; Kostopoulos 2003 and 2008; Trubeta 1999). Given the ambivalent

96 On the symbolism of blood and its transformation, see Linke (1999).

way they were treated, minorities in Greece stayed in an 'ethereal light'. Slavs, Albanians, and after 1923 increasingly also Turks, remained a challenge for the national narrative, but now they were discussed, or concealed, in a new framework that was strongly marked by the bilateral relations between their nation state of reference and Greece. The changing interstate relations in the Balkans put national doctrines to the test and caused alterations in previous perceptions of homogeneity insofar as they were projected over the factors that were disturbing the desired homogeneity.

The attempts to provide anthropological evidence regarding the racial affinity of the so-called 'refugees' of Greek origins, who arrived in the Greek state from Asia Minor and other territories of the former Ottoman Empire during the 1920s, should be seen in light of the shifts in the relationship between the Ottoman past and the Turkish national state. The racial anthropological arguments shifted from initially raising questions of their possible affinity to Asiatic races to ascribing them a genuine Greek character. The integration of the so-called 'refugees' into the Greek national narrative did not occur automatically, at least not from the very beginning. Their arrival challenged Greek society not only because of their considerable size (approximately 1.5 million individuals), but also because it raised the question of their possible negative impact on national homogeneity. Their 'otherness' was materialized on the grounds of their social calamity but also the fact that they came from the Orient, and a state that was now hostile towards Greece. In the 1930s they were treated in the anthropological debates under the aspect of an 'Asiatic factor' that could disturb Greek homogeneity and the racial consistency of the Greek *fili*. They were conceived as a vehicle to bring former Orientalistic representations into the state territory, making these representations current and epitomizing them within a certain population category. This assertion was, however, also widespread beyond the anthropological discourse. While the autochthonous population considered and treated them as foreigners for many decades to come, there were also demands on exploring their 'biological quality' and comparing them with those of the autochthonous Greek population, as for instance those raised by the physician, liberal politician and convinced eugenicist Apostolos Doxiadis (1874–1942) in the inter-war period (Doxiadis 1928b: 97). (Ironically enough, Apostolos Doxiadis himself was born in what had previously been Ottoman territory, the Bulgarian town of Asenovgrad.) From the anthropological point of view, in principle the theory of the 'strong nucleus' of the Greek race provided a means for restoring the disrupted affinity of the oriental 'refugees' to the Greek *fili* as soon as it postulated the Hellenic/Greek element to be predominant over any Asiatic influence. However, it was felt that additional anthropological investigations might provide sound evidence for this theoretically asserted affinity, and Koumaris consequently used the occasion of a presentation of results from blood examinations at the Greek Anthropological Society to underline the necessity of conducting comparative investigations with 'old' and 'new' Greeks in order to determine racial similarities.[97] In the aftermath of World War II, he 'repaired' their deviance on the grounds of anthropological findings that proved the belonging of the 'refugees' to the Greek racial community. He claimed to have found a haematological relation (described as an 'obscure bond') between the 'Anterior-Asiatics' and other Greeks by means of comparative blood examinations. His finding was that there was a high percentage of blood group O among refugees, which, he argued, indicated a Mediterranean influence, thereby demonstrating the racial affinity between the 'old' and 'new' Greeks (Koumaris 1948).

The concerns for the racial descent of "refugees from Asia Minor" arose because they came from a territory that had previously belonged to the former Ottoman Empire, but currently to a hostile state, namely Turkey. The affinities between the peoples of the former Ottoman Empire, and in particular the Greco-Turkish racial/*filetic* bonds, were put on a new footing in the aftermath of the Treaty of Lausanne (1923) which forced a population exchange between Turkey and Greece and also institutionalized minorities in the two countries. A possible detection of racial affinities in these 'refugees' to the Orient would objectify their bonds with a hostile state and consequently justify Turkish nationalist claims on Greece. The exaggerated Greek–Turkish antagonism dictated and strengthened the premises of national homogeneity.[98]

---

97 Papamarkou and Sklepa, *PGAS* (28 October 1941): 28–9. Stavropoulos, *PGAS* (28 October 1941): 30–1. In the discussion following the two presentations, Koumaris remarked that the blood affinities of 'old Greeks' and 'new Greeks' should be studied.

98 Opposing voices to Turkish–Greek hostility and the population exchange also used the argument of descent in condemning the violent separation of peoples who had hitherto been bound by historical and other ties. So did, for instance, Symeon Iosifidis, a physician of Greek origins from Caesarea. Iosifidis asserted that the forced population exchange and Kemalist nationalism violently divided two 'sibling peoples'. He assumed a *filetic*, psychic and social affinity of Greeks and Turks that might be temporarily hidden due to the violent depressions, but which could nevertheless emerge to the surface at any time.

His view was expressed in his article titled "Turkey, Islamism and Civilization" that appeared in a periodical addressed to 'refugees': Iosifidis (1925). In the General State Archives (Athens), Iosifidis's Archive (1923–39), [File K 134, φ.δ.], and in a serialized article

In view of this development, the distinct emphasis that the 1956 winner of the 'Koumaris award on the Greek race' had laid on the Turks is impressive; indeed, they were even referred to in the title of his awarded treatise: "The Aegean, Cradle of the Aryans and Hellenism: Cultural Antecedence and Radiation, Racial Unity, the Racially Negligible Intrusion of Slavs and Turks".[99] The prize winner was a military man, General Xerxis Livas (1899–1965), and his study was an attempt to elaborate a comprehensive model to explain the singularity and racial homogeneity of Hellenism. The emphasis on the Turks was exceptional within both the Greek anthropological circle and the official minority discourse of that period, in which silence usually reigned with regard to the presence of Turks. Even though unspoken, the Turks referred to by Livas were embodied in the Muslim minority in the northern border region of Greek Thrace; a minority (the only one officially recognized in Greece) that had been established by virtue of the Treaty of Lausanne (Trubeta 1999). Livas's particular concern for the Turks needs to be seen in the light of the exacerbation of Greek–Turkish relations in the mid-1950s which signalled the (re)start of a long period of repression for the Muslim minorities in Greek Thrace (Trubeta 1999: 69–100). On an ideological level the strategy Greek nationalists pursued in coping with the Muslims in Thrace was based on a distinction between the three largest sub-groups of the minority, namely the Turkophone Turks, the Slavophone Pomaks and the Roma. In the proliferating, populist and often anthropological-like literature the descent of the sub-groups was treated with respect to their possible affinity to the Greek *fili*. Until the late 1990s, when a new generation of Greek scholars in the humanities started to unveil the nationalistic nature of the minority question, professional and amateur researchers concealed the presence of Turks whereas they were anxious to uncover the secret Hellenic descent of the Pomaks that was allegedly hidden behind their violent Islamization (cf. Trubeta 1999; 2008). An early example of this is the PhD thesis by Nikolaos Xirotiris (1971), one of the last members of the GAS and currently a professor of physical anthropology at the Department of History and Ethnology of the Democritus University of Thrace. Xirotiris sought to draw conclusions for the Hellenic descent of the Pomaks on the basis of

titled "Filetic Affinities between Greeks and Turks" which appeared in the weekly newspaper *Attention* [*Προσοχή*], the magazine of the 'refugees' in Greece, Iosifidis (1936).

[99] A summary of this study was printed in the *PGAS* (29 November 1956): 26–43 with an English summary. The entire treatise was published in the same year and a second edition appeared in 1963; Livas (1963/1956).

their blood groups, along with supplementary ethnological and historical evidence, oral traditions, customs and so on. This was not the first time that Xirotiris had investigated members of the Muslim minority by blood examinations. In 1967 (the year that he joined the GAS), he delivered a paper with the title "Negroes of Thrace and Blood Groups of Women in Northern Greece".[100] His research subject (the 'Negroes') referred to a small sub-group of the Muslim minority in Thrace whose members are Turkish-speakers and whose origins are assumed to be African because of their black skin. His paper contained statistics on their blood groups, without coming to any conclusions on affinities; nor did he outline issues of descent.

Without any doubt, the affinities of Greeks to Albanians and Slavs (also Vlachs or Aromanians, though to a lesser degree)[101] rather than to Turks, preoccupied anthropologists, scientists, intellectuals, politicians and even the Greek public for a long period of time. Once the former Ottoman peoples had become embedded in the minority question of antagonistic states, the debate surrounding the Slavic and Albanian influence over Greece changed in becoming dependent on the outcome of the interstate relations. From the 1930s onwards, Albanians and Slavs (similarly to the Turks) had been thought of as physically absent from Greek state territory, either because of acculturation and absorption, or simply because there was a general silence concerning their presence for all their discriminatory treatment; and this rendered them a phantom minority. Keeping pace with this tendency, the focus of anthropological reasoning had shifted from the hitherto emphasized acculturation to the negation of any correlation with foreign peoples and races. In this context, religion, and Islam in particular, was reinterpreted as an interruption of the Hellenic character of some Muslim communities. The argument was that forced Islamization had caused a temporal de-Hellenization that should be restored.

After the founding of the Albanian state, most of the Albanians who stayed in Greece were Christians and their elites were widely acculturated. Even though resident in distinct communities, from a Greek nationalist point of view they were not regarded as posing any immediate threat to the sovereignty of the state. The situation was different in the case of the

100 See Xirotiris, *PGAS* (12 November 1967): 24–6.

101 Dendias expressed the same views, *PGAS* (1 May 1948): 10–27. On the Vlachs or Aromanians see Kahl (2003).

Albanophone Muslim population, known as Chams (or Čames), who were resident in Epirus, in the Greek–Albanian border region. The relation of this border population to Albania, Greece and also to Turkey arose as a matter of urgency in connection with the population exchange between Turkey and Greece. During the Treaty of Lausanne negotiations, the question as to whether this population should also be exchanged caused some controversy; it was eventually decided that it should be excluded (Kretsi 2002a). Chams were again in the spotlight during World War II, when they were regarded, and treated, as collaborators with the occupying forces. They were persecuted and displaced during and after the war.[102]

Usually, in the Greek anthropological discourse no distinction was made between Christians and Muslims when speaking of the contested 'Albanian element' in Greece. An explicit reference to the Chams before the GAS in 1948[103] may well be the only exception. The lecturer portrayed them as an 'authentic Greek population' whose de-Hellenization had occurred in very recent times during the course of their Islamization. Despite different languages and religions, Greeks and Muslim Albanians living in Thesprotia (a department of Epirus) were related by blood and marriage, the affinity of which, the lecturer argued, could be confirmed by local Albanian and Greek families who had common ancestors.[104] But how many Muslim Albanian families had remained in Thesprotia after World War II? The official census of 1951[105] recorded 77 individuals in Thesprotia and a total of 127 in Epirus,[106] whereas in 1920 they had numbered 20,319[107] and 21,800[108] respectively. The large majority of these "undesired compatriots", as they were aptly designated by the historian Giorgos Margaritis (2005), had been persecuted and expelled or had left their homes to avoid repression.

The 'disappearance' of minorities from the national narrative concerned the Slavs or Slavophone inhabitants in Greek Macedonia more than all other minorities. The Slavs were associated with the still topical

[102] See Kretsi (2002b); Margaritis (2005): 133–216. The issue of the Chams was not least associated with the property question; see Kretsi (2003).

[103] Dendias, *PGAS* (1 May 1948): 10–27.

[104] Dendias, *PGAS* (1 May 1948): 10–27, 19.

[105] Greek National Statistical Service, *Results of the 7.4.1951 Population Census*: Αποτελέσματα απογραφής πληθυσμού της 7/4/1951, vol. 2, Athens, 1961, pp. 212–47. Unpublished data, here cited in Kostopoulos (2003): 71.

[106] National Statistical Service of Greece, *Results of the 7.4.1951 Population Census*: Αποτελέσματα απογραφής πληθυσμού της 7/4/1951, vol. 2, Athens, 1961, pp. 212–47.

[107] Data from the Ministry of Foreign Affairs, cited in Kostopoulos (2003): 71.

[108] Unpublished data, here according to Kostopoulos (2003): 71.

conflict concerning the status of Macedonia and therefore continued to be a hotspot for Greece and its Slavic neighbours during the 20th century, and also into the 21st century (Karakasidou 1997; Voss 2003). From the late 1930s, during the dictatorship of Ioannis Metaxas, this conflict (known also as the 'Macedonian question') was exacerbated when repression against the Slavophone inhabitants of Greek Macedonia increased. And after Greece had been drawn into World War II the tension with Bulgaria intensified due to the Bulgarian occupation of eastern Macedonia and Thrace (1941–4). As the historian Philip Carabott argues, in the prevailing atmosphere of anti-communist hysteria during World War II, Slav-speakers in Greek Macedonia were generally associated with the communists,

> because of the public acceptance by the KKE (Communist Party of Greece) in early 1925 of the Comintern-ruling on the right to self-determination of all the *ethnotites* (nationalities) of Greek Macedonia and Western Thrace, which was changed in March 1935 to equal national and political rights for all "national minorities" within the country (Carabott 2003: 155).

The echo of these developments can be traced in the internal debates of the Greek Anthropological Society, in which the relationship between Slavs and Greeks was still discussed, albeit under new conditions. The focus of the arguments had shifted from earlier assertions that Slavs had been absorbed into the strong nucleus of the Greek *fili*, to explicit statements on the 'Macedonian question' arguing for the exclusively Greek character of Macedonia. The prevailing assertion continued to be that the Slavic racial influence upon Greeks was minimal, if it existed at all, and that Slav physical and psychic characteristics were not detectable in the Greek population. More extreme views were not missing, for example, that Slavs could not have any influence upon Greeks since the Slavs were inferior in terms of civilization.[109] General Xerxis Livas pointed to the heart of this contentious matter by asserting that since the Slavs were not a distinct population group in Greece, the 'Macedonian question' had no basis in fact.[110] Already, in his prize-winning treatise, Livas had earlier asserted that the Balkans and Asia Minor constituted a unified space and the influence of Turks (and Slavs) on this region was insignificant. Livas himself, as well as most of those involved in the debates of the GAS, sought to corroborate these assertions by drawing on historical, cultural and linguistic

---

109 Dendias, *PGAS* (1 May 1948): 10–27.
110 Livas, *PGAS* (12 March 1965): 21–7.

arguments. The most prominent historical approaches included those by the historian Konstantinos Amantos, an expert on issues related to Slavs in Greece.[111] Ethnological observations also formed the basis on which Anna Triantafylidou, the president of the Lyceum of Greek Women [Λύκειο Ελληνίδων], struggled to corroborate the Greek character of Macedonia, which she outlined at a meeting of the Greek Anthropological Society in 1945.[112]

When these views were expressed in the circle of the GAS, the notion of *fili* had passed through a long period of transformation that corresponded to historical developments that greatly influenced the perception and representation of Greece as a large, imagined, community. The anthropological discourse partially operated as a mirror of the national doctrine, keeping the main axioms of the latter but transcribing them in racial terminology. Given that there was never a professional group of trained anthropologists in the period under consideration, the protagonists in elaborating racial anthropological concepts of the Greek *fili* included Clon Stéphanos, but the most significant was Ioannis Koumaris. A further physical anthropologist by training, who entered the stage in the final years of the Koumaris era, and who was concerned with Greek racial descent, was Aris Poulianos. He embraced the same line of arguments in his efforts to buttress the Greek national doctrine, and he also assumed the autochthony and racial homogeneity of the Greek population.

7.7 *Aris Poulianos on the Racial Origins of the Greeks*

"The origins of the Greeks" was the subject of Aris Poulianos's study that paved his entrance into the anthropological scene and also ensured that he received the Koumaris Award for studies on the Greek race in 1962,[113]

---

[111] Amantos, *PGAS* (22 February 1926): 10–31.
Further publications by Amantos include the monographs: Amantos (1920); (1923); (1934); (1952).

[112] Triantafyllidou, *PGAS* (27 June 1945): 25–31.

[113] *PGAS* (2 April 1962): 20–31. In this brief summary of the award-winning study Poulianos presented his central conclusions and working methods. See also Poulianos, *PGAS* (12 March 1965): 27–38.
Since Aris Poulianos was in Moscow in 1962, the prize was awarded in his absence. He received it personally at a celebratory meeting at the Academy of Sciences in Moscow by Christos Maltezos, a member of the board of directors of the Greek Anthropological Society, on the occasion of Maltezos's journey to Moscow to attend an International Health Organization seminar.

even though he did not hold an academic position. The 13th revised edition, printed in 2004, carries the complete title *The Origins of the Greeks: An Ethnogenetical Study—Mathematical Evidence for the Descent of the Greeks* (Poulianos 2004). This was his PhD thesis, supervised by Professor Georgy Frantsevich Debetz, and which he submitted in 1960 at the Lomonosov University in Moscow.[114] The theme of his research was the racial descent of Greek nationals, and his hidden programme was the intention to prove the racial anthropological homogeneity of the Greek population; the challenge he was faced with, in doing so, was to conform to the views of communist theoreticians, historians and politicians on the Greek nation.

The main arguments of Poulianos's study were the racial homogeneity of the Greeks and their autochthony in the Greek territory. His views conformed to Koumaris's racial theory and his line of argument reveals the contemporary national state of affairs in treating the minority question in Greece. But Poulianos went a step forward in corroborating the autochthony of the Greeks and claimed to have found anthropological evidence for the beginning of all human life in Greek territory. In contrast to the above-mentioned recipients of the racial anthropological discourse, Poulianos was an anthropologist by training. His research was widely used and cited in Soviet academia and kept pace with developments in physical anthropology. In terms of working methods he was at the cutting edge given that he included population genetics which, next to anthropometry, played an important role in such anthropological research,[115] as well as utilizing supplementary historical and linguistic sources.

He drew the sample for his anthropological investigation from Greek political refugees in the Soviet Union;[116] those who had lived in Soviet exile after the end of the Greek Civil War and the defeat of the leftist front (1949). He began his research in 1956 in Tashkent, Uzbekistan, where a large number of Greek Civil War refugees resided, and he later continued his work in other Soviet republics. Reflecting retrospectively on the circumstances that facilitated his research, he remarked that "one could say that the Civil War, which drove thousands and thousands of people into Central Asia, also had something positive: [it provided an opportunity] to

---

114 The first edition of his thesis was published in 1960 in Russian and also in Greek (Poulianos 2004/1960).

115 Cf. also a further paper presented to the Greek Anthropological Society: Poulianos, *PGAS* (12 November 1966): 43–52.

116 Details on how he managed to cope with his demanding project are included in the fifth edition of his study and are quite informative with regard to the circumstances under which he had conducted his research (Poulianos 2004: 13–18).

begin research on the biological study of the Greek population" (Poulianos 2004: 14). The fact that these individuals originated in different Greek regions made them a representative sample, all the more so because they included Slavophones, Vlachs or 'refugees' from Asia Minor and other erstwhile Ottoman territories (or their descendants). For the purposes of comparison, Poulianos included in his sample a group of Russians as well as Uzbeks and Georgians who were to be compared with the Slavs from Greek Macedonia and the Greeks from Asia Minor respectively. Summarizing, he drew the following central conclusions: the Greek population was essentially autochthonous to the Greek state territory; all types he examined were part of the East Mediterranean branch of the large European race; the Greek population was anthropologically and racially homogenous; the influence of Slavs, Albanians and other peoples upon the Greeks was negligible; the Anterior-Asiatic racial type was detected neither in his sample of the Greeks from Asia Minor nor in the rest of the Greek population;[117] and contemporary Greeks were descendants of the ancient peoples who, in the course of history, adopted the Greek language and culture. On the whole, and all external influences notwithstanding, the basic Greek anthropological type had been sustained and perpetuated into the present. Poulianos believed he had proved "the biological continuity of the Greek population for dozens of millennia" (Poulianos 2004: 15).

But Poulianos was uncomfortable with these conclusions because he felt that they did not coincide with the views of the Marxist historian Ioannis Kordatos (1891–1961), and the former leader of the Communist Party of Greece Nikos Zachariadis (1903–71). (The latter was at that time a political refugee living in the Soviet Union where he remained until he committed suicide in 1971.) Poulianos's problem was that, according to the Marxist view, the continuity of the Greek people was a matter of *historical* (not of biological) character and its starting line coincided with the *Byzantine* period. His supervisor helped him to solve this dilemma. Debetz encouraged him to choose a supplementary sample of between 100 and 150 Greeks who had arrived in the Soviet Union only very recently. On the basis of this new sample, Debetz himself conducted a comparative investigation at the Racial Department of the Institute for Ethnography at the Academy of Sciences in Moscow. Debetz's findings affirmed Poulianos's conclusions, and the latter was able to complete his dissertation,

[117] Poulianos, *PGAS* (2 April 1962): 29.

even though his conclusions were not congruent with the Greek Marxist historiography.

Those involved in the Greek anthropological discourse were not indifferent to the fact that Aris Poulianos came from Soviet academia; his Soviet academic affiliation was considered in the evaluation of his findings when his study was awarded the 'Koumaris Award". The members of the awarding committee ascribed a dual symbolic significance to their decision: from a political point of view, they welcomed a study that came from "behind the Iron Curtain", which could be understood as a contribution to world freedom.[118] Furthermore, they acknowledged the national importance of the prize-winning study which had been conducted in a Slavic country but nevertheless corroborated Greek national concerns. Yet reservations were not entirely absent either, as is clear from the remarks made by General Xerxis Livas, who was asked to make the laudatory speech. Livas expressed his reservations concerning the insufficient strength of Poulianos's assertions regarding the Greek origins of the Slavic speakers in Greek Macedonia.[119] But Poulianos outperformed his critics with respect to his national aspirations. If Ioannis Koumaris had asserted the autochthonous character of the Greek race in the region of the Aegean Sea and the current Greek state, Poulianos believed that he had detected the oldest traces of human life on the European continent, in and around the cave of Petralona; a cave of archaeological importance that had been discovered by chance in 1959 in the village of Petralona in the northern Greek prefecture of Chalkidiki. A year later, a skull was found there as well.[120]

The national implications of his investigations, but also his conflicts with the Greek authorities, attracted a great deal of public attention. After arriving in Greece from the Soviet Union, Poulianos started to carry out excavations in the cave, but soon he was forced to interrupt his activities during the military junta period, from 1967 to 1974. As he noted in the introduction to the fifth edition of his dissertation, he himself was

---

118 On the occasion of the presentation of the prize, the president of the Greek Anthropological Society disclosed to the members that, on 1 January 1955, he had addressed an appeal to Soviet anthropologists asking for their commitment to ending the cold war and opening the 'Iron Curtain'.
See statement by Koumaris in *PGAS* (2 April 1962): 11.

119 Xerxis Livas's statement in *PGAS* (2 April 1962): 12–15.

120 http://www.aee.gr/english/6petrlona/petralona.html (last accessed: December 2012).

detained and interrogated by the Intelligence Service in 1967,[121] probably because of his previous residence in the Soviet Union. His researches were also restricted due to a conflict with Spiridon Marinatos, the head of the Archaeological Office in the Ministry of Culture during the dictatorship (and a well-known member of the Greek Anthropological Society).[122] Only after Marinatos's death, which coincided with the end of the dictatorship (1974), could Poulianos resume his investigations.

On the basis of his findings, Poulianos saw confirmation of his hypothesis that Greece had played a crucial role in anthropogenesis. His conclusions as to the age of the skull discovered in 1960 also differed from all previous estimations by other anthropologists.[123] Unlike all the other anthropologists who had examined the skull and classified it as Neanderthal, Poulianos claimed that the 'Petralonian man' belonged to a species older than the Neanderthals. He named the remains of the masculine skeleton 'Archanthropus' (i.e. archaic *Homo sapiens*) and classified him as an upright walking 'Homo erectus';[124] the first ancestor of humankind. His assumption was reinforced by the discovery of further skeletal remains, tools and other artefacts in and around the cave area. On the basis of this evidence Poulianos questioned the mainstream scientific hypotheses that Africa was the continent of anthropogenesis, claiming instead that 'Archanthropus' had not migrated but was autochthonous to the region of Petralona, and therefore that humankind was born in South-Eastern

---

[121] See the biographical notes on the official website of the Anthropological Association of Greece that he founded in 1976: http://www.aee.gr/english/2apoulianos_biogr/apoul_biogr.html (last accessed: December 2012). His biography has been published by his wife, Dafni Poulianou: Poulianou (2006a) and (2006b).

[122] His conflict with Spiridon Marinatos was carried out very publicly during the 1970s. See, for example, some notes in the Athenian press: "Anthropological research and the continuation of the conflict between Mr. Poulianos and Mr. Marinatos" ["Αι Ανθρωπολογικαί έρευναι και η συνέχεια της διαμάχης κ.κ. Πουλιανού-Μαρινάτου"], *Βραδυνή* (3 July 1972); "New Statements by Poulianos" ["Νέες δηλώσεις του Πουλιανού"], *Νέα* (3 July 1972); "Does the skull discovered in Crete belong to man or to deer? Mr. Poulianos responds to Marinatos" ["Σε άνθρωπο ή σε ελάφι ανήκει το κρανίο που βρέθηκε στην Κρήτη; Ο κ. Πουλιανός απαντά στον κ. Μαρινάτο"], *Ακρόπολις* (2 July 1972); "Severe Response by the Archaeological Office on the Question of Petralona Cave" ["Οξυτάτη απάντηση της Αρχαιολογικής για το θέμα του σπηλαίου Πετραλώνων"], *Βήμα* (30 June 1972); "Severe Response of Mr. Sp. Marinatos to Mr. Poulianos" ["'Εντονος απάντησις του κ. Σπ. Μαρινάτου εις τον κ. Πουλιανόν"], *Εστία* (29 June 1972).

[123] E. Breitinger, "Der Neandertaler von Petralona", Presentation at the Seventh International Congress of Anthropological and Ethnological Studies, Moscow 1964 (cited in Poulianos 2004: 222). Cf. Kannelis and Savvas (1964).

[124] The conclusions of these studies were published in *Anthropos* [*Άνθρωπος*], the journal of the Anthropological Association of Greece, between 1974 (the journal's founding year) and 1980. Cf. Poulianos (2004): 222–79.

Europe, and specifically in Greece. Additionally, he was convinced that he had discovered in the Petralona cave the oldest traces of fire and the oldest known tools (at least until that day) that had ever been used by the human species.[125]

In Poulianos's anthropological narrative, Greece was homogenous in terms of the racial composition of its population which was autochthonous in the territory of the current Greek state and even featured a racial anthropological continuity from antiquity up to present. And Greece was not only the cradle of all European civilization; it was also the birthplace of all humankind. Poulianos was not concerned with developing a general racial theory, or with finding a link with social developments in society. He was exclusively concerned with racial/physical anthropological investigations which conformed to the predominant national idea. Therefore, Ioannis Koumaris remained the only Greek anthropologist by training who sought to define Greek communality at large by racial anthropological means.

125 See a series of articles by Poulianos in the journal *Anthropos*: Poulianos (1976); (1974); (1978). For a summary account of the findings in the Petralona cave and Archanthropus, see Poulianos's monograph in the series "Library of the Anthropological Association of Greece": Poulianos (1982).

CHAPTER EIGHT

THE EUGENIC CONCEPT OF *FILI*

Los que falsificaron la firma de Dios
They forged the signature of God
(Viriato Sención 1992)

> Eugenics is the science which deals with all influences that improve the inborn qualities of a race; also with those that develop them to the utmost advantage. The improvement of the inborn qualities, or stock, of some one human population will alone be discussed here (Galton 1904).

When Sir Francis Galton (1822–1911) presented his proposal for improving British society by means of controlling human reproduction before the British Sociological Society in 1904, he was making history (Galton 1904 and 1909/2004). His lecture marked the birth of the 'well-born-science' or *eugenics*, as Galton himself named his concept, borrowing from the Greek suffixes *ευ-* (well) and *-γένος* (heritage/born). Human breeding by selective procreation was the core idea behind this concept. The novelty he introduced was not just the concept per se, he had first coined the term in the 19th century, but at that time it had not attracted the level of public interest that it would subsequently attain, and even before the term eugenics had been imprinted, intervention into human procreation in order to control reproduction had been implemented at least in North America by doctors who practised the sterilization of paupers and individuals labelled 'feeble-minded' (Black 2003; Bruinius 2006). However, it was only after Galton had made his proposals that eugenics started to be widely embraced, not only in Great Britain but in many other countries as well, resulting in an international movement that grew in strength as the 20th century progressed (Kühl 1997).

Today there is good deal of literature that explores the different eugenic currents and which brings to light their specific national as well regional characteristics. The historiography of eugenics has also raised questions on the diverse points of view from which eugenicists considered their concerns (e.g. Weindling 2011). The eugenic phenomenon can be approached from different perspectives: as a science of human heredity, as a vision of a perfect humanity, an ideology addressed to the improvement of a single society, a social movement or as a political programme for immediate intervention into social relations. Regardless of whether it is a long-term target or a project for immediate implementation, eugenics has always been ideologically justified on the grounds of the collective good, the welfare of the nation, race, society or humanity at large; simultaneously, it has always been tied to social politics applied in single countries. In global terms eugenic thinking bases itself on a deeply technocratic understanding of progress, according to which evolution should be calculated by human intervention rather than being left to chance. Prognosis and prevention then emerge as central tasks for politics and its allied sciences. Eugenic ideas were lauded, and to some extent applied, in an international environment and in single countries in which either the social welfare state was being founded, or its establishment demanded by elites.

The main purpose of this chapter is to disclose the particular character of eugenics in Greece and to identify the ways in which eugenicists conceptualized *fili* as a collectivity. How was *fili* conceived and ranked in eugenic projects? To what extent did the eugenic and racial hygienic ideas expressed in the anthropological discourse intersect with or differ from each other? How is the eugenic conception of *fili* related to the prevalent Greek national ideology, and which semantics do heredity and ancestry obtain in this? What were the social questions that eugenic proposals sought to address and were these questions and answers imprinted in the eugenic usages of *fili*? In dealing with these questions particular emphasis will be placed on the main social actors involved in the eugenic and racial hygienic movements; the major targets and propositions of the most important currents in Greece; and their possible impact on policy-making. The professional careers of leading actors in Greek eugenics will be considered in relation to their contribution in propagating eugenic ideas and establishing eugenic measures or, conversely, to preventing the implementation of such measures in Greece. Generally, the Greek eugenic discourse will be approached as a field in which eugenic ideas and practices were negotiated by the actors involved. Crucial to negotiation were action and reaction, the approval and disapproval of eugenics. Hence, in addition to the different views expressed by the proponents of eugenics, I will also consider those who challenged the ideas of eugenic science as well as those actors whose proposals to cope with social issues went beyond eugenics. The central argument of my analysis is that, at least in the first decades of the 19th century and especially in the interwar period, eugenic thinking in Greece was embedded in socio-political reformist projects that advanced the establishment of welfare institutions.

Adherents of racial hygiene were exceptional among Greek eugenicists and sympathizers of eugenic ideology. In the post-war period, eugenics became closely interwoven with genetics, thereby conforming to an international reformist eugenic movement to which Greek racial anthropological discourse did not find any link.

8.1 *Eugenic Representations of Hellenic Antiquity: A Matter of European Self-Consciousness*

The assertion that the origins of the eugenic idea could be traced back to Hellenic antiquity was circulated among eugenicists at the beginning of the 20th century. The proper initiators of this historical relationship were European eugenicists (from the German-speaking countries and the National Socialists in particular) who regarded eugenic measures as having first been applied in ancient Sparta, where infanticide was practised. Proponents of this idea included the German racial anthropologist, human geneticist and eugenicist Fritz Lenz (1887–1976) who formulated this assertion in a series of lectures on eugenics and ancient Greek philosophy,[1] and Alfred Ploetz, who had portrayed Lycurgus and Plutarch as "conscious racial hygienists" (Ploetz 1895: 5–6). The argument for the ancient Hellenic origins of eugenics was also embraced in Switzerland by proponents of euthanasia. In 1923, Alfred Hauswirth, the head doctor and representative in the Berne canton, suggested the use of euthanasia for 'madmen' and 'idiots' countrywide, an idea that he justified on the grounds that euthanasia was also practised in ancient Sparta (Huonker 2002: 66).

The idea of the ancient Hellenic origins of eugenics was embraced by several Greek eugenicists as well. Apart from Ioannis Koumaris,[2] Moisis Moiseidis (1880–?), the gynaecologist and author of numerous books on eugenics, even went so far as to dedicate a Greek-language monograph to this subject, *Eugenics and Childbearing in Ancient Greece*.[3] The professor of criminology Konstantinos Gardikas outlined the same view in a 1934 German-language article, published in a prominent journal of criminology in Germany (Gardikas 1934c). Gardikas placed the origins of eugenics

---

1 According to Stavros Zurukzoglu (1896–1966), who was in fact opposed to this correlation, in a review by St Zurukzoglu that appeared in the journal *Medicine* [*Ιατρική*], 1 May 1925: 32–3.

2 Koumaris (1931b) and (1931c).

3 Moiseidis (1925). This publication was based on a study honoured by the Medical Society [Ιατρική Εταιρεία] in 1924.

in the 'heroic age' and in the era of Aristotle, without however taking a stand on contemporary debates about eugenic policies.[4] A further advocate of the alleged ancient Hellenic origins of eugenics was the professor for hygiene Konstantinos Moutousis (1891–?). In his inaugural lecture as chair-holder for hygiene at the University of Athens,[5] Moutousis incorporated Orthodox Christianity along with Hellenic antiquity into his eugenic narrative. He asserted that eugenics, in its practical implementation, should be based on the combination of the ancient Hellenic ideas of Plato and Aristotle, Christian principles and current scientific achievements (Moutousis 1933: 873). This combination would ensure the creation of a modern state, a 'Polity' [Πολιτεία] that would rest upon antique and Christian ideals and scientific rationality (Moutousis 1933: 872).

This correlation of eugenic thought with Hellenic antiquity appears to have corresponded with the self-perception of Europeans as emerging from an ancient high civilization. Even an opponent of the correlation of eugenics to antiquity, as Zurukzoglu was, did not abandon the 'classic sources' when it was a question, if not of corroborating, then at least of embellishing his theses. Zurukzoglu introduced his German volume on the *Prevention of Genetically Diseased Offspring* with passages from texts by Euripides and Plutarch, with which he tried to vindicate his argument that controlling childbearing by eugenic means was necessary for humanitarian reasons (Zurukzoglu 1938: 7). It seemed that referring to the cradle of European civilization elevated the rank of eugenic ideology, despite the fact that contemporary Greece was viewed negatively in National Socialist ideology. Even in the 1930s, when German Philhellenism declined or became clearly disconnected from contemporary Greece, Hellenic antiquity continued to be an ideal for German racial ideology (Marchand 1996: 341f.). Finally, the association of racial ideas and policy with Hellenic antiquity and Sparta by the National Socialists was also relatively typical (Chapoutot 2008). The antiquity cult was in reality forced to the margins in the overall Greek eugenic movement.

The incorporation of eugenics into a primordial myth of (either European or Greek national) origins was based on an exclusive interest in the present; drawing on the past ultimately served to update current ideologies. A similar tendency can be observed among Jewish eugenicists in the

[4] Greek criminology has not yet been subject of research from the eugenic perspective.

[5] The lecture was delivered on 9 November 1933 and was about "Eugenics and Hygiene in Modern Societies": Moutousis (1933).

inter-war period who saw the history of the Jewish nation as a "eugenic experience on a large scale".[6] Moses was deemed by Jewish eugenicists to be the first to implement the eugenic ideal. According to Kamila Uzarczyk, some advocates of Jewish eugenics saw the Bible and the Talmud as containing eugenic legislation (Uzarczyk 2007: 289).

Today scholars point at the initial forms of eugenics (so-called 'proto-' or 'primitive eugenics') that emerged prior to the proper period of modern eugenics, namely in the late 19th and above all in the 20th century.[7] Certainly, the idea of human breeding via selective reproduction and its implementation within a restricted community is not a novelty; nor did it first emerge in modernity. The historian A.G. Roper (1992), who has addressed the subject of 'Ancient eugenics', has shown that infanticide was practised (mostly for religious reasons) in several historic cultures worldwide. But the strongest tendency among current researchers is to approach eugenics as a specific form of bio-power that emerged at the end of the 19th century.

In line with this latter approach, I will argue by means of the Greek paradigm that at the core of the eugenic projects was the intention to manage social inequality in modern society through birth control, with the population being treated as biological capital and poverty assessed as social pathology.

8.2 *The Social Question and Overpopulation: Eugenics for Governing Social Stratification*

Eugenic proposals had flourished on the basis of the processes that accompanied the industrialization and the modernization of societies. They posed one option among several others that addressed social differentiation in the interior of societies and which looked for a method to cope with social inequality while advancing the establishment of welfare institutions at the same time. Eugenicists attempted to meet this challenge by giving priority to the regulation of human reproduction and evaluating the members of society as more or less useful for the whole. In eugenic terms this was translated into an inherent conflict between the

6 Hans von Pezold (1932) and Gerson Lewin (1934); cited in Uzarczyk (2007): 288.

7 See for instance Roper (1992); cf. the report on the conference "Proto-Eugenic Thinking before Galton" held in Washington, 25–7 September 2008: http://www.ghi-dc.org/files/publications/bulletin/bu044/bu44_083.pdf (last accessed September 2012).

biologically superior and inferior parts of the population. In this chapter I will argue that, at the heart of the eugenic projects, in Greece like in other European countries, there was an intention to meet the challenge of the so-called social question and that of overpopulation, both of which required governance of social stratification and the distribution of resources in a modernizing society during a period in which social welfare institutions were gradually beginning to emerge.

The 'social question' arose in the 19th century as a discursive term to designate a cluster of diverse social phenomena caused by the Industrial Revolution on the European continent. Rather than describing a single phenomenon, the 'social question' became a means for interpreting a radical new epoch that departed from former conditions and styles of living for both individual and collective actors. Proposals to cope with this new challenge therefore encompassed the philosophical, intellectual and political prospects of society. The 'social question' became a research field of the social sciences, and it was also treated from a philosophical point of view (e.g. Stein 1897) or even specified as a 'moral question' (e.g. Ziegler 1891). Yet it also became a field of struggle for political action, especially that connected with the rising socialist movement. Although the phenomena associated with the 'social question' were to be observed in particular societies, both the issue as such and the proposals for its solution possessed universal outlooks as well. The rise of the social question went hand in hand with the creation or/and growth of urban centres and the modernization of societies and social institutions, especially at the turn of the 19th century and during subsequent decades; this certainly occurred to varying degrees in each country. A stratum of lumpenproletariat was in formation as well as a labour stratum in an insecure social situation whose members oscillated between the middle classes and the paupers. The question of the peasants in the countryside was posed with respect to their possible contributions to the modernization of societies.

In Greece the 'social question' was first formulated when industrialization took hold in the last quarter of the 19th century in connection with the precarious living and working conditions of the lower social strata and the emerging working class. In the early decades of the 20th century the discourse of the social question acquired a new quality as a consequence of the growth of the working class and growing poverty in the urban centres.[8]

---

[8] On the transformation of social stratification and the accompanying modernization of Greek society and state institutions during the inter-war period, see Mavrogordatos and Chatziiosif (1988); Moskof (1988); Pizanias (1993); Alexiou (1994).

These developments became even more urgent after 1922, when the 'refugees' arrived from the former Ottoman territory. Social and state institutions were unprepared and unable to meet this challenge as welfare institutions had not existed for long, and their development was thus accelerated, especially after 1922, to include an institutionalized public healthcare system, a system of employment insurance and, related to this, a system of health and social insurance for the working classes (Papastefanaki 2011). The 'social question' had been equated to a broad extent with the 'labour question' by various institutions and political agencies (Liakos 1993: 462). In addition to successive Greek governments, political parties of all wings, unionists and labour organizations—such as the Labour Centre of Athens [Εργατικό Κέντρο Αθηνών], founded in 1910—but also the International Labour Office, were all involved in this undertaking.

The range of proposals made to deal with the 'social question' reflects the different objectives Greek elites pursued and the visions to which they were committed. In this wide platform of propositions it is possible to recognize patterns of political and societal roles that are indicative of the position these elites concede to the lower social strata. The proposals ranged from technocratic projects for partial infrastructural reform up to socialist projects that aimed at a fundamental reorganization of political relations by means of revolution. Within this kaleidoscope of diverse options, a shared point among those who were preoccupied with the social question was the factual incongruity between the urgent needs of a large part of the people on the one hand, and the insufficiency of the existing economic and infrastructural resources to meet these needs on the other. A point marking the crucial difference of the diverse proposals was whether the growing numbers of those among the lower social strata were conceived as objects in the implementation of state policy, or as subjects of action.

Socialist thinkers as well as leading socialist and communist political figures underlined the central role of the working class in solving the social question.[9] From their perspective, the labour class had an active role to play in the shaping of a new political and societal order. Thus the question with which these socialists and communists were concerned was whether the liberation of the working class could, or indeed should, be realized by means of democratic reform, or whether a class revolution was

9 Such figures were, for example, Georgios Skliros, Nikolaos Giannios (1885–1958) and Pantelis Pouliopoulos (1900–43).
See Skliros (1907); Giannios (1914); Pouliopoulos (1934).

necessary. In spite of fundamental differences on points of criticism and long-term targets, socialists, communists, as well as a significant number of liberals, emphasized the immediate need for improving the living and working conditions of the weakest members of the population. In doing so, the former also addressed their criticism to the deficient or paternalistic means by which the Greek state and international organizations struggled to cope with the precarious living conditions of the weak social classes (Liakos 1993: 499–502).

Other scholarly elites and technocrats in the inter-war period, such as economic experts, increasingly addressed the focus of their propositions to economic and infrastructural reforms which were to be implemented in state policy. For example, the director of the Chamber of Commerce in Epirus saw a possible solution to the precarious situation of large parts of the population in the promotion of domestic industry (Garoufalias 1930; cf. Graikos 1930). Others prioritized population policies and demanded the rationalization of existing resources by means of restricting their access to recipients from certain, selected sectors of society. The circle of the latter introduced the issue of the 'problem of overpopulation' and shifted the focus to the natality question: Greek demography should be managed through the control of births and the rationalization of human reproduction. From their point of view, economic and social progress was contingent on the birth rate in certain strata of the population. This drew on a way of thinking which interpreted the social question in terms of the biological reproduction of the population and which distinguished population categories according to assessments of their biological quality.

The rhetoric of overpopulation and the need to regulate both the quantity and quality of the population mirrored a flourishing European discourse that linked the rationalized distribution of resources with the biological reproduction of the population. This problem was embedded in population policies and was widely discussed in international networks of experts and politicians. In several European countries, in fact, Malthusian theories concerning restrictions on reproduction were about to be replaced during this period by a preoccupation with the decline in the national birth rate, or the "ghosts of decline in the birth rate" in the words of Ursula Ferdinand (2007). During the first decades of the 20th century, methods were developed for 'thinking about population', which resulted from a desire to improve the biological substance of the population, on an increasingly international scale (Mackensen 2002; Ferdinand 2007). Birth control as a means for attaining both optimal size and the best quality with regard to the population was broadly considered an imperative in population policy. Concepts were elaborated due to a fear of degeneration, which was thought to threaten national societies and humanity in its entirety. Ideas and experiences for coping with the population question were exchanged internationally. Yet shared concerns and joint activities notwithstanding, each country was primarily concerned with its own population. International meetings of experts gave new impetus to national discourses on how to deal with the population issue on the local level. In Greece, a comprehensive account of the international state of affairs on 'thinking about population' and population policy was provided by the director of the National Statistical Survey of Greece [*Εθνική Στατιστική Υπηρεσία της Ελλάδος*] in the Greek daily press,[10] on the occasion of the International Congress for Statistics held in Rome in 1931, in which he outlined the anxieties of other European countries, like Germany and France, which complained of a declining birth rate in the inter-war period. This problem included both quantitative and qualitative aspects, namely the development of the overall size of the population as a feature of the quantitative power of a state and differences in fertility between parts of the population that were thought to contribute to social progress to varying degrees.

The inter-war Greek case was not congruent with the paradigm of other European countries with respect to the trend in the birth rates. In Greece a remarkable population growth could be observed, due primarily to the relatively high birth rate among the rural population. Moreover, in the wake of the 1912–13 Balkan Wars, and especially after 1922, the Greek population grew substantially as a consequence of the annexation of new territories and the arrival of 'refugees' of Greek origin.[11] At the same time, infant mortality, although considerable, was believed to be manageable. In view of these developments, a segment of those engaged with the population question agreed that the country was overpopulated. This assertion was based mainly on the idea that domestic resources were insufficient for nourishing the entire population because of major deficiencies in the organization of society and economy. This problematic situation deteriorated further in the face of increasing urbanization, which went hand in

[10] See Michalopoulos's article on "The Decline of Births" published over four parts in the Athenian newspaper *Free Forum* [*Ελεύθερον Βήμα*]: Michalopoulos (1931).

[11] In 1907 the Greek population was estimated to be 2,631,952; in 1920 it increased to 5,016,889 and in 1928 to 6,204,684.

hand with unemployment in the cities, as well as the decline of agrarian production in the countryside, despite agrarian reforms.[12]

In light of this situation, the adoption of the population discourse of other European countries by Greek scholars and politicians appears somewhat paradoxical at first glance. Yet the vehicle for transmitting this discourse into Greek reality was the issue of degeneracy, which emphasized the quality rather than the quantity of the population: "*Quality* may not be sacrificed for *quantity*; all the more when recent statistics undoubtedly bear testimony to evident dysgenic indications in our fili" (Moiseidis 1932: 26). This statement was made by the physician and publisher of the medical journal *Health* [*Υγεία*], Moisis Moiseidis, who published prolifically on eugenics and provided the earliest treatises on this issue in the Greek language.[13] Although Moiseidis sympathized with Malthusianism,[14] he did not argue for an outright decline in the birth rate but for a selective reduction of births, particularly in those cases "where the financial means are insufficient for sustaining and nourishing the descendant and a faulty *quality* is foreshadowed in the poor health of the offspring".[15] Regulating the reproduction of the domestic population by applying state policies that sought to maintain an optimal balance between quality and quantity was at the core of eugenic thinking.

In eugenic rhetoric, the issue of fertility was deemed to pose a problem extending to Greece from a dual perspective. Despite the annual growth in population, a decline in the Greek population was allegedly likely to take place over the long term, and given the shortage of resources and the higher birth rate among the poorest classes, the biological quality of the population would inevitably degenerate. Essentially, eugenicists proposed the redistribution of societal resources, interlinking the 'social question' with the 'population question'. They suggested reducing the percentage of the needy in the population by restricting their biological existence rather than by changing the conditions of inequality. The emphasis on the 'quality' of the population and the omnipresent ghost of degeneration effectively shifted from social care to managing inequality through birth control and eugenic selection. In addition to the required establishment

---

[12] On the agrarian reforms in the period 1917–40 in relation to overall economic development in Greece, see Kostis (1988).

[13] His monographs include Moiseidis (1912); (1925); (1932); (1934) and many others.

[14] See his monograph on Malthusianism: Moiseidis (1932): 26.

[15] Moiseidis (1933): 35. Cf. Moiseidis (1932).

of infrastructure, a revision of the institutional framework for family and marriage was also demanded in order to realize a social transformation.

In the journal *Archive of Economic and Social Science* [*Αρχείον Οικονομικών και Κοινωνικών Επιστημών*], the head physician for the Piraeus municipality suggested solving the problem of overpopulation by developing the agrarian and industrial economic sectors (Chatzivasileiou 1925). This could not, however, solve the problem of the degeneration, the causes of which lay within social stratification. Taking as his point of departure the assertion that variations in fertility rates in different social classes were responsible for an increasingly degenerate populace, Chatzivasileiou pointed to the lurking threat of the qualitative decline in the Greek *fili* which, in his opinion, stemmed from two causes: first, disease that plagued the poorest strata of the population; and second, low natality rates in the upper ('the leading') social classes that resulted from an intentional abstention from childbearing. The latter was the result of a new model for the modern family with fewer children "that caused the degradation of the family in moral terms; both the will as well as the sense of responsibility for the future of the nation and the *fili* do not exist any longer" (Chatzivasileiou 1925: 258). Drawing upon arguments from Fritz Lenz and Sebald Rudolf Steinmetz (1862–1940), Chatzivasileiou advocated the regulation of births rather than their overall reduction. In concrete terms this meant decreasing the birth rate in certain parts of the population while simultaneously ensuring lower child mortality rates through better healthcare. As he argued, precarious living conditions, poor nourishment, and diseases like malaria and tuberculosis, were frequent among 'refugees' from the former Ottoman territories as well as among the lower social classes—both threatened Greece with degeneration.[16] The head physician of the municipality of Piraeus also voiced his views before the Greek Anthropological Society, where he advocated positive eugenics, arguing for the promotion of children from gifted families rather than the restriction of births because of fears about hereditary illnesses,[17] and that the hidden threat to the Greek *fili* lay in losing the gifted and leading families rather than in diffusing severe hereditary diseases. For his part he advocated care for both the quality and quantity of births. He was convinced that the more

16 In another article the same author approached hygiene by means of anthropometry: Chatzivasileiou (1929a). On social legislation and social hygiene see Chatzivasileiou (1929b).

17 See Chatzivasileiou's statement at the anthropological society: Chatzivasileiou, *PGAS* (27 December 1945): 57–9.

children born to a family, the greater the probability that there would be geniuses among them. Chatzivasileiou disclosed to the Greek Anthropological Society (in 1945) that he had previously attempted to persuade state institutions to establish an individual health card for pupils, a kind of individual register of the illnesses of pupils as well as their families. However, his proposal did not meet with approval.[18]

If the above-mentioned experts can be classified as moderate adherents of eugenics who mostly approved positive measures, there was also a hard core of eugenicists who proposed negative measures and regarded the regulation of reproduction among different social classes to be a means of eliminating poverty and its side effects. Moiseidis, for instance, was convinced that birth control would not only prevent degeneration, but would contribute further to solving the problem of social inequality by reducing the impoverished population through restrictions on their biological reproduction (Moiseidis 1932: 26). In other words, he suggested that the fight against poverty would effectively be won through the elimination of the poor. This was a matter of gaining mastery of human evolution by political intervention (against natural selection), either in order to solve urgent social problems or to facilitate the proliferation of the fittest members of society with the broader aim of creating a perfect society. This fundamental eugenic principle radiated from the projects of eugenicists like Panagiotis P. Panagiotou, who, while asserting the functionality of natural selection, did not abandon the claim to intervention in human evolution by institutional politics. Panagiotou asserted that infant mortality ensured the survival of the fittest children and that natural selection occurred among mentally disabled and asocial individuals.[19] Even when their physical constitution was adequate, the death rates among mentally disabled individuals were remarkably high: "bums, beggars, drunkards, prostitutes and criminals have only a few chances to reach a ripe age".[20] But in comparison with their high birth rate, in Panagiotou's estimations, the deaths of individuals of 'inferior value' that resulted from natural

[18] Cf. information communicated by the physician Georgios Kosmetatos in a 1946 meeting of the Greek Anthropological Society that he had also advised state institutions to establish this measure but his suggestion similarly met with indifference.

Kosmetatos, *PGAS* (30 December 1946): 49–54.

[19] Panagiotis Panagiotou expressed his views after the World War II before the Greek Anthropological Society. Panagiotou, *PGAS* (30 December 1946): 69–73.

He would later become a member of the board of directors of the Greek Eugenic Society.

[20] Panagiotou, *PGAS* (30 December 1946): 71.

selection were incongruously low. This was because current progress in medicine enabled the prolongation of the lives of 'defective individuals'. In his opinion, medical care should be granted conditionally to such individuals:

> We understand and excuse [*sic*] the prolongation of their lives insofar as either the individuals under consideration possess other valuable qualities, or their disabilities have been repaired after medical intervention so that their lives have become congruent with the life of the community.[21]

Panagiotou developed an explanatory pattern for social relations in which social stratification, profession and intelligence were interconnected in a deterministic way. He assumed that members of different professional groups featured different degrees of intelligence and mortality rates. This was because intelligence, which he deemed to be decisive for the choice of profession, was allegedly transmitted by parents to their offspring. Such families belonged to the high social strata of society.

> From a eugenic point of view, the different mortality rates between professional groups depend absolutely on the degree of hereditary differences between these groups. The correlation of profession and intellect comes clearly to the surface when we take into consideration that intellectually inferior individuals shoulder the lower salaried professions.[22]

Panagiotou believed that hereditary factors determined the intelligence of individuals and that these factors even prevailed over the social environment. In fact, he regarded the social environment as such to be a hereditary condition rather than an external and changeable factor of human life. Even when environmental conditions were altered for the benefit of individuals, he asserted that the intellectual capacity of the latter remained unchanged since this was determined by the social environment in which individuals were born. He interpreted professional illnesses through the same lens of biological determinism. The mortality of members of professional groups was allegedly indicative of the way natural selection operated in order to erase intellectually inferior individuals from the community. Panagiotou was an adherent of both negative and positive eugenics. In his opinion, eugenic measures should support the reduction of natality among inferior individuals and increase the birth rate among those who held superior hereditary traits. And in a contradictory argument, he also

21 Panagiotou, *PGAS* (30 December 1946): 70.
22 Panagiotou, *PGAS* (30 December 1946): 71–2.

asserted that a further concern of eugenics was to reduce mortality rates independently of intelligence.

Generally speaking, eugenicists thought that intelligence and social position were linked together by biological ties and in the form of hereditary traits that were transmittable to the next generations. Social background was essentialized and perceived as a biological quality inherent in individuals and social strata. This is the very assertion on which the father of eugenics, Sir Francis Galton, elaborated his concept. On the grounds of pedigree studies Galton concluded that heredity was the primary factor in advancing the creation of the leading strata, including the intellectual, political and other elites. This view was embraced by those who considered biology and heredity to be the driving force of history, based on the assumption that the main protagonists in historical progress (or better, the only authorized protagonists) were elites, a view that was voiced also in the humanities. While addressing the relationship between biology and history before the Greek Anthropological Society in 1945[23] the prominent historian Konstantinos Amantos outlined the crucial role played by heredity in the creation and reproduction of elites. Amantos axiomatically presumed that 'leading families' produced numerous leading personalities. Thus the reconstruction of genealogies, or the "family history of great men", could, in his opinion, assist and facilitate the interpretation of historical developments. In this undertaking, eugenics was important insofar as the preservation and transmission of the qualities that these elites possessed should not be left to chance but advanced by strategies of reproduction. Marriage strategy was deemed instrumental to ensuring the perpetuation of gifted families by producing adequately gifted progeny. At the same time, Amantos believed that there was no guarantee that the geniuses who might emerge from the "conjunction of biologically powerful natures" would necessarily commit their genius to the progress of humanity and history: "Sometimes it occurs that the coupling between biologically powerful natures provides us with men of genius who dominate and thrill us either for good or for bad."[24] Alexander the Great and Julius Caesar were mentioned as two paragons of such disparate careers of geniuses. The former exemplified a genius who used his gifts for civilizing the world, whereas the latter used his to destroy civilizations. With respect to Adolf Hitler, Amantos felt incapable of making a personal state-

---

23 Amantos, *PGAS* (27 December 1945): 48–55.
24 Amantos, *PGAS* (27 December 1945): 54.

ment since, in his opinion, Hitler was a case study for biologists and physicians rather than for historians. Interestingly enough, in the discussion that followed Amantos's presentation before the GAS, the only criticism that was made came from a biologist who pleaded for a reassessment of the limits of heredity in determining human action. Instead of embracing biological determinism, the biologist argued, it was necessary to take into consideration further factors of social significance that would be able to advance hereditary dispositions, such as the social environment and family, among others.[25]

The point at issue, however, was not so much whether these factors should be taken into consideration, but rather the significance that should be ascribed to them with respect to their instrumentality in designing concepts of humanity and society. Beyond the macro-level of history, the adherents of eugenic ideas engaged in detailed discussions regarding the role that diverse social strata could play in achieving their final goal, namely a biologically superior society and the perfect humanity. In the process of the remarkable transformations that occurred in European societies, particularly in the period after World War I, social stratification was characterized by high mobility. The proliferation of both the low social classes and the middle strata, along with the creation of welfare institutions, was an integral part of the modernization of societies. The lower and middle social strata were centrally positioned in the eugenic projects with regard to the optimal system of governance. In the Greek case a coherent eugenic programme to deal with social stratification was elaborated by the paediatrician and director of the Paediatric Polyclinic of Athens, Apostolos Doxiadis. His case is worth being examined at length since he also was politician in liberal governments under Eleftherios Venizelos and therefore had the opportunity to influence politics through his official position. He made public his ideas on developing a eugenics-based population policy in several Athenian newspapers and journals.[26]

---

[25] Vlasidis, *PGAS* (27 December 1945): 56.

[26] The liberal daily *Free Forum* [*Ελεύθερον Βήμα*], for example, published two of his serialized articles, the first entitled "Problems of Life" in autumn 1930, and a second some months later entitled "Biological Problems". In the former series he deals with issues such as the relationship between biology and the church; the reduction or regulation of births; overpopulation, etc. Doxiadis (1930a); (1930b); (1930c); (1930d); (1930e); (1930f); (1930g); (1930h).

In the series on "Biological Problems" he deals with the following issues: heredity and environment; Mendel—the father of the science of heredity; heredity; the mystery of life; ascertaining the sex of embryos; human evolution; the decay of humankind; the optimal

Doxiadis argued that the disparity in the reproduction rates between the 'upper, leading strata' and the 'lower ones' was the real reason for the imminent degeneration of the Greek race. He estimated that of the 220,000 individuals born in Greece every year, between 100,000 and 120,000 died, resulting in an annual net surplus of 100,000 people. Nevertheless, the overall incongruence in the reproduction of the social classes adversely affected the general biological quality of the population. He considered the decline of births "in those families who govern the world today and the increase in those who—rightly or wrongly—learn to subordinate themselves" to be responsible for the degeneration, which "leads humanity to the edge of a dark and abominable abysm" (Doxiadis 1933: 45–6). The long-term persistence of this development would ultimately endanger the Greek *fili* itself.[27] His views did not escape some criticism. His assertion that humanity generally and Greek society in particular were degenerating due to the proliferation of the low social strata, for example, was challenged by the counterargument that the lower social classes needed social care rather than restrictions on birth (Saratsis 1931).

Social care was, however, a focus of Doxiadis's activity. Together with his involvement in the state apparatus,[28] he was also committed to the protection of children, combining professional expertise with political activism. One of his major achievements was to found the Patriotic Foundation for the Protection of Children [Πατριωτικό Ίδρυμα για την Προστασία του Παιδιού], of which he was also the head.[29] His field of interest was not restricted to caring for children but extended to more general issues regarding the family and the population question. He elaborated concepts of both family and population policy, based on the rationalization of human reproduction through eugenic birth control starting from the assumption that, like all humanity, Greece was also threatened by degeneration. From his point of view, control over reproduction should be an integral part of a larger eugenic programme that was urgently needed to prevent imminent degeneration. But such radical intervention could also achieve the ultimate goal: the creation of a Greek population of superior

---

number of children in a family: Doxiadis (1930i); (1931a); (1931b); (1931c); (1931d); (1931e); (1931f); (1931g); (1931h).

[27] Apostolos Doxiadis argues much the same point in his column "Biological Problems" in the newspaper *Free Forum* [*Ελεύθερον Βήμα*]: Doxiadis (1930i); (1931a); (1931b); (1931c); (1931d); (1931e); (1931f); (1931g); (1931h).

[28] Apostolos Doxiadis was minister of health (1922–4), deputy minister of hygiene (1928–9) and a senator (1932–3) in liberal governments.

[29] Theodorou and Karakatsani (2011) have explored the views and activities of Apostolos Doxiadis as well as eugenic ideas among Greek paediatricians.

biological quality. Indeed, in Doxiadis's thinking, the possible implementation of eugenic measures on family, childbearing and marriage issues was treated as a pressing priority:

> The goal of practical eugenics is to preserve the hereditary capital of families who own it to improve their living conditions and to exterminate hereditary stigmas that burden so many families today. How may this goal be realized?...[By] providing guarantees for the health of the couple about to get married and evaluating their ethics; by excluding from reproduction all organisms that may endanger the health of the race due to mental or somatic defects; by facilitating marriage by the state through the founding of agencies that will provide all necessary guarantees for the precise evaluation of people and things (Doxiadis 1933: 47).

Like many of his fellow scholars and politicians at the time, both in Europe and elsewhere, Doxiadis conceived the state as a living organism; a collective subject which possessed body and spirit. At that time this was an influential idea according to which legislation, policy, economy, culture and all sectors of society were thought of as organic forces which constituted the state and its idiosyncrasies (Lemke 2007: 20). That Doxiadis had embraced this conception of the state is clear from his serialized article "Social Biology–Biological Politics" (Doxiadis 1928a). What Doxiadis meant by 'biological politics' was what Foucault has termed "state control of the biological...[namely] the acquisition of power over man insofar as man is a living being" (Foucault 2003: 239–40). Doxiadis grounded his approach on the programmatic assertion that the objective of state politics was life as a whole. From his perspective, 'biology' was a key notion for understanding history and life because it "teaches us about the history of humanity". State politics should comply with the principles of 'social biology' (i.e. "the science of the conditions of life" should form the basis of society). The conclusions of 'social biology' were deemed to provide the basis upon which 'state biological politics' should define the optimal size of the population as well as guarantee its 'quality', defined, in his words, as "the value of each individual, that is, the eugenics of a people" (Doxiadis 1928b).

Doxiadis was well informed regarding the developments in the international eugenic movement and he also visited the Second Hygiene Exhibition in Dresden (from May to October 1930)[30] from where he sent a report

[30] No official Greek delegation was present at the Hygiene Exhibition in Dresden in 1930. In the following year the Ministry for Hygiene did send delegates. See *Internationale Hygiene-Ausstellung Dresden 1930*, Amtlicher Führer, Verlag der Internationalen

to the Greek press (Doxiadis 1930a). Doxiadis buttressed his assertions with arguments and practices from international sources, such as the efforts of Margaret Sanger (1879–1966) to link birth control and eugenics in the United States.[31] In practice, he proposed a human engineering project with the aim of providing answers to the perceived problem of overpopulation, which he considered to be closely linked to social stratification. In order to counteract the alleged danger of degeneration, Doxiadis suggested a rational modus operandi for the selective biological reproduction of social classes. He elaborated a multi-layered concept of stratification that linked profession, social position and natality. According to his model there were four social classes in Greece, each with different fertility rates: the ('leading') upper class consisting of bankers, scientists and officials who tended to have two children; the middle class, members of which tended to have three or four children; the petit bourgeoisie class of merchants and tradesmen with four to five children; and, finally, a class of workers and farmers who had five or as many as seven children.[32] The state needed to promote an increase of natality in the higher classes (the bearers of 'hereditary capital'),[33] and the reduction of the birth rate in the lower social classes. The latter measure seemed to him to be premature, at least at that point in time. On the other hand, reducing the birth rate in the lower classes alone would not solve the problem since while this might lead to an improvement in the population in terms of quality, the population as a whole would decline in size due to the low birth rate in the upper classes.[34] In effect, the danger of degeneration and, finally, the collapse of the Greek *fili* could not be avoided. In order to counteract this threat, Doxiadis suggested advancing the role of the middle class, which he considered the most appropriate means to balance the instability of the two extremes: the biologically superior but individualistic upper classes and the biologically inferior but prolific lower classes.

The middle class in Greece that Doxiadis referred to had been growing since the turn of the 20th century as a result of industrialization, urbanization, private entrepreneurship and the founding of small companies and

---

Hygiene-Ausstellung, Dresden 1930; *Internationale Hygiene-Ausstellung Dresden 1931*, Amtlicher Führer, Verlag der Internationalen Hygiene-Ausstellung, Dresden 1931 (2nd edition).

[31] See Doxiadis (1930c). On Margaret Sanger's campaign for eugenic birth control, see Rosen (2004): 154–7.

[32] Doxiadis (1930g) on the overpopulation in Greece.

[33] Doxiadis (1933) on eugenics.

[34] He deals with the question of "reduction or control over births?" in Doxiadis (1930d); (1930e).

factories. In the 1920s the Greek economy experienced a second wave of industrialization, and the period from 1923 to 1938 is regarded by economic historians as the breakthrough with regard to industrial capitalism in the country, even though it was based mostly on the founding of small companies.[35] A significant part of these strata lived in risk, being threatened with unemployment and insecurity due to fluctuations in the labour market and economy. However, Doxiadis not only addressed the *existing* middle class in Greece, but moreover embraced an abstract idea of the middle class as a guarantor of social cohesion. In doing so, he drew lines to distinguish the middle of society from the bottom, and demanded the promotion of the former to higher positions. Although this idea was formulated in eugenic terms, the role Doxiadis ascribed to the social middle was consistent with the double orientation in advancing the middle class more generally, as described by Nobert Elias: "The gates downward should remain barred; the gates upward should open" (Elias 1993: 23).[36] In Doxiadis's words, negative eugenics discarded the unused and damaging elements of the social body. In terms of positive measures, he demanded state support, primarily to address the needs of the middle class, for example by establishing institutions for promoting children from such families. He himself, as director of the Foundation for Protecting Children, was concerned with advancing the situations of families and children in need.

Doxiadis essentially proposed an efficient, rationally organized system of social control over childbearing. This was a programme of human economy within which the value of human life was estimated in terms of economic costs and thus of the costs and benefits for the community. He had in mind a long-term vision as well as a programme for immediate implementation, both of which would be based on state interventionism:

> I do not know if it will ever happen that, like the tax inspector, there will also be a family inspector who, after intensive study and medical certification, will determine the number of children permitted to every married couple (Doxiadis: 1930e).

---

[35] Papoulia (2007); Agriantoni (1986). On the transformation of social stratification in Greece during the inter-war period, as well as the accompanying modernization of society and state institutions, see Mavrogordatos and Chatziiosif (1988); Moskof (1988); Pizanias (1993); Alexiou (1994).

[36] For a recent theoretical approach to the middle class, see Fischer (2007).

In the short term, however, Doxiadis believed that a series of partly institutional measures aimed at reducing child mortality (through the protection of maternity and childhood) would be more realistic. This project became the focus of his Patriotic Foundation for Protecting Children.[37] Nonetheless, he continued to argue that childbearing needed to be controlled and selectively reduced, and that anyone who could threaten the health of the community should be excluded from reproduction (Doxiadis 1930d). Further proposals included the establishment of agencies to control the health of bridegrooms, ending any propaganda advocating large numbers of children among 70–80 per cent of the population (coinciding with the two lower social classes) and supporting social strata whose low birth rate was a result of financial constraints (Doxiadis 1931h and 1933: 47).

The middle level of society was thought to be a worthy recipient of investment with state and social resources since this was conceived as a (potential) guarantor of quality in terms of both biological capital and moral values. Although at risk of poverty, the middle level of society was at least capable of being educated with ethical values and thus represented hope for the *fili*. Doxiadis, like most eugenicists, distinguished between those poor who were carriers of moral values and deserved state protection, and those who lived like parasites in the social body.[38] While the latter should be excluded from the right to reproduction and childbearing, state support should address the vulnerable parts of the middle class. In the eugenic narrative, which utilized biological and medical terminology for depicting social circumstances, poverty was evaluated according to middle-class moral values. Social calamity, marginalization and their attendant phenomena including poor health were interpreted in terms of order and disorder and described as social disease.

The following chapter addresses the moral evaluation of the social condition, and especially the question of health and disease, and argues that this widespread approach among Greek eugenicists drew its core idea from the hygienic discourse and bio-power that arose in Greece in the 19th century. Eugenicists maximized the moral connotations of hygiene, but moreover, they endorsed the repressive character of the suggested measures to cope with individuals who were thought to be bearers of social diseases.

---

[37] See his articles on the protection of mothers and care for children, Doxiadis (1930d), (1930e).

[38] Socialists such as the British Fabianists were not excluded from this line of argument; see Niemann-Findeisen (2004).

8.3 *Social Disease, the Promiscuity of the Poor and Eugenically Coloured Hygiene*

Erst kommt das Fressen,
dann kommt die Moral

Feed is the first thing,
morals follow on

Bertolt Brecht, *Die Dreigroschenoper*
(The Threepenny Opera), 1928

The rise of the Greek state in the 19th century and the attempts to establish modern state and societal institutions also led to a transformation in the perception of disease and poverty. Both disease and poverty were conceived in terms of social order and disorder, and indeed of social pathology. 'Social disease' is the proper category for describing social disorder using medical terminology. This is a manifestation of the medicalization of society, meaning "a process by which non-medical problems become defined and treated as medical problems, usually in terms of illness and disorders" (Conrad 2007: 4). The terminological couplet 'health–disease' becomes a cornerstone for a holistic worldview and at the same time of a reductionist view of society in which a wide range of social phenomena are approached and explained by medical terms. The couplet 'health–disease' is equivalent to the antithetical couplet of order vs disorder and connects the moral and social norms of the prevalent value system. From the medicalized point of view, society, metaphorically represented as an organism, is perceived as a *topos* of morality, as a moral totality (Conrad 2007: 52). 'Social disease' is defined so widely as to include any kind of supposed social abnormality deemed to threaten not only individual members of society, but the desired order and the 'collective body', the "man-as-species" (Foucault 2003: 242), with biological degeneration, or indeed, ethical decadence.

Social disease was a widely used term in the European discourse during the inter-war period. It acquired its semantics in the context of bio-power and, as such, it became an integral part of the population question. Social disease did not threaten the individual but the collective body, and it did not threaten it by death in the sense of interrupting life; the threatening death was "something permanent, something that slips into life, perpetually gnaws at it, diminishes it and weakens it" (Foucault 2003: 244). Ultimately, degeneration is resident within the realm to which the bio-power has access: the realm of life. The threat of degeneration provides reasons for intervention in favour of life; but only of the life that is evaluated as worthy of support.

In inter-war Greece, under the term 'social disease' were subsumed epidemics and endemic diseases, malaria, trachoma, prostitution, venereal diseases (above all syphilis), hereditary propensity to criminality, insanity, alcoholism, drug addiction, epilepsy, leprosy, tuberculosis, professional diseases and, to a lesser degree, cancer.[39] A pathological lifestyle was ascribed to carriers, or potential carriers, of so-called social diseases, which were claimed to be hereditary. Venereal diseases, criminality, and insanity in particular, either interlinked or separately, were the symbols *par excellence* of supposed hereditary social pathology. 'Hereditary syphilitics' were thought to be a true danger to society since "most invalids, disabled, as well as the most degenerate and criminal individuals, proceed from them" (Drakoulidis 1933: 56). This claim was advanced by the dermatologist, venereologist and psychoanalyst Nikolaos Drakoulidis (1900–85), best-known as a novelist under his *nom de plume* 'Aggelos Doxas'.[40] Drakoulidis believed that syphilis, alcoholism and drug addiction among parents was the main reason for juvenile delinquency, since children of addicted parents, "when they survived, suffer from epilepsy, idiocy, insanity; [they] are criminals, cachectic and degenerate" (Drakoulidis 1933: 59). The vehicle for transmitting such social pathologies to the next generation was of a moral nature: it was the promiscuity, perversion and uncontrolled sexuality of the parents. Another advocate of the same view, the psychiatrist Georgios Vlavianos, justified his proposition for limiting the sexuality and fertility of criminals on the grounds that "the majority of them feature a psychopathology, and they are either congenital criminals or suffer from a morbid increase in sex drive that causes certain crimes" (Vlavianos 1939: 11).

The violation of moral and social norms was deemed to have a negative impact on the bearing of sane offspring. Moisis Moiseidis claimed that sexual excess, promiscuity and the abuse of alcohol, all of which were prevalent among workers at the end of the working week, had negative consequences for their offspring, and asked the rhetorical question:

> How many workers and burly villagers have cachectic children because they combine alcohol and tobacco with sex? Children who are conceived

[39] See, for instance [Anonymous] (1925); Vathis (1925); Drakoulidis (1933); see also the entry "Marriage" in the *Great Greek Encyclopaedia*, written by Drakoulidis [no year].

[40] Nikolaos N. Drakoulidis (1900–85) studied medicine in Athens, Vienna and Paris, specializing in venereal diseases and psychoanalysis. He introduced the field of sexology into Greek medicine.

> on Saturday (Legrain's *enfants du samedi*),[41] the day when the workers draw their wages, are mentally degenerate (Moiseidis 1933: 30).

Supposed carriers of social diseases, or individuals thought to be prone to social disease, were charged with an ascribed overall abnormality and were situated in the lower social classes. In a teleological interlinking of illness with low social strata and delinquency, an ophthalmologist referred to the dissemination of trachoma with the following words:

> the criminal, as is well-known, … is lecherous or simply deranged; further, he has additional defects and, very often, is dirty. Crimes can be observed too often in the lower social strata where dirtiness is also more usual (Charamis 1926).

In the eugenic rhetoric, pauperism as an extreme manifestation of poverty was generally linked to social pathology or more directly to "social degeneration", as the psychologist Georgios Sakellariou asserted in a paper on "Criminality from a Psychological Viewpoint" which he presented at the Greek Anthropological Society.[42]

The consideration of poverty as a form of social pathology was neither a novelty of the inter-war period nor an argument exclusive to eugenicists. Rather, it surfaced in the 19th century simultaneously with claims to rationalize social organization in order to meet the challenge of the modernization of society. However, poverty was not always ascribed negative connotations; occasionally it was also used by those lobbying for the rights of the paupers as a vehicle for political criticism against the failure of state policy to care for the lower social strata. Such views are witnessed in Greek publications from the second half of the 19th century, as for example in an article entitled "Paupers" which was published in 1874 in the *Economic Review* [*Οικονομική Επιθεώρησις*].[43] The author termed pauperism a 'pathology' and described the precarious situation of the poor as a disease that required therapy and prevention. Another article on "Epidemics and Population", which appeared in the same journal, addressed the diffusion of epidemic illnesses among the lower social classes and

[41] Apparently he meant Paul-Maurice Legrain (1860–1939). According to Prestwich (1997), Paul-Maurice Legrain was the most prominent, yet least representative figure in the late 19th-century French temperance movement. He founded the first popular temperance organization and was France's leading expert in the treatment of alcoholics. Cf. Person (2005): 135.

[42] Sakellariou, *PGAS* (27 January 1932): 3–24, 9.

[43] Pappalouka Eftaksias (1874). The *Economic Review* [*Οικονομική Επιθεώρησις*] was published from 1873 to 1888.

discussed Malthusian tenets from a critical point of view ([Anonymous] 1874). Several other contributions in the same journal used similar terminology, thus betraying the Europe-wide diffusion of the biological holism of that period and the perception of society as an organism that could be afflicted with illnesses. *Economic Review*, the journal in which these articles appeared, is regarded by historians as the ideological organ of the intellectual current of positivism in 19th-century Greece (Moskof 1988: 153f.). The journal's editor, Aristeidis Oikonomos (1835–90), was a judge and lecturer in law at the University of Athens, as well as being a member of a group of utopian socialists, the Saint-Simonists (Moskof 1988: 153f.). The influence of Saint-Simonism[44] upon Greek intellectuals grew considerably in the period after the adherents of Saint Simon's utopian socialism had moved to the Greek town of Nafplio (the first capital of the burgeoning Greek state from 1828 to 1833) in 1832 as protégés of the Greek 'French Party' and its founder, the politician Ioannis Kolettis (1774–1847).[45] As a member of the Saint-Simonist circle, Aristeidis Oikonomos was concerned with the social question in Greece, which he regarded as closely connected with the labour question as it emerged in the country's first wave of industrialization. In a proposed agenda of social policy he prioritized the health of the people, which he viewed as a factor in the rising pauperism, or the risk of pauperizing, the working and living conditions of the lower social classes due to poor nourishment, child labour, professional diseases, their susceptibility to epidemics, and so on. In order to respond successfully to the challenges posed by the transformation of Greek society, Oikonomos thought that—in addition to the self-organization and political activities of the working and lower social classes themselves[46]—it was necessary to implement appropriate state policies that would advance economic and technical infrastructure.

The creation of state welfare institutions was a general demand among diverse sections of the elite in Greece who argued that there was a need

[44] On the Saint-Simonists and their ideology, see Picon (2003).

[45] Moskof (1988: 148).

On François Graillard (1792–1863), the intellectual and Philhellene adherent of Saint Simon (1769–1825) who fought on the Greek side during the war of independence from the Ottoman Empire, see Dimakopoulou (1979).

[46] Oikonomos prompted the population to found self-help organizations and promoted them by founding associations like the Society of the Friends of the People [Εταιρεία Φίλων του Λαού] in 1865, which advanced a broad programme of political education. Four years later, he created a further foundation called the 'Association of the People, Self-Help', an ephemeral organization like the others that preceded or were to follow it, according to Moskof (1988): 154.

for the provision of healthcare to the population. State attempts to create health institutions were supported by an emerging, professional group of physicians with an academic education who were also involved in establishing a health insurance system for the population. In a lengthy process (which was sooner or later to take place throughout Europe), health began to be seen as a public good. Healthcare passed (or was meant to pass, according to the claims of the emerging political and scholarly elite) into the jurisdiction and responsibility of the state. In the course of creating modern health institutions in 19th-century Greece, the perception of illness changed from individual suffering to a threat for the collective body that could effectively be treated by means of social surveillance (Korasidou 2004/1995 and 2002). Indeed, the state's first attempts to treat the health of the people concentrated on creating an alliance between police and doctors in order to inspect and control the dissemination of illnesses.[47] Parallel, private initiatives involving medical associations and philanthropic organizations worked on coping with poor health as a side effect of social calamity.

In this period, it was not only the semantics of health that was subject to transformation, but also of poverty, in such a way that both became interwoven with issues pertaining to health. The landscape of pauperism changed with respect to both the parts of the population to which the poor themselves belonged and the agencies that lobbied for them. In the last three decades of the 19th century the poor population in the few urban centres in the country included, along with the beggars (the 'old paupers'), the badly paid and unemployed workers, women, adolescents and children. Athens in particular was becoming an industrial centre and a magnet for rural–urban migrants who arrived from several parts of continental Greece and the islands, and from Crete, which was still (until 1886) under Ottoman rule (Korasidou 2004/1995: 66; Agriantoni 1986: 193). Changes also occurred with respect to the lobbyists for the poor. In the second half of the 19th century secular philanthropic agencies had begun to challenge the exclusivity of the church's philanthropic care. Poverty was now often considered a kind of social injustice, and the status of workers, and especially of the female poor and workers (Salimpa 2004: 169), was equated with pauperism per se. Simultaneously, liberal

[47] See also Ruisinger (1997) and an account of the first measures concerning public hygiene and social welfare from the founding of the Greek state to the end of Otto's era: Rozou (1976).

intellectuals strove to provide explanations for poverty, arguing, in some cases, that social inequality was a necessary condition for a functional social order.[48] In contrast to such liberal assertions, the Saint-Simonist Aristeidis Oikonomos believed that pauperism resulted from industrialization and the new economic conditions, and that it should be treated by social policies aimed at benefiting both the paupers and overall economic development.

Along with living and working conditions, health gradually became a matter that was taken up by lobbyists for the rights of the lower social strata—the part of the population that was most susceptible to poor health. Whether they were private medical initiatives, political activists or philanthropic organizations, all agreed that healthcare should be passed over to state jurisdiction. Health ceased to be a matter of individual well-being and became a societal resource embedded in the sphere of 'public or social hygiene', notions inaugurated in Greece in the same context as the beginning of the current of industrialization. The shift from the individual to the collective level justified the establishment of the term 'social hygiene' (often used synonymously with 'public hygiene' or 'hygiene of the people') which, in the 20th century, would become the terrain on which the rhetoric of social disease flourished.

Particularly in the inter-war period, experts emphasized that the scope of social hygiene extended to preventative health policy (Oikonomopoulos 1922; Charitakis 1929); hygienists complained about the confusion caused by interpreting social hygiene as being "restricted exclusively to fighting infectious diseases" (Chatzivasileiou 1925: 265). Instead, they argued that social hygiene should include a comprehensive statistical picture of national demography in terms of birth and mortality rate, and that it should be concerned with registering the living conditions of the population.[49] This consideration illustrates that the perception of the population in inter-war Greece had entered the age of bio-power, where it was comprehended and treated as a political, scientific, biological matter; a matter of power (Foucault 2003: 245). Social hygiene as a technique for the regulation of human life was assumed to be in a position to shape the fields of social and economic policy in the interests of preventing

---

[48] Even in the *Economic Review* [*Οικονομική Επιθεώρησις*], liberal views that justified inequality as a natural law were present. See for instance a treatise on *Rich and Poor* which argues in this way: Valsamakis (1875).

[49] See, for example, Panagiotakou (1928). On the scope and tasks of social hygiene and the respective institutions, see also Charitakis (1929), (1928a) and (1928b); Poliadis (1936).

disease and, more generally, any pathology thought to be immanent among the population. Hygienists and those concerned with the health of the population advanced what is usually termed 'social technology', serving the implementation of bio-politics. The birth rate, the mortality rate, longevity, "together with a whole series of related economic and political problems ... become biopolitics' first objects of knowledge and the targets it seeks to control" (Foucault 2003: 243).

This is the framework within which Greek sympathizers of eugenics (who proliferated in the first decades of the 20th century) elaborated their proposals. In practice, hygienists and eugenicists shared a desire to combat and prevent social diseases, and indeed to control population developments through scientific and legislative means. They essentially used the same vocabulary and appeared to pursue the same ends, namely the common welfare of the community. The borderline between hygienic and eugenic claims was often so blurred that their advocates frequently used the two notions interchangeably. In the Greek discourse, the prevalent usages of the terms 'eugenics' and 'social hygiene' suggested that the former was often thought of as an ideal target placed in the distant future, whereas the latter meant a programme for immediate operation in order to resolve urgent problems. Eugenic ideas were adopted in social hygienic programmes, so that a kind of "eugenically coloured hygiene", in the words of Béatrice Ziegler (2002), eventually arose from their merger. Yet crossing the fluid boundaries between eugenics and hygiene was as easy as switching from care to repression. Eugenicists assessed the 'value' and 'biological quality' of individuals and even raised them to become a condition for awarding care. The conditional award of healthcare was the most fundamental difference between eugenicists and social hygienists.

Stavros Zurukzoglu attached a great deal of importance to the distinction between hygiene and eugenics. He pleaded for this distinction on several occasions[50] and specified it in his limited number of Greek publications as follows: while hygiene protected and promoted every individual, eugenics sought to anticipate the proliferation of "hereditarily negative elements that counteract the protection they enjoy from individual hygiene".[51] He argued that "hygiene protects and promotes any individual irrespective of its value in terms of biology and culture", while "eugenics aims to safeguard

50 For instance, he did so, in reviewing the book by Moiseidis (1925). Zurukzoglu criticized the author's failure to make any such distinction. The review appeared in the journal *Medicine* [*Ιατρική*] 3, no. 1 (May 1925): 32–3.

51 Zurukzoglu, *PGAS* (11 May 1925): 14–40, 16.

and proliferate, for the most part (if not exclusively), those who hold true biological and cultural value".[52] The negativity of eugenic selection was based on a presumed positive norm, aimed at promoting physically and mentally capable people in what Zurukzoglu described as a "teleological attempt to maintain the nation, humanity, and civilization".[53]

In health–disease discourses, it was not only health that became a public resource—the public itself was viewed as the source for the successfulness of eugenic and hygiene agendas. Both hygienists and eugenicists attempted to influence public awareness through the broad dissemination of their ideas. The popularization of hygienic and eugenic tenets posed a challenge that could be met by means of propaganda, in the original terminology, through 'enlightenment campaigns'. This is typical of the 'medicalization of society' wherein medical power diffuses into society and health and disease are not exclusively a matter for the medical profession but demand the mobilization of the people (Conrad 2007: 4). In Greece, physicians delivered their expertise directly to the public while several institutions and agencies provided them with forums for presenting their ideas. Lectures in broader, non-medical circles and publications in the popular press were the most frequently used means for spreading such ideas. The Athenian philological club, Parnassus [Παρνασσός], was one such institution that often hosted promoters of eugenic ideas, lectures on related subjects and hygiene exhibitions. One agency dealing with the "fight against social disease" advancing the dissemination and popularization of hygienic and eugenic ideas in inter-war Greece was the National Council of Greek Women [Εθνικό Συμβούλιο Ελληνίδων]. This feminist umbrella association encompassed numerous organizations and enjoyed a great deal of influence in the middle-class women's movement. Indicative of the importance the council attached to the role of social hygiene was the establishment of a Department for Social Hygiene, an internal commission that was engaged with questions of hygiene and eugenics. The stated aims of the council's activities were:

> to combat and reduce the diseases which plague Greece, to improve the health and eugenics of our race [in original: *ράτσα*], to teach mothers how to maintain the physical and mental vitality of their children... to combat

---

[52] Zurukzoglu (1925e): 222. This chapter, like the lecture delivered at the anthropological society, is based on the author's German-language book; Zurukzoglu (1925a).

[53] Zurukzoglu, *PGAS* (11 May 1925): 14–40, 16.

> ignorance and superstition and to provide general hygienic educational propaganda.[54]

Additionally, the National Council of Greek Women suggested a programme of preventive hygiene that extended to propaganda films, lectures to be held in factories and other workplaces and the establishment of information centres for the public. It also organized a series of lectures which were held in spring 1932 at the Athenian philological club Parnassus, the Archaeological Association and in residential districts of Athens.[55] The first series covered "social diseases such as tuberculosis, syphilis, malaria, trachoma, and drug addiction" (Drakoulidis 1932a: 11). Further lectures on hygiene in marriage and childbearing followed, in which prominent scholars, most of whom were advocates of eugenics, disseminated their views on eugenic birth control and restrictions on reproduction. In 1932 the Department for Social Hygiene established the Library of Social Hygiene, a book series under the editorship of Nikolaos Drakoulidis.

Efforts to popularize and spread eugenic ideas on reproduction and fighting social disease typically formed part of an overall concept that included demands for passing legislative measures. Eugenicists supported the transformation of external obligations (to be enforced by state institutions) into values internalized by the population. In doing so, eugenicists joined other critics of state policy in Greece, who complained that the welfare and healthcare systems had serious infrastructural deficiencies. At the very least, policies for preventing diseases by means of improving living and working conditions, ensuring proper nourishment and so forth would benefit the state itself, since the expenses of healthcare would be reduced while the working classes would be able to work better and for longer hours (Oikonomopoulos 1934; Doxiadis 1931h). This provides evidence of the perception of illness as a question of population policy and thus, as Foucault puts it, illness was thought of as a permanent factor which "sapped the population's strength, shortened the working week, wasted energy, and cost money, both because they led to a fall in production and because treating [it] was expensive" (Foucault 2003: 244).

---

54 See the programmatic opening text of the Library of Social Hygiene by Nikolaos Drakoulidis (1932a).

55 The two districts were Petralona and Vyronas. According to information in the journal *Hellene* [*Ελληνίς*], "Department of Hygiene. Lectures on the Social Disease" ["Τμήμα Υγιεινής. Διαλέξεις κοινωνικών νοσημάτων"], *Ελληνίς* 12, no. 6–7 (1932): 131–2; and 14, no. 11 (1934): 231.
See also Drakoulidis (1932b).

Eugenicists in particular shifted their focus to evaluating the population, and demanded a selection of those who most deserved to benefit from welfare institutions. In a number of efforts to establish a health system and a system of occupational and social insurance, the interventions suggested by eugenicists for saving the demographically endangered country and protecting the social or national body from diseases were considered and discussed as possible solutions. Needy individuals appeared to constitute a burden on the development of the welfare system. As the psychiatrist Simonidis Vlavianos argued:

> the ill and cachectic products of ... criminal marriages [between alcoholics, lunatics, the feeble-minded and idiots] are a terrible burden for the social body in terms of both their appearance as well as material expenditure (Vlavianos 1939).

Sympathizers of eugenic ideas proclaimed their concern for the common good, and their humanitarian rhetoric (and their professional status as experts) made them legitimate partners in the fight for promoting the welfare of the population. However, their agendas extended the existing 'enlightening campaigns' to measures of a repressive character that criminalized social behaviour, poverty and even illness: "The fight against common social diseases is the care of all current criminologists and legislators",[56] as Spiridon Ardavanis-Limperatos wrote in his column "Eugenics and Social Hygiene" in the medical press (Ardavanis-Limperatos 1935a). The repressive character of the suggested measures was deemed to be justified by the sanctity of the long-term target: an ideal world. "The salvation ... [and the] welfare of the peoples lay in the hands of the legislator", claimed the professor for hygiene Konstantinos Moutousis in his inaugural lecture at the University of Athens in 1933 (Moutousis 1933: 872).

With regard to the eugenic agenda, there were not only campaigns of public enlightenment, but also negative eugenic measures to be applied by the state politically. The most frequently discussed of these negative eugenic measures included establishing premarital medical examinations as a precondition for allowing or sanctioning marriage, and the sterilization of the potential carriers of social disease. The debates among scholars were often highly controversial.

---

[56] This is an article on eugenics, social hygiene and the healing of criminality: Ardavanis-Limperatos (1935a).

8.4 *Scholars on Eugenic Politics, Sterilization and Prenatal Health Examinations*

Attempts to control reproduction and marriage were based on the conviction that the purpose of the family was to constitute cells in a healthy society, nation, race and humanity, providing the collective body with superior human capital. One practical way to meet this goal was by establishing appropriate legislative measures that allowed or enforced the medical certification of the health of bridegrooms prior to marriage, and the sterilization of individuals affected by social disease. Physicians occupied a prominent position in debates on this issue, since they were deemed to possess expertise and 'objective knowledge' about health and the prevention and combating of illnesses.

One of the earliest written references to eugenics, and to premarital health examinations in particular, is found in a 1917 essay entitled "Eugenics" (Kairis 1917) which was based on a lecture of the same title that had been delivered at the Sinaia Academy in Athens. During this presentation the lecturer expressed doubts as to both the necessity and the possibility of establishing this specific eugenic measure in Greece, a view that contrasted with that of Moisis Moiseidis (1922), who had promoted this idea in his monographs on eugenic issues and possible practical measures, including what may well be the first ever Greek monograph on eugenics.[57] During the ensuing years, references to eugenic premarital examinations proliferated. Aside from medical publications,[58] possible restrictions on reproduction and marriage were also mooted in the Athenian daily press and in populist journals. The liberal journal, *Labour* [*Εργασία*], for example, provided a permanent forum for debates on public hygiene and the population question, and while eugenic thought did not predominate, it did occasionally find a place for expression. For instance, *Labour* also published papers presented at the First Panhellenic Congress for Hygiene, held in Athens in April 1930; one of these papers (Moiseidis 1930) specifically addressed premarital medical examinations. One of the most fervent proponents of eugenic birth control, Nikolaos Drakoulidis, contributed to familiarizing the Greek public with this issue through numerous articles and treatises. His ideas on marriage, birth control and prenuptial health

57 Moisis Moiseidis was co-editor of *Health* [*Υγεία*], a journal of *Practical Medicine* (as its title) that was supported by the Greek Red Cross and aimed to reach "all social classes".

58 See [Anonymous] (1936). See below for a discussion of the legislative institutions.

certification were published widely in the Greek popular press and newspapers.[59] He also wrote the entry on "Marriage" in the *Great Greek Encyclopaedia* [*Μεγάλη Ελληνική Εγκυκλοπαίδεια*] where, among other matters, he addressed the necessity for state control over reproduction by means of prenuptial health certification (Drakoulidis [no year]). Still later, another instance where these ideas were aired was a lecture on "Health of the Bridegroom and the Medical Certification for Marriage", delivered by an air-force physician, to the Chamber of Industry and Commerce in Thessaloniki (Ntoulas 1936).

Even though adherents of eugenics were generally sympathetic to the idea of establishing medical certification as a means of ensuring healthy offspring, opinions varied significantly with respect to how this could be implemented in practice. Points of contention included whether such certification should be voluntary or obligatory, and further, which measures should be applied in cases of diagnosed disease or disability. The options discussed were manifold, including a legal prohibition on marriage, voluntary abstention from marriage by couples and voluntary abstention from childbearing, contraception, voluntary abortion and even sterilization. Sometimes the measures proposed combined a number of different options. For instance, the psychiatrist Simonidis Vlavianos (1933) endorsed a ban on marriage by law, as well as voluntary sterilization for the insane, feeble-minded, alcoholics and those suffering from tuberculosis and syphilis. According to a rather moderate tendency, health propaganda should persuade couples to have themselves examined voluntarily and, if deemed medically pertinent, to abstain from marriage and procreation. Broadly speaking, a strong tendency among Greek adherents of eugenics gave top priority to winning over the public through argumentation, rather than by trying to enforce a ban on marriage by legal means. Several physicians participating in the eugenics debates underlined the social imperative of choosing spouses according to eugenic criteria and thereafter creating ideal conditions for conceiving children, or even developing 'codes of marriage' for choosing a spouse (Moiseidis 1933: 20–8). However, consultation and voluntarism alone were deemed insufficient without institutional action. Drakoulidis, for instance, suggested the establishment of a broad-based network of agencies, charged with the organization and management of educational propaganda for the Greek population, and the foundation of a

[59] See Drakoulidis (1932c) and especially Drakoulidis (1933) in which the author provides a list of his publications on this issue in the Greek press.

eugenic society to coordinate these agencies (this measure was not, however, adopted in practice at that time).

A major point of contention that was raised by both moderate and hard core eugenicists concerned the extent to which institutional conditions in Greece were appropriate for effectively implementing measures like the prenuptial health certification given the infrastructural deficiencies in the country. On the other hand, however, the implementation of eugenic measures was seen as a challenge, the response to which would provide an impetus to the establishment of welfare institutions, such as the newly formed occupational and health insurance system, the social care system, and so on.

The controversy among proponents of eugenic measures increased when sterilization was at issue. A large part of those who approved of sterilization in principle as a means of promoting a healthy collective body were, at the same time, sceptical about its possible implementation in Greece not only because the necessary infrastructure was unavailable, but also because public opinion might be opposed to it. This scepticism surfaced in the journal of the feminist organization, the National Congress of Greek Women, when Moiseidis's book on eugenic sterilization was reviewed. On the one hand, the anonymous reviewer underlined the convenience of such a measure, especially in his contemporary period, "when millions of people are becoming unemployed and yet those who are healthy and physically and mentally fit, are barely able to ensure their livelihood".[60] Yet, on the other, the reviewer considered the implementation of eugenics to be inopportune, on the grounds that "the public consciousness that would allow such an intervention into individual life still failed".[61]

The majority of the defenders of eugenic sterilization favoured voluntary rather than forced sterilization with the assent of the individuals concerned or their relatives. Advocates of forced sterilization included the physician and chair of hygiene at the University of Athens, Konstantinos Moutousis, who dedicated his inaugural lecture as chair (on 9 November 1933) to eugenics.[62] Moutousis argued that the implementation of eugenics by states would result in the improvement and perfection of 'human material' while contributing to the maintenance of humanity as whole.

---

60 Review in *Hellene* [*Ελληνίς*] 15, no. 3 (March 1935): 69–70.

61 Review in *Hellene* [*Ελληνίς*] 15, no. 3 (March 1935): 69–70.

62 Moutousis (1933). His speech was published several days later in the medical review *Clinic* [*Κλινική*] 47, (25 November 1933).

The positive effects of what he called 'eugenic hygiene' included a reduction in the costs associated with 'the ill or infirm', 'epileptics', the 'mental a-normal', 'psychopaths', 'criminals' and other 'disabled individuals' (Moutousis 1933: 867). Regarding practical measures, Moutousis suggested that, along with the public promotion of families who were bearers of 'good qualities', compulsory sterilization should be imposed for those who were thought to be "hereditarily burdened with a pathological disposition" (Moutousis 1933: 872). His statement caused an immediate reaction from the church, which condemned both forced sterilization and sterilization in general in its official journal *Ecclesia* [*Εκκλησία*].[63]

While Moutousis's statement may have been inconsistent with the ideas promoted by mainstream Greek eugenicists, who advocated voluntary sterilization, it is nevertheless remarkable because it was voiced only a few months after the passing of sterilization laws in the Third Reich, the notorious 'Law for the Prevention of Genetically Diseased Offspring' [*Gesetz zur Verhütung erbkranken Nachwuchses*, 14 July 1933]. The enactment of this legislation in Germany also led to an intensification of the debates in Greece; moreover, it had a decisive impact upon opinions about sterilization and caused much uneasiness. These ambivalent and controversial opinions among eugenicists both worldwide and in Greece were discussed by Moiseidis (1934), whose own attitude was also ambiguous. Although he accepted the core principles of eugenics, he was ultimately unable to side with any one of the fronts "[without] running the risk of confrontation with the others" (Moisideis 1934: 19). Despite his initial scepticism, Moiseidis sided with the 'moderate eugenicists', advocating voluntary and remedial sterilization with the assent of the persons involved on condition that it would contribute to combating degeneration and protecting the *fili* (Moisideis 1934: 67–8).

Moiseidis's anxiety was not without foundation. In the Greek medical and daily press several articles commented on, and typically condemned, the German legislation. For example, immediately after the legislation had come into effect, the Greek medical press announced protest actions by German scholars against this 'barbaric measure' ([Anonymous] 1934). Some weeks earlier the psychiatrist Simonidis Vlavianos had characterized the enforcement of sterilization in Germany as a "product of an absolute dictatorship" (Vlavianos 1933). At the same time, the author of the article pleaded for voluntary sterilization and the banning of marriage for those

[63] The brief response had the title "Forced Sterilization": [Anonymous] (1933).

who suffered from tuberculosis and syphilis, for alcoholics, the insane, the feeble-minded and idiots. He argued that these individuals should be excluded from reproducing by means of voluntary sterilization and a ban on marriage, in order to prevent the birth of children who would burden the health system and end up in correctional institutions, since many such children became "paranoiac criminals" (Vlavianos 1933).

The issue of the sterilization of criminals was also addressed on the occasion of the German legislation and its possible adoption in Greece in the Athenian press (Katopodis 1934). The columnist, who approached this theme from a juristic point of view, rejected the introduction of sterilization in Greece on the grounds that, while the measure might be feasible for a country with a large population and developed social hygiene programmes, this would not be the case in Greece, which had a small population and an underdeveloped infrastructure.[64] Keeping up with the broader criticisms emanating from the Greek public,[65] the author argued that priority should be given first to the modernization of the correctional system, the improvement of terms and conditions of detention, education of the personnel employed in the correctional institutions and the study of criminality rather than the immediate implementation of such a measure. The diagnosis of a hereditary predisposition to criminality, a conceivable reason for sterilization, would be possible only after conducting investigations and compiling reliable statistics.[66] A similar criticism addressed to the deficient organization and inadequate infrastructure of the correctional system had been made by the psychologist Georgios Sakellariou two years earlier. At a session of the Greek Anthropological Society[67] Sakellariou discussed the wide variety of crimes in Greece and

---

64 Katopodis (25 February 1934).

65 The same author, Giorgos Katopodis, published further articles in the journal *Labour* [*Εργασία*] in which issues of criminality and reforms of the Greek correctional system were discussed, especially in 1933. Cf. the view of the socialist Platon Drakoulis (1858–1934): Drakoulis (1927).

66 This article reflected the contemporary state of affairs in the Greek judicial system as well as in the studies of criminology in which a series of reforms were announced. See also Gardikas (1934d). Gardikas reported that he had invited contributions from Greek and foreign scientists, among them also Gustav Aschaffenburg (1866–1944), the editor of the above-mentioned journal, who spoke about "The Significance of the Psychopaths for the Society and the Law, Environment, Sterilization" ["Die Bedeutung der Psychopathen für die Gesellschaft und das Recht, Anlage und Umwelt, Sterilization"]. The precarious situation of prisoners in Greece and the reforms in progress were reported by Dimitris Karanikas in the same journal: Karanikas (1934).

67 Sakellariou, *PGAS* (27 January 1932): 3–24. See also Sakellariou's study on criminality in Greece from a psychological and social point of view: Sakellariou (1932).

the role the 'mental capability' of criminals played therein, distinguishing between hereditary and environmentally determined criminality. He proposed the improvement of the correctional system by means of reforms and the better organization of the police, rather than the sterilization of criminals. Sakellariou's objection pertained to both the severity of this measure and the problematic situation of the Greek correctional system, which inhibited the diagnosis of the hereditary predisposition to criminality.[68]

Radical proposals to restrict the marriage and reproduction of criminals by forced sterilization came from the psychiatrist and docent at the University of Athens, Georgios Vlavianos, who prioritized state intervention rather than educational campaigns and reforms. On the occasion of a lecture he gave during the hygiene exhibition in 1938 at the Athenian philological club Parnassus, Vlavianos argued in favour of establishing a premarital health certificate by law, in order to impede the reproduction of mentally deficient offspring (Vlavianos 1939). Further measures could extend to committing those affected to an asylum and the imposition of compulsory sterilization. In particular, Georgios Vlavianos considered the sterilization of mentally disabled individuals, and especially criminals, to be necessary. He claimed that 75 per cent of criminals were afflicted with "congenital perversion" and a "pathologically heightened sex drive" (Vlavianos 1939). He regarded sterilization as more humane than internment for the individuals concerned, and as being economically beneficial for the Greek state, since the latter would profit through the reduction in the number of asylum inmates as well as the expenses of combating criminality. In his opinion, restrictions on marriage, commitment to asylums and sterilization should not be applied exclusively, but rather concurrently. Thus, if a couple, after medical examinations, was deemed to be unfit for bearing healthy children, then both measures should take place: the marriage should be forbidden and the couple should either be committed to an asylum or sterilized. Vlavianos was convinced that a ban on marriage would not preclude defective individuals from bearing children. Hence a ban on marriage without intervening in procreation would encourage childbearing outside the institution of marriage, meaning that reproduction would escape state control.

[68] See an account of eugenic views in psychiatry and criminology by Kazolea-Kavolari (2002).

The debates on eugenic measures and their possible implementation did not remain confined to periodicals and lectures in the public or in associations. There had been several occasions to enact eugenic legislation in Greece, and in some cases this indeed took place. Proponents of eugenics, physicians, jurists and other scholars offered their expertise to the state, either voluntarily or on request. Yet antagonism among professional groups concerning who had the requisite authority to advise on state policy was by no means absent. Greek state institutions and the legislative body seriously considered the implementation of such measures and even tried to impose them on several occasions, although usually with little success.

8.5 *Eugenic Birth Control by State Institutions*

In the above, the shifting semantics of illness from individual suffering to an issue concerning the collective social body has been witnessed in 19th-century Greece, a shift that signifies the first attempts to implement bio-policies in the country. At the core of this discourse and its related policies was the close association of sexuality, marriage and childbearing and their regulation by the state. Early legislative restrictions on marriage in order to avoid unhealthy progeny were implemented in the very first years of the founding of the Greek state and during the rule of King Otto. A royal decree addressed to the police from 1837 prescribed the means for treating individuals affected by communicable diseases such as typhus, plague, cholera, smallpox, venereal disease, some ophthalmologic diseases and leprosy.[69] By virtue of this decree, the police, with the assistance of doctors, were charged with identifying diseased individuals and isolating them or committing them to doctors. An exception was made in the case of the Hansen disease, for which the legislator issued a specific instruction advising restrictions on marriage and childbearing for the affected individuals. The police were also asked to prevent marriages of men and women who suffered from leprosy in order to avoid their proliferation. This exceptional measure concerned an illness that was not fatal, but the individuals affected would continue to live diseased lives that were thought to threaten the healthy collective body with infection. Later, in

[69] Police regulations from 31 December 1836 (12 January 1837) by King Otto. According to the habilitation treatise by Vasileios Rozos which deals with early measures for social and healthcare in Greece: Rozou (1976): 101.

the 20th century, lepers were prohibited from marrying by legislation that was enacted in 1920[70] and which remained in force until 1955[71]—it would appear that it was subsequently abrogated after it had been discovered that leprosy was not a contagious disease.

Restrictions on marriage for reason of illnesses had already been addressed by the Greek parliament in the 19th century. A somewhat coherent proposal was shaped in 1870 by a deputy from the island of Tinos (Zallonis 1870), a physician by training, who requested the imposition of a legal ban on marriage for midgets, as well as for individuals suffering from tuberculosis, leprosy, elephantiasis and cachexia caused by syphilis. He recommended that couples who were about to marry should have their health assessed by two physicians, and that they should also submit a health certificate issued by a judge. In his 16-page petition, the deputy also addressed one problem that might emanate from such legislation; namely, whether marriage restrictions violated individual rights. His own answer was clear: "individual freedom is inviolable insofar as it does not harm the state". It is uncertain whether parliament indeed discussed this petition, as there is no reference to it in the minutes of the parliamentary debates from that period.

Much later in the 20th century, however, similar proposals cropped up concerning eugenic restrictions on marriage and childbearing. These appeared on the agenda of the legislative body when health legislation was being debated. In 1919, during a parliamentary debate on the draft law "On the Establishment of Medical Practices, Hospitals, Convalescent Homes and Sanatoriums against Tuberculosis",[72] a deputy from Cyclades (S. Papavasileiou) recommended imposing a ban on marriage for those suffering from tuberculosis. He argued that tuberculosis was a hereditary disease, typically a side effect of syphilis. In his opinion, the wording of

[70] See Law No. 2450 for 1920 which prohibited marriage among lepers; in *Official Governmental Gazette* [ΦΕΚ] 182 A'- 18 February 1920 [Νόμος 2450 "Περί μέτρων προς περιστολήν της λέπρας"].

[71] Marriage restrictions for lepers were abandoned in 1955 by virtue of Law 3369/1955 on "Measures for Combating Leprosy", *Official Governmental Gazette* [ΦΕΚ] No. 258 A' 23 September 1955 [ΝΔ 3369/1955 "Περί μέτρων καταπολέμησης της Λέπρας", ΦΕΚ 258 A' 23/9/1955].

[72] See the debate of the 20th Parliamentary Session of 17 December 1919, reported in documents of the Parliamentary Debate on the Articles of the Law Draft on the Founding of Medical Offices, Hospitals, and Sanatoriums. *Gazette of Parliamentary Debates* [ΦΕΚ], No. 20, 17 December 1919: 287–292 [A' συζήτησις επί των άρθρων του σχεδίου νόμου "Περί ιδρύσεως αντιφυματικών ιατρείων, νοσοκομείων, αναρρωτηρίων και σανατορίων". Πρακτικά Ενάρξεως των Εργασιών της Δ' Συνόδου της Κ' Βουλευτικής Περιόδου. Εφημερίς των Συζητήσεων της Βουλής Συνεδρίασις 20ή της 17ης Δεκεμβρίου 1919, 287–92].

the draft law under discussion was insufficient for eradicating both the causes and the spread of tuberculosis. The deputy argued that the most effective measure would entail legal marriage restrictions for consumptives by means including the health certification of couples prior to marriage or even their sterilization. His archetypal model for doing so was the sterilization law passed in the American state of Indiana in 1907, abridgments of which he had read during the 1919 parliamentary session. While advocating sterilization in principle, he expressed scepticism about passing a similar law in Greece on the grounds that the public was likely to disapprove of the measure. The ensuing parliamentary discussion illustrated the diversity of opinions about eugenics among Greek politicians. Another deputy, Michail Katopodis, whose professional background was in medicine, disputed the hereditary character of tuberculosis, and declared himself opposed both to marriage restrictions for those who suffered from tuberculosis and to sterilization for any reason. But Konstantinos Alavanos, a deputy and physician, advocated the opposite, pleading for both the introduction of a medical certification system for bridegrooms and for compulsory sterilization. In responding to Papavasileiou's request, he also underlined the state's need to strengthen intervention in health institutions and limit the actions of private health agencies.

The proposed eugenic restrictions on marriage and childbearing did not pass in that parliamentary session. Yet the same issue appeared anew on the agenda of the Ministry of Hygiene and Welfare [Υπουργείο Υγιεινής Προνοίας και Αντιλήψεως] some years later. As the medical press reported,[73] the then minister addressed a letter to the Holy Metropolis of Athens [Ιερά Μητρόπολις Αθηνών], which queried the possibility of enforcing medical certification as a condition for allowing an Orthodox marriage. If the church would agree, he asked for it to lead the way in defining the diseases that were to be considered impediments to marriage.[74] The eugenic idea in the foreground of this initiative was stated explicitly in the minister's letter: "One of the main concerns of the State Health Office is the study and possibly the implementation of eugenics."[75] The minister signing the letter indicated that the ongoing dysgenic tendency among Greeks, the infant mortality rate and the dissemination of venereal

73 [Anonymous] (1925).The correspondence between the ministry and the church was published by Nikolaos Drakoulidis (1933): 65–7.

74 The church was authorized to issue marriage licences in Greece prior to the establishment of a civil marriage code in 1980.

75 Correspondence reprinted by Drakoulidis (1933: 65).

diseases (mainly syphilis) all demanded the immediate establishment in Greece of "a series of hygienic measures for protecting Greek heritage [in original: *γένος*] from dysgenics and promoting eugenics that are the basis of every hygienic goal ... as in the rest of the civilized world". The church responded immediately to the letter, approving the minister's proposals.[76] Nevertheless, a law demanding prenuptial health certificates was not passed on that occasion either. Even so, the initiative attracted the attention of the medical press, which emphasized the enormous importance of this matter for society, even going so far as to suggest convening a commission to examine the possible implementation of marriage certificates ([Anonymous] 1925).

Thereafter, state institutions continued searching for a way to draft a law of international standards. Immediately following the church's positive response, the Greek government contacted the German Ministry of the Interior [Reichsminister des Inneren] in order to request advice, based on the German experience of legislative measures to safeguard the health of offspring.[77] In response, the German ministry informed the Greek government about the lack of any recent domestic legislation on this matter.[78] Indeed, in the 1920s the German government itself had searched for and adapted texts concerning laws on marriage restrictions passed in other countries, including Brazil, Mexico, Turkey and the Scandinavian countries.[79] German institutions were especially interested in the Turkish law, "Le réglementation du marriage" that had been passed in 1926 and established health examinations prior to marriage.[80]

While the negative answer from the German state temporarily interrupted the Greek government's attempts to impose prenuptial health certification by law, marriage restrictions for eugenic reasons were to appear once again on the agenda of the Greek national legislative body in 1930. This happened during the debate on the revision of the Civil Law, in the

[76] See the church's answer on 4 November 1925, reprinted in Drakoulidis (1933): 67.

[77] Das Auswärtige Amt, Berlin. Dem Herrn Reichsminister des Inneren. Abschrift VIZ 9993, Griechische Gesandtschaft, Berlin W8, den 15. Januar 1926 (Request from 18 December 1925). BA, R 1501/109380, Blatt 146.

[78] Reichsminister des Inneren, An das Auswärtige Amt, Berlin 15. Januar 1926. Betreff: Gesundheitszeugnisse für Ehebewerber'. Bundesarchiv, R86 N. 2372. On the establishment of premarital health certification in Germany, see Kesper-Biermann (2007).

[79] The German Federal Archive [Bundesarchiv] testifies to the correspondence between such state institutions worldwide. See, for instance, BA, files R. 86; R. 1501.

[80] It was the Law No. 37 (Stanboul 6.2.1926), "Le réglementation du marriage". The text of this law with related correspondence between German ministries can be found in BA 1501/109380.

third liberal government of Eleftherios Venizelos.[81] The revision commission was mainly composed of jurists, all of them prominent professors of law. One member of the revision commission, a professor of private international law, Georgios Maridakis (1890–1979),[82] pleaded for a legal ban on marriage for lepers,[83] epileptics and those suffering from either tuberculosis or syphilis.[84] Although his proposition was finally rejected by 11 of the 15 members of the revision commission, the debate brought to light controversial views on the possible implementation of eugenic measures in Greece, as well as those pertaining to the idea of eugenics in general. The members of the revision commission of the Civil Law agreed on the importance of eugenics to ensure the health and welfare of society and the nation. But beyond consensus on this point, their opinions differed remarkably with respect to the practical means for achieving this goal. Controversial matters included whether or not the diseases in question were indeed of a hereditary nature, and whether radical measures such as a ban on unhealthy marriages were appropriate and necessary. Furthermore, members of the commission were sceptical about the feasibility of the suggested eugenic measures, especially given the lack of technological and scientific infrastructure in Greece and severe deficiencies in the healthcare system. In addition to these reservations, the professor for civil rights Georgios Balis (1879–1957)[85] underlined the risk of the proposed restrictions becoming "repressive and socially dangerous".[86] The economist Kyriakos Varvaresos (1884–1957) added to this criticism the argument

81 The first Venizelos government lasted from 1910 to 1916, the second from 1918 to 1920 and the third from 1928 to 1933.

82 Georgios Maridakis (1890–1979) later became a member of the International Arbitration Tribunal (The Hague 1940) and the European Tribunal for Human Rights (Strasbourg 1959).

83 The reference to leprosy here seems paradoxical since there was already a law in existence from 1920 that prohibited lepers from marrying.

84 The proceedings of this meeting were published in the *Gazette of the Debates on the Revised Civil Law*: [*Εφημερίς των Συζητήσεων του Νέου Αστικού Κώδικος*], Issue 5, 12 March 1931 (Meeting no. 6, on 29 December 1930): 82–96; (Maridakis's proposal is on p. 91).

85 Georgios Balis (1879–1957) was one of the most important experts on civil rights in Greece. In 1938 he was tasked with the revision of the Civil Law. His concepts formed the basis for the Greek Civil Law that remains in force today.

86 Revising Commission of the Civil Law, 6th session, 29 December 1930: *Gazette of the Debates on the Revised Civil Law* [*Εφημερίς των Συζητήσεων του Νέου Αστικού Κώδικος*] 5 (12 March 1931): 92.

that prohibiting marriage on the grounds of poor health was nothing but a Malthusian principle that restricted individual freedom.[87]

Varvaresos thus voiced an ethical dilemma in the proposal, one which derived from the incompatibility of care for the community and the protection of individual rights. On the one hand, it was necessary to safeguard the health of the *fili* and the people, yet on the other restrictions on marriage for health reasons would inevitably violate individual rights. His statement was countered by the constitutionalist Alexandros Svolos (1892–1956)—a highly prominent figure from the liberal wing of the socialist movement who was later actively involved in the resistance during World War II.[88] In Svolos's opinion, the dilemma could be resolved if everyone accepted that the most important priority was Greece's common good. He assumed that the highest ideal, the good of the collective body and future generations, prevailed over individual welfare, which in turn legitimized any restriction on individual freedom. Svolos was the only member of the commission who, in addition to premarital health examinations, proposed sterilization, in keeping with the American paradigm. He did so despite being familiar with the fact that critics of sterilization characterized it as a "tyrannical and barbarous measure".[89] Still, for Svolos, the Civil Law was the most appropriate framework for imposing such a measure in Greece, after careful examination and consultation with experts.

The final rejection of the proposed premarital health certification resulted in protests from those physicians excluded from decision-making processes, who claimed that they should have been consulted. One of them was Nikolaos Drakoulidis (1933) who reported that some physicians,

---

[87] Kyriakos Varvaresos (1884–1957) studied law in Athens, as well as economics, political science and statistics in Germany. After completing his studies he was appointed head of the Department for Statistics in the Greek Ministry of the National Economy in 1911. In 1932, he became minister of economics in the government of Eleftherios Venizelos and Alexandros Papanastasiou. Varvaresos wrote numerous monographs including, among others, *The Theory of Population*: Varvaresos (1912).

[88] Alexandros Svolos (1892–1956) studied jurisprudence in Athens and later became a professor at the National University in Athens (1923). He was also an outstanding politician from the moderate left wing. Under the dictatorship of Ioannis Metaxas, Svolos was persecuted and exiled. During the Second World War, he actively participated in the resistance and the government-in-exile. Among other activities, he was president of the Socialist Party in the 1940s, the Union of the People's Republic and later (in 1953) was a founding member of the "Greek Union for Human and Civil Freedoms" [Ελληνική Ένωση υπέρ των ελευθεριών του Ανθρώπου και του Πολίτου].

[89] Minutes of the Revising Commission of the Civil Law, 6th session (29 December 1930), *Gazette of the Debates on the New Civil Law* [*Εφημερίς των Συζητήσεων του Νέου Αστικού Κώδικος*] 5 (12 March 1931): 94.

in trying to reverse the decision, had addressed a series of protests to the Highest State Hygienic Office [Ανώτατο Υγειονομικό Συμβούλειο του Κράτους]. Although claiming some success, no measure seems to have been implemented by state institutions that would corroborate this. However, this failed attempt was not the last effort to impose medical certification on Greek bridegrooms through legislative measures, as the state was to make further attempts to pass a law to enforce premarital health certificates during the post-war period.[90] In 1958, for example, the Greek parliament debated a very similar proposal that still failed to become law. Four years later, in 1962, the Ministry for Social Welfare submitted a draft law for implementing such a certification procedure to the Faculty of Law at the University of Athens, but the faculty never even considered the proposal.[91]

Attempts to put restrictions on marriage into practice failed several times, even when the legislative framework for doing so existed. This, for instance, was the case with a 1937 law passed during the dictatorship of Ioannis Metaxas on combating trachoma and 'hereditary syphilis'.[92] By virtue of this law the minister was authorized to demand premarital health certificates from the inhabitants of areas considered to be at risk of trachoma and syphilis. In fact, this legislation was not applied.[93] The most illustrative example regarding the inconsistent implementation of legislation was the 1920 law that banned marriage for those suffering from Hansen disease. The leprosarium on the island of Spinalonga near Crete[94] was exempt from this prohibition, with the Autocephalous Church of Crete taking responsibility. In 1935 the Autocephalous Church, yielding to pressure from the internees of the leprosarium and in order to prevent cohabitation, decided to allow and bless marriages in the leprosarium

90 Information according to the professor of law Panagiotis Zepos (1908–85) who reported on such efforts by the state during a 1965 public debate of the Greek Society of Eugenics and Human Genetics [*Κοινωνική Επιθεώρησις. Μηνιαίον Όργανον Ενημερώσεως του Υπουργείου Κοινωνικών Υπηρεσιών. Πρακτικά της Ελληνικής Εταιρείας Ευγονικής και Γενετικής του Ανθρώπου*] 4, no. 9–12 (Sept.–Dec. 1976): 5–29, 17–20.

91 According to Panagiotis Zepos (1908–85) during the 1965 public debate of the Greek Society of Eugenics and Human Genetics.

92 Law No. 651/1937 on Combat of Trachoma and Hereditary Syphilis [N. 651/1937, "Περί καταπολεμήσεως των τραχωμάτων και της κληρονομικής συφιλίδος"].

93 According to the jurists Panagitis Zepos and Faidon Vegleris in the public debate on premarital health certification at the Greek Society of Eugenics and Human Genetics [*Κοινωνική Επιθεώρησις. Μηνιαίον Όργανον Ενημερώσεως του Υπουργείου Κοινωνικών Υπηρεσιών*] 4, no. 9–12 (Sept.–Dec. 1976): 5–29, esp. 20, 23.

94 The leprosarium on the island of Spinalonga was founded in 1903, in the period of the autonomous Cretan state (1896–1913), and was abolished in 1957.

(Savvakis 2008: 138). The legislative restrictions on marriages were abrogated in Crete, as in the whole country, in 1955.[95]

For all intents and purposes, the Greek state repeatedly failed to establish eugenic birth control on a large scale despite the claims made by physicians, scholars, politicians and social agencies. Infrastructural deficiencies—usually referred to in the Greek eugenic discourse as 'backwardness'—do not provide a sufficient explanation for this outcome when considered in isolation. One could even reverse the position by arguing that a national eugenic project would be a step towards modernization and counteracting backwardness, even supporting infrastructural development. This was exactly the line of argument pursued by several Greek eugenicists who justified their demands for regulating reproduction by eugenic means as a challenge for promoting welfare institutions and overall progress in the country; by exterminating 'unfit individuals', eugenic state politics would promote the welfare system and advance society's fittest members. The adherents of eugenic views that found entrance into state and social institutions were mainly those of a liberal persuasion; however, proponents of the German model of racial hygiene were not absent, although they were rather marginal in the eugenic circles and much more so in Greek academia as a whole.

8.6 *Greek Racial Hygienists and the Transient Appearance of Stavros Zurukzoglu*

Racial hygiene was a specifically German model of eugenics whose denomination and first conceptualization go back to Alfred Ploetz (1860–1940). In the German-speaking countries the eugenic idea was outlined most clearly at the end of the 19th century by Ploetz's programmatic book *The Capability of Our Race and the Protection of Weak Individuals* [*Die Tüchtigkeit unserer Rasse and der Schutz der Schwachen*] which appeared in 1895. In the book Ploetz conceptualized a 'racial hygienic model' of eugenics at the core of which stood the universal concept of race, in contrast to the predominately nation-oriented eugenic concepts of Great Britain and other countries. In the subsequent period, the German eugenics

[95] Marriage restrictions for lepers were abrogated in 1955 by virtue of the law 3369/1955 on Measures for Combating Leprosy, *Official Governmental Gazette* [ΦΕΚ] No. 258 A', 23 September 1955 [ΝΔ 3369/1955 "Περί μέτρων καταπολέμησης της Λέπρας", ΦΕΚ 258 Α' 23/9/1955].

movement pursued what Paul Weindling calls a *Sonderweg*.[96] Anthropology and race theory were instrumental to the conceptualization of this eugenic current and its particular association with the questions of population and hygiene. "For the composite term *Rassenhygiene* first involved the component of 'race' as the intersection of anthropology with varieties of race theory: physical, psychological and cultural. The second part of this double-barreled term, 'hygiene', primarily referred to the science of public health and prevention of pathogens" (Weindling 2011: 37). The crucial element that characterized the peculiar feature of German racial hygiene in comparison to other eugenic currents, at least during the National Socialist era, was that it became the ideological basis of the racial state (Burleigh and Wipperman 1991); that is, a state type based on biological totalitarianism whose ideological fundaments were provided by scholars. Further features of the distinct character of racial hygiene were its close association with the expansionist plans toward the East, the intention to encompass German *völkisch* elements and the intersection of ethnicity, disease and social welfare. These issues stood on the top of the agendas of the adherents of racial hygiene in German-speaking countries as well as their recipients elsewhere. Although starting from the same point of departure, racial hygienists in Germany, Central and South-Eastern Europe had developed varied conceptions for racial hygiene but also for race (Weindling 1989 and 2011). With respect to its perceptions, Paul Weindling asserts that "German racial hygiene was a forceful presence: as a model, as a collaborative enterprise, and as an expansionist politics. What is emerging is a fuller understanding of health, welfare and population issues and how these intersected with larger state health and welfare organizations" (Weindling 2011: 54).

In Greece, racial hygienic conceptions were rare; the term itself was hardly applied by the adherents of eugenics at all. Racial hygiene was probably embraced by only a handful of eugenicists. The few exceptions included the anthropologist Ioannis Koumaris, the chair-holder of hygiene at the University of Athens, Kontantinos Moutousis, and the medical statistician, bacteriologist and specialist on social and racial hygiene of European stature, Stavros Zurukzoglu. All of these individuals had an educational background in Germany; the latter made a career in Switzerland and his appearance in Greece was ephemeral. But each of them eventually came to interpret both racial hygiene and race in different ways. Ioannis

96 Weindling (1989). On Austria, cf. Baader, Hofer and Mayer (2008).

Koumaris set racial hygiene at the top of his anthropological programme, but his ideas were out of step with the mainstream eugenic discourse of his contemporaries. A comparison between the racial hygienic proclamations by Koumaris, as depicted in the previous chapter, with the demands of the mainstream eugenicists demonstrates that, while the latter were concerned with how to modernize society through the use of social welfare institutions and control over the population, Koumaris's main preoccupation was the avoidance of miscegenation that would be caused by emigration. He also represented Greece as a racial state, and translated any legislative measures on restrictions to marriage into racial state politics. The goal that he believed would be achieved by this was the advancement of the Greek race. In contrast to Koumaris's ontologization of race and the metaphorical usage of hygiene in the sense of cleanliness, Stavros Zurukzoglu defined racial hygiene against a strong theoretical background but also advocated (and greatly contributed to) the implementation of eugenic measures, and sterilization in particular, in a modernist social framework (Weindling 2011: 36). His definition of race was also strongly contextualized in the racial hygienic discourse, drawing in particular on Alfred Ploetz.

Despite the prominent position he had held in Switzerland, Stavros Zurukzoglu's influence upon the Greek eugenic movement was almost entirely irrelevant and his presence in the Greek scientific scene was transient, although it did not go unnoticed. He delivered an extensive introduction to eugenics at the first meeting of the Greek Anthropological Society in 1925,[97] and in the same year he also published some book reviews in the Greek medical press. Even these few statements demonstrate a quite different overview of eugenic issues than those of the mainstream Greek eugenicists. More importantly, it was obviously the view not of a recipient of eugenic ideas but of a theoretician who had formed the eugenic discourse in the racial hygienic fashion. When he delivered his lecture "On Eugenics" to the anthropological society he was at the beginning of his career in Switzerland. At that time he was based in Athens, where he stayed for a year from autumn 1924 in order to conduct research on the Mediterranean countries.[98]

---

[97] Zurukzoglu, *PGAS* (11 May 1925): 14–40. A further paper was presented by Zurukzoglu some years later at the anthropological society (Zurukzoglu himself was absent at that session and Koumaris read the paper): Zurukzoglu, *PGAS* (3 December 1929): 80–4.

[98] In Ziegler (2002).

Zurukzoglu was educated and had his first professional experience in Germany and Switzerland. After studying medicine in Berlin, Berne, Munich and Geneva, he earned his doctoral degree in 1921 in Berne, with a dissertation on bacteriology and hygiene. He worked in Munich (from 1922 to 1924) with leading scientists such as Max von Gruber (1853–1927), the professor of hygiene and bacteriology, and Ignaz Kaup (1870–1944), professor of social hygiene, eugenics, genetics and medical statistics. Six years later he acquired his *venia legendi* at the Bacteriology Institute of the University in Berne.[99] In Berne he was associated with the Institute for the Study of Contagious Diseases [Institut zur Erforschung der Infektionskrankheiten] which was headed by Georg Sobernheim (1865–1963), the professor of hygiene and bacteriology. From 1927 onwards Zurukzoglu was Sobernheim's first assistant, and conducted experimental studies at his institute as well as at the pharmacological and dermatological institute. His research focused on eugenics, population studies, alcoholism and criminal biology. Zurukzoglu became an important figure in the scientific scene in Switzerland where he was also engaged as political adviser.

His year-long stay in Athens (1924–5) coincided with the founding of the Greek Anthropological Society, which he immediately joined. The main part of his lecture at the first meeting of the anthropological society in Athens was extracted from his German monograph on *Biological Problems of the Racial Hygiene and the Civilized Peoples* [*Biologische Probleme der Rassenhygiene und die Kulturvölker*] published the same year in Munich (Zurukzoglu 1925a). This book did not go unnoticed in Greece. It was reviewed in the medical press by Athanasios Tsakalotos (1885–1956), a physician who was noted for his studies on Lamarckism and the history of science.[100] On the occasion of the book presentation the reviewer also referred to Zurukzoglu's lecture at the Greek Anthropological Society[101] and underlined some reservations that had been expressed in the discussion following the lecture. Indeed, the anatomist Georgios Sklavounos (siding openly with the Lamarckists) assumed that, in opposition to Zurukzoglu's assertion, acquired traits could in fact be inherited.[102] This

---

[99] See the biographical data on Zurukzoglu in Fischer (1933): 1732. His professional career is the subject of Ziegler's lecture (2002).

[100] See review in Tsakalotos (1926). Tsakalotos's publications included an early treatise on Jean-Baptiste de Lamarck (1744–1829) and his agency: Tsakalotos (1909/1911); (1931).

[101] Parts of this lecture were published in three instalments in the medical periodical *Medicine* [*Ιατρική*]: Zurukzoglu (1925b); (1925c); (1925d).

[102] *PGAS* (11 May 1925), discussion: pp. 40–3. See more on Zurukzoglu's criticism of Lamarckism in Zurukzoglu (1925b): 30–3. Furthermore Sklavounos wanted to encourage

partial dissent did not detract from the importance Tsakalotos ascribed to Zurukzoglu's book. Given the rising interest and the "recent remarkable activities in eugenics and social hygiene" by Greek scholars, Tsakalatos argued that "a translated publication of such a brief and clear-sighted treatise [which] bears witness to the expertise of the author would be welcome in our country" (Tsakalotos 1926). Indeed, Zurukzoglu's expertise was clearly revealed not only in the book but also in his lecture at the Greek Anthropological Society. His presentation, an extensive depiction of the scope and tenets of eugenics, was in tune with the newest developments in the field in addressing the major questions and controversial points in the contemporary discussion.

In fact, Zurukzoglu's approach differed on several points from the prevalent eugenic trend in Greece. First, he used race, 'vital race' [*Vitalrasse*] in particular, as his central operative notion. A further difference in his approach, deriving from the former, was that he disapproved of the idea of *degeneration*, and instead introduced *regeneration* as a central facet of his eugenic concept. With this preference he essentially opposed both evolutionism and cultural pessimism. He regarded eugenics as an autonomous science, emancipated from its original connection with Darwinism and evolutionism. Zurukzoglu was convinced that the absolute adjustment of an organism to its natural environment was impossible—evolution had its limits. This presumption smoothed the way for the intervention into human evolution by the state and its policies, the latter of which should be based on scientific knowledge. In contrast to the general spirit of decadence that penetrated his contemporary intellectual circles and the eugenic movement in particular, Zurukzoglu embraced the more optimistic perspective of regeneration. In doing so, he appeared to stand close to an ideology of progressivism and to share a belief in the perfectibility of the human race. Accordingly, degeneracy could eventually be permanently overcome.[103] This conviction and the notion of the 'vital race' formed the basis for his eugenic, and clearly racial hygienic, conception. The idea of the 'vital race' had been first elaborated by Alfred Ploetz (1895) to describe the 'sound parts' of a human race, encompassing the carriers of the proper qualities of this race that should therefore be promoted and cultivated. In the 20th century the concept of 'vital race' was

Zurukzoglu to develop his approach, while he also urged him to admit that Weismans's theory was in the meantime obsolete (discussion pp. 40–3). He might as well have 'carried coals to Newcastle', however, since Zurukzoglu was critical of this theory.

[103] For the impact of this idea on Jewish eugenics, see Presner (2007).

embraced and cultivated by German racial scientists during the period of National Socialism and at the Kaiser-Wilhelm-Gesellschaft in particular (Schmuhl 2003: 29).

Zurukzoglu specified the main task of racial hygiene as being "the study of the conditions for the maintenance of the proper flow of a 'vital race' and furthermore the best possible formation of its cultural abilities" (Zurukzoglu 1925a: 141). Against this background he distinguished between two types of degeneracy: hereditary and cultural degeneracy. The latter (and most important) type emerged insofar as a race became incapable of producing culture due to the lack of a leading stratum. From his point of view a 'vital race' was threatened once its ability to reproduce itself culturally had disappeared, long before it degenerated physically. The *diminution* of cultural ability could result in hereditary degeneracy (Zurukzoglu 1925a: 109). Zurukzoglu translated the importance of elites for the perpetuation of an entity ('nation' in the Greek lecture, 'Volk' in the German monograph) into racial terms, arguing that, in productive cultural nations, the appearance of a temporary deficit in 'superior intellects' (apparently meaning the lack of elites) did not necessarily cause degeneracy but produced temporary decimation. At the bottom of his approach was essentially a nominative culture and class; nevertheless, the term he proposed for describing the phenomenon of degeneracy was borrowed from biology. He suggested introducing the term *Geneoktonie* (in the German-language book; *γενεοκτονία* in the Greek lecture). This term was derived from the Greek word *γενεά* (meaning *Geschlecht*, generation, kin) and the infix *-κτονία* (meaning murder/homicide). Thus, *Geneoktonie* means "the murder of a generation".[104] In his lecture at the Greek Anthropological Society he defined *Geneoktonie* as the "elimination of a generation (or an important family) before achieving its optimal reproduction".[105] *Geneoktonie* might be related terminologically to *genocide* (*γενοκτονία*), which was introduced into the Greek language prior to its establishment in the international discourse after World War II. Objectively, this term was applied in Greece at the beginning of the 20th century to describe the persecution of Armenians in Turkey.

The core idea behind *regeneration*, which was essential to his racial hygienic approach, was that even though the emergence of talents and geniuses might be delayed for a period, the ability of a populace to

104 Zurukzoglu (1925a): 113; cf. Zurukzoglu, *PGAS* (11 May 1925): 35.

105 Zurukzoglu, *PGAS* (11 May 1925): 35.

regenerate was not terminated. Thus this populace would not necessarily disappear. Once the elites were recreated, they could successfully regenerate and the threat of '*Geneoktonie*' would be avoided. Zurukzoglu believed that elites should be recruited from all social classes. With this assertion he differed from mainstream eugenicists generally and Greeks in particular, the majority of whom advocated the promotion of the upper and middle social classes. Indeed, in his German-language book, Zurukzoglu openly opposed this "aristocratic current among racial hygienists" and pleaded for the advancement of the best elements of society, comprising all social classes (Zurukzoglu 1925a: 107). A distinction between inferior classes of workers and peasants on the one hand, and superior upper classes on the other would lead to "the lumpenproletarianization of the population" (Zurukzoglu 1925a: 108). It is worth mentioning that the same view (unusual for the Greek eugenic scene) was expressed a decade later by Konstantinos Moutousis. In his previously mentioned inaugural lecture in 1933 as chair for hygiene at the University of Athens, Moutousis underlined the importance of the workers and moreover the peasants, together with the upper classes, for the eugenic improvement of the Greek *fili* (Moutousis 1933).

The idea of regeneration was a method for coping with the social and population question. Certainly, Zurukzoglu and eugenicists generally saw themselves faced with the challenge of managing the contradictions that arose out of the intention to control and rationalize the reproduction of the population while maintaining both its optimal size and quality. Zurukzoglu argued that if the upper classes alone were promoted then their possible '*Geneoktonie*' would have fatal implications for "cultural ability" and thus the existence of the race as a whole, all the more so since the upper social classes, considering their "social endogamy", had a greater propensity for biological degeneracy than those in the lower strata (Zurukzoglu 1925a: 169). Furthermore, he argued, the upper classes were prone to mental diseases, as was typical for gifted individuals and geniuses more generally. On the other hand, "fully-valuable working and cultural types" could be found among members of the working classes who live in favourable conditions (Zurukzoglu 1925a: 109). On the basis of pedigrees, Zurukzoglu argued that it was possible to find among peasant families a good number of "valuable individuals" who had achieved social advancement. Even if such families would temporarily cease to bring forth distinguished individuals, they could still do so in the future. Zurukzoglu was thus a proponent of marriages between members of different social strata. Regeneration, nevertheless, should not be left to chance. It should be controlled and managed in focused efforts and practical eugenic measures in order to advance culturally valuable elements.

References to Greece were rare in his approach. Only in lectures at the Greek Anthropological Society and in the published versions of these did he make use of his racial hygienic theory in order to take a stand on racial views that applied specifically to the Greek case. His regeneration thesis was instrumental in rejecting the theories of Gobineau and his adherents, according to which ancient Hellenic civilization had decayed as soon as the blond race had disappeared. He countered this model with the argument that a 'nation' was able to regenerate by producing leading personalities and strata, geniuses and intellectuals ('superior intellects') as soon as conditions allowed for doing so.[106] In contrast to Gobineau's assumption of the superiority of the blond race and its alleged exclusivity in creating culture and civilization, Zurukzoglu argued that every advanced race was able to develop culture, and that primitive races (as ancestral Germanic and Hellenic peoples once were) would be able to do so in the future. Besides, he argued, the blond race was well represented in Greece—to the same degree that it was present in southern Germany, central France and Italy.

In his lecture to the Greek Anthropological Society specific references to the Greek case were confined to the above-mentioned passage on Gobineau and a further remark that eugenic measures were necessary in Greece. The major part of this presentation addressed the subject of applied eugenic politics in general. With respect to Greece he complained about the lack of interest in eugenics by both the state and scholars. Suggestions on concrete measures to be applied in Greece were formulated in a popularized version of the same lecture.[107] This version addressed practical eugenic measures rather than theoretical racial issues, and included an additional section entitled "Practical Applications" (Zurukzoglu 1925g: 266–7). Here he presented detailed proposals for implementing eugenic measures in Greece in order to combat epidemics, endemic diseases and infant mortality. Although he insisted on the distinction between general hygiene (the care for the individual) and eugenics (care for the commonality) and sided with the latter, he suggested a combination of individual hygiene and eugenics as a compromise for the present. He advocated granting individual care after evaluating each individual and ascertaining their quality.

106 Zurukzoglu, *PGAS* (11 May 1925): 37.

107 See Zurukzoglu (1925e); (1925f); (1925g).

The practical measures suggested in this text included the internment of "damaging individuals" in special institutions where "they could be treated in a productive way" (Zurukzoglu 1925g: 266). He also took a stand on controversial questions of his time such as sterilization, control over childbearing and the banning of marriage for eugenic reasons. Although Zurukzoglu basically favoured the sterilization of "hereditarily degenerate individuals", he opposed forced sterilization, which he considered unrealistic for "scientific as well as social reasons" (Zurukzoglu 1925g: 266). At that time he called for voluntary sterilization on condition that either the persons concerned or their relatives agreed with the procedure. With respect to marriage between 'degenerate individuals', he was still inclined, in 1925, to approve voluntary abstinence from childbearing rather than the prohibition of marriage. In his monograph he was explicit in expressing his disagreement with the suggestion made by the Society for Racial Hygiene [Gesellschaft für Rassenhygiene] in Berlin for establishing compulsory prenuptial health certification. Instead, he regarded educational campaigns, such as providing bridegrooms with appropriate brochures, to be more expedient, at least at that point in time (Zurukzoglu 1925a: 155). More generally, he attached a great deal of importance to popular education. He approved of the promotion of social hygiene by raising awareness about its importance among the working class, who were the most vulnerable to venereal disease, tuberculosis, malaria and alcoholism.

In contrast to the Greek psychiatrist Simonidis Vlavianos, who considered alcoholism a major threat to the Greek society,[108] Zurukzoglu claimed that alcoholism had not yet reached Greece (Zurukzoglu 1925g: 267). Alcohol consumption in Switzerland became an area of expertise for him, especially after 1927 at which time he documented the so-called 'alcohol question' in Switzerland by order of Jean-Marie Musy (1876–1952), the head of the Swiss Federal Council [Bundesrat]. The documentation was to serve as the basis for new alcohol legislation. In the following period he also edited a series on the alcohol question in Switzerland.[109]

---

[108] Several years later, the psychiatrist Simonidis Vlavianos raised alcohol as a central issue, as argued in the *Medical Newspaper* [*Ιατρική Εφημερίς*]: see for instance, Vlavianos (1931); (1934). On Simonidis Vlavianos's contribution to the development of psychiatry in Greece, see Kazolea-Tavoulari (2002): 93.

[109] See Zurukzoglu (1935). In the introduction (p. 8) it is mentioned that he contributed to the drawing up of the law banning the consumption of alcohol.

Zurukzoglu's involvement in the Swiss scientific and political scene was decisive for his attitude towards eugenic measures. In the 1930s his position hardened, especially after the passing of the sterilization laws in the German Reich. In a volume that critically examined and evaluated the notorious German law on the prevention of genetically diseased offspring, Zurukzoglu, as editor and the main contributor, examined the extent to which a certain German law was applicable in Switzerland (Zurukzoglu 1938). While he had earlier advanced voluntary sterilization after consulting with the individual concerned or their family, he now approved enforced sterilization "when a family history would lead to the conclusion that progeny of low-value could be anticipated with the utmost probability" (Zurukzoglu 1938: 46).

In the case of Switzerland, in order to safeguard valuable offspring he first of all recommended restrictions on sexuality and individual freedom, for instance internment in asylums, a ban on marriages, and so on. He suggested that each individual case needed to be examined separately in order to choose the most appropriate and effective measures to cope with it. Prevention instead of sterilization could serve as an alternative treatment for those suffering from hereditary diseases, provided that the patients were conscious of their deficits and cooperative with the authorities (Zurukzoglu 1938: 54–5). His view bore the deep imprint of the Swiss system of control over socially deviant individuals in which charity played a crucial role. Recent research has brought to light the fact that sterilization was also applied in Switzerland in the aftermath of World War II, and that it was even supported by the charity system and practised in charity institutions (Hauss and Ziegler 2009; Ritter 2009). What is particularly impressive in Zurukzoglu's 1938 edited volume is the totalitarian perspective on the organization of society by the contributors: individuals are caught up in a control system in which doctors, statisticians, bureaucrats, asylums, social workers and accompanying social charity associations collaborate in estimating the value of individuals and treating them accordingly. Zurukzoglu's own contribution to this volume is an especially insightful source of information on eugenic measures, in particular sterilization, that were implemented in his time, not only in Switzerland but worldwide. Notably, there is no mention of Greece.

It is possible that Zurukzoglu's interest in and impact upon the eugenic movement in Greece was marginal. It is just as likely that he did not neglect his sympathizers in the Greek academic arena, given that he was proposed as chair for public and social hygiene at the University in Athens in 1939.[110] Zurukzoglu did not accept the candidature, probably because of his ambitions in Switzerland. Indeed, only a few months previously, in 1938, the Swiss Ministry of the Interior had appointed him director of the Department of Statistics in recognition of his contribution to the legislation on the ban on alcohol. Despite his position as political adviser, Zurukzoglu was anxious to develop his university career in Switzerland as well. However, he never fully achieved this goal.[111] The historian Béatrice Ziegler interprets Zurukzoglu's unsuccessful academic career as the result of an overall failure to establish social hygiene and eugenics as a teaching discipline at the medical faculty of the university in Berne.[112] Nevertheless, this does not mean that eugenics was rejected at this university. On the contrary, Ziegler concluded that eugenic research and teaching had met with tacit acceptance. Zurukzoglu continued to research and publish his racial and eugenic views even in the post-war period; his research also continued to be tightly linked with politics. The topics he dealt with, along with racial issues, concerned addiction, alcoholism in particular, and the system of health insurance.[113] Zurukzoglu's racial hygienic concepts were clearly linked with the Swiss social problematic and were interpreted appropriately by state and societal agencies. In this way his work became relevant for policy-makers in Switzerland.

A possible acceptance of the candidacy for the chair of hygiene at the University of Athens by Zurukzoglu would likely mean the introduction of racial hygiene in Greek academia. However, this opportunity was instead given to the adherents of the German version of eugenics, Ioannis Koumaris and Konstantinos Moutousis, and also ended unsuccessfully; in the case of Koumaris because of the limited scope of his teaching on physical anthropology with a focus on palaeontology, whereas in the case of Moutousis this was due to professional antagonism at the faculty. As chairholder for public and social hygiene at the University of Athens (in 1939), Moutousis was anxious to establish a laboratory for hygiene that would

---

110 In: Historical Archives of the Athenian University, *Minutes of the Meetings of the Professors of the Medical Faculty*, 20 October 1938–3 October 1939, vol. 18, pp. 102–19, 117 [Συνεδρία Ιατρικής Σχολής, Σάββατον 4 Μαρτιού 1939, Πρακτικά Συνεδριάσεων καθηγητών Ιατρικής Σχολής, 20/10/1938–3/8/1939, Τμ. 18, σ. 102–119].

111 Stavros Zurukzoglu started his unsuccessful attempt to acquire a position at the university in Berne in 1932, according to Ziegler (2002) on the basis of his personnel file at the University of Berne.

112 Ziegler (2002).

113 Zurukzoglu (1956a); (1956b); (1963).

be integrated with his chair.[114] This laboratory was designed to conduct research on all topics related to social hygiene, including eugenics and racial hygiene. As models for his project he acknowledged the German institutes for hygiene and racial hygiene which he admitted having visited and with whose research he had become familiar. The laboratory to be established would, like the German model, contribute to keeping the field of social hygiene at the Athenian university at the cutting edge of affairs.[115] Social hygiene in Germany encompassed theoretical and empirical research related to issues of social disease, anthropometry and racial measures.[116]

The main problem involved in implementing the project appears to have been the fact that there were two chairs for hygiene in the same medical faculty: one for public and social hygiene headed by Moutousis, and another for hygiene, epidemiology and medical statistics held by Professor Alivizatos. Due to the similarity of subjects, there was a certain rivalry between the two chair-holders. The proceedings of the meetings of the medical faculty reveal that the points of contention included questions as to whether the epistemic field of social hygiene was merely theoretical or if it should encompass experimental and empirical research that would require a full laboratory, as Moutousis proposed. At a meeting of the medical faculty,[117] the epistemological and empirical content of hygiene was specified with respect to its relationship to hygiene, public or/and social hygiene, statistics, social disease and eugenics. However, the most important source of contention seemed to concern the competition between the two professors and chair-holders, rather than epistemological questions. Finally, in German academia, which was Moutousis's

114 The project and its main targets were outlined at a meeting of the medical faculty of the University of Athens.

Meeting of the Medical Faculty on 19 May 1939: Historical Archives of the Athenian University, *Minutes of the Meetings of the Professors of the Medical Faculty* (20 October 1938–3 October 1939), vol. 18, pp. 223–64.

115 Meeting of the Medical Faculty on 19 May 1939: Historical Archives of the Athenian University, *Minutes of the Meetings of the Professors of the Medical Faculty* (20 October 1938–3 October 1939), vol. 18, pp. 223–64.

116 The medical faculty supported the establishment of this laboratory. The present study did not follow the activities of the laboratory. Meeting of the Medical Faculty on 19 May 1939: Historical Archives of the Athenian University, *Minutes of the Meetings of the Professors of the Medical Faculty* (20 October 1938–3 October 1939), vol. 18, pp. 223–64.

117 Meeting of the Medical Faculty on 19 May 1939: Historical Archives of the Athenian University, *Minutes of the Meetings of the Professors of the Medical Faculty* (20 October 1938–3 October 1939), vol. 18, pp. 223–64 [Συνεδρία Ιατρικής Σχολής, 19 Μαΐου 1939, Πρακτικά Συνεδριάσεων καθηγητών Ιατρικής Σχολής, 20/10/1938–3/8/1939, Τμ. 18, σ. 223–64].

archetype, demography and statistics were also sciences allied to social hygiene at that time (Ferdinand 2009). The competition concerned issues related to authorization over the head of institutions such as the Museum of Hygiene, or to shaping the content of the field of hygiene.

Without finding an entrance into the university and having gained only a few adherents in Greece, racial hygiene ultimately remained a marginal current in the Greek eugenic movement. The collaboration between the proponents of racial hygiene seemed to be weak and somewhat infrequent.[118] Similarly marginal was the rank of racial hygienic proposals to the then contemporary concerns of Greek society, both prior to and after World War II.

8.7 *Beyond and against Eugenics: Still a Subjective Decision*

The context in which bio-power and the hygienic discourse emerged and operated in the 19th and especially the 20th century was certainly pivotal to the rise of eugenics both generally and in Greece in particular. However, the eugenic phenomenon per se was not an unavoidable result of this context, much less an inevitable phenomenon. After all, not only adherents but also opponents, critics and sceptics of eugenics existed in the same context and, eventually, there was a large majority who thought and acted beyond eugenic frames of reference. A coincidence of favourable conditions facilitated the rise of eugenics, but this does not absolve certain actors of having been responsible for their choice. Peter Weingart's conclusion that eugenicists during the National Socialist era "were not forced into their positions, censored, or otherwise abused by a totalitarian regime" (Weingart 1999: 165) is all the more applicable in non-totalitarian regimes such as Greece during the inter-war period when eugenics flourished. Finally, with the exception of the German racial state, the most fervent adherents of eugenics were liberals—the Scandinavian countries being a shining example in this respect. Likewise, in Greece,[119] the most influential currents of eugenics were politically liberal. Generally

[118] It is not without interest that the chair-holder of physical anthropology, Ioannis Koumaris, was present at that meeting at the medical faculty; he did not, however, take a stand in the discussion.

[119] Research focusing on eugenics, especially during the dictatorial regime of Ioannis Metaxas (1936–41), has not yet been carried out. To date my own studies give no indication that the eugenic movement intensified during that period, or that it had any specifically fascist character.

speaking, liberal eugenic models proved to be more durable and effective in their targets and their implementation than models linked with totalitarian regimes.

The challenge faced by actors in both totalitarian and liberal political systems was that they were provided (the former obviously to a lesser degree than the latter) with a choice between diverse options, and that they had to carry the consequences of their decisions. In turn, being aware of the possible consequences and deciding to take their chances was vital in mobilizing or inhibiting action.[120] Opponents of eugenics were well aware that their attitudes were potentially dangerous to their careers and networking, first and foremost because of the risk of professional marginalization. Since eugenics had represented a vehicle for the modernization of society (Turda 2010), especially before World War II, those who opposed eugenics risked being characterized as outdated. Inherent to the process of modernization advanced by those eugenicists was the emergence of new professional groups, whose members endeavoured to establish themselves as authorities in managing developments in society and making an impact upon state policy. If in the 19th century evolutionism had broken away from the fatalism of religion in favour of rationality, eugenics carried the mark of progress, and of a kind of evolution that was not left to nature but became subject to calculation by human handling through scientific, political and administrative means. In this process, eugenics was acting like many other ideologies that blame the victim and which are:

> developed through a process that shows every sign of being valid scholarship, complete with tables of numbers, copious footnotes, and scientific terminology. Ideologies are quite often academically and socially respectable and in many instances hold positions of exclusive validity, so that disagreement is considered unrespectable or radical and risks being labeled as irresponsible, unenlightened, or trashy (Ryan 1976: 22).

Certainly, modernization, illness, poverty and social calamity were not interpreted exclusively through the lens of eugenics. The presentation of the following cases is intended to sharpen the contours of the diverse and contradictory tendencies that formed the landscape in which eugenic projects emerged or were communicated in the Greek case.

---

120 This quality distinguishes human from non-human agency and, as suggested by Actor-Network-Theory, evidences the need to re-examine theoretical concepts of agency. Non-human agency (if it can even be properly denominated as 'agency' at all) is not optional but is a possible result of the concurrence of diverse factors.

Opposition to eugenics was often uttered in a far quieter voice than approval. Disagreement with the introduction of health examinations by law was, for instance, expressed by the neurologist and psychiatrist Konstantinos Papastratigakis (1891–1929)[121] in his 1925 review of a book entitled *On the Frequency of Psychopathic Phenomena in Greece*.[122] The author of this book, Michail Gianniris, was a psychiatrist and director of the psychiatric asylum Dromokaiteion [Δρομοκαΐτειον] in Athens. The reviewer criticized Gianniris who, among other measures, had pleaded for the prevention of psychopathic phenomena by legally enforcing psychiatric examinations in order to verify the mental and physical health of couples prior to marriage. From the reviewer's point of view, the implementation of such measures was "absolutely impossible".

A prominent opponent of eugenic control over reproduction was the bacteriologist, hygienist and professor for 'Social Biology' at the Panteion Institute in Athens, Dimosthenis Eleftheriadis (1885–1964).[123] Besides his academic career, Eleftheriadis was also for a time a sanitary inspector for the Department of Hygiene of the Ministry of the Interior, and for a short period the secretary general in the Ministry of Hygiene (March–August 1936).[124] Eleftheriadis contributed a chapter on "Heredity in the Family" to the volume on the *Hygiene of Marriage*, published by the National Council of Greek Women (Eleftheriadis 1933). Here he expressed his moderate but explicit opposition to the "obsession with hygiene and the healthy body culture" that characterized the inter-war period. In his own words: "In recent years there has been a great deal of discussion about the improvement of humankind by means of gymnastics or hygiene. Unfortunately, all of them have no relation to reality" (Eleftheriadis 1933: 84). Although he did not question the importance of heredity upon individuals, family and society, he did express some moderate scepticism regarding the effectiveness of control over childbearing by means of premarital medical

---

[121] Konstantinos Papastratigakis (1891–1929) studied in Athens and France; he taught psychiatry at the Military Academy for Hygiene [Ανωτέρα Σχολή εκπαιδεύσεως υγειονονομικών αξιωματικών]. In 1927, he became professor for psychiatry at the University of Athens. His publications included Papastratigakis (1915); (1929).

[122] The review was published in *Medicine* [*Ιατρική*] (1 May 1925): 33–4.

[123] Dimosthenis Eleftheriadis (1885–1964) studied medicine in Greece and Germany, specializing in hygiene and bacteriology.

[124] This was during the conservative governments of Konstantinos Demertzis (1876–1936) and Ioannis Metaxas (before the dictatorship of 4 August 1936). Konstantinos Demertzis (1876–1936) died in April 1936, one month after his appointment as prime minister, and was succeeded by Ioannis Metaxas. Dimosthenis Elefteriadis was deposed from office at the Ministry of Hygiene after Ioannis Metaxas's putsch on 4 August 1936.

examinations. Drawing arguments from biology, he claimed that medical examinations were unable to expose hereditary pathological features when these only existed in a latent form. Moreover, "by means of athletics and hygiene we may improve to some extent the phenotype but not the genotype of the people" (Eleftheriadis 1933: 84–5). Eleftheriadis later radicalized this critical position. Immediately after World War II he spoke out emphatically against birth control by eugenic means. An article in the Athenian press reported his view on medical certificates for bridegrooms as well as the question of degeneration (Stathatos 1948). In Eleftheriadis's opinion, hereditary diseases were not so pervasive that they could result in the degeneration of a whole society. He himself stood against all eugenic measures and characterized the prenuptial health certification especially as being:

> graver and more unrealistic than sterilization, not so much due to the difficulty of its implementation (about which I may make only the worst and most disgraceful predictions), but rather because neither are such measures necessary nor is science able to exterminate or even reduce hereditary diseases…Most of the proponents of prenuptial health certification are Neo-Malthusians masquerading as eugenicists. I do not need to discuss the economic, demographic or, above all, ethical consequences of such measures as I consider these self-evident (Stathatos 1948: 1).

In his textbook, *Lessons of Social Biology*, a subject he taught at the university, Eleftheriadis also opposed eugenic birth control, advising people to "distrust those who claim to desire [their] happiness" (Eleftheriadis 1948: 171). This warning was well founded considering the fact that proposals for handling the social issue and its side effects by eugenic means were too often justified as a commitment to the common good, as a form of humanitarian ideology. The eugenicists' claim to humanitarianism reflects the vicious cycle of ambivalences that characterizes modernity: removing and at the same time creating ambivalence (Bauman 1991). More specifically, it brings to light a basic ambivalence in modern humanitarianism: on the one hand, modern social welfare is based on the universal idea of equality; on the other, the provision of social welfare in reality occurs selectively. This contradiction is fundamental to modern forms of social inequality; nevertheless, granting social welfare either selectively or universally is still a decision made by actors. It also depends, among other factors, on which relationship with the state and its institutions is ascribed to individuals or collective subjects in need; they are either objects to be governed by state politics or are social actors and citizens to be protected by state politics.

The latter consideration was the case of the venereologist Petros Apostolidis (1896–1988) who treated venereal disease beyond the scope of eugenic ideology and the discourse of 'social diseases'. Petros Apostolidis practised as a physician in Ioannina (in Epirus) from 1923 to 1967.[125] He was a socialist and was known as 'the red mayor of Ioannina', after serving in this office for a mere four months, from December 1944 to March 1945, the period during which Ioannina was under the control of the National Liberation Front (EAM). In his two-volume memoir Apostolidis describes his experience with patients who suffered from contagious diseases and castigates state institutions for their ignorance (Apostolidis 1981 and 1983; cf. Dalkavoukis/Karamitsos 2008). He wrote:

> Wage-earners, workers, villagers, young and stoutly-built people come to my medical practice. They suffer from syphilis, tuberculosis, and other contagious diseases. They need no more than a little aid to convalesce and return to production; these people are national capital. But they are poor and helpless (Apostolidis 1983: 153).

In several places in his memoir Apostolidis provides revealing observations about the dissemination of syphilis in the region in which he practised during the inter-war period. In a report submitted to the ministry he depicted in detail the situation and the medicine required for curing the patients. Indeed, the ministry sent him medicine and instructions to start with the therapy. However, he felt unable to do so due to the lack of the necessary infrastructural assistance:

> But how could I systematically administer a therapy to people who live in twenty-four different villages quite distant from one another when I possessed a single medical practice?...I was inwardly in agony lest, giving injections irregularly, I would inflame their sleeping illness instead of healing them. I depicted my fears in a report addressed to the Ministry. Did anybody ever read it? I doubt it (Apostolidis 1983: 134–5).

Like those who were involved in the eugenic discourse, Apostolidis ascribed a crucial role to state institutions in combating contagious diseases. He nevertheless differed from eugenicists in that, from his point of view, the responsibility of the state was not in the lack of legislation intended to eliminate or punish poor individuals and poor health, but the lack of state

[125] Daklavounis and Karamitsos (2008). The text is based on a paper presented at the conference in Athens and other cities in the 1940s which was held in Aegina (Greece), in June 2007.

care for their healing. Apostolidis underlined the importance of state care and his own powerlessness to help his patients due to the lack of state support: "What is the use if I or other doctors do not accept money for the consultation? The patients need medicine and nursing and they find them nowhere." And, he stated, "I side with them" (Apostolidis 1983: 153).

Apostolidis's lone commitment to socialism can hardly explain the fact that he considered venereal disease outside the eugenic rhetoric of 'social disease'. Socialist adherents of eugenic ideas were not a rarity; on the contrary, British Fabianists and several German and Austrian socialists had made striking contributions to eugenics (Niemann-Findeisen 2004; Baader 2008). With respect to Greece, the above-mentioned constitutionalist Alexandros Svolos was a resolute socialist, and simultaneously an advocate of eugenic sterilization and eugenics generally. Comparing the reasoning of these two Greek socialists, Svolos and Apostolidis, one can recognize the different priority that each attached to individual welfare vis-à-vis the common good. This difference is not least a matter of elitism that obviously characterizes Svolos's attitude but seems to be missing in Apostolidis's case. In eugenic discourse, the relationship between the individual and the community gradually became a central question. Sympathizers of eugenics (regardless of their political ideology) answered it in favour of the community, even though this meant discrimination against or even the elimination of individuals. This priority reveals the limits of modern ideologies committed to the liberation of the people. Furthermore, it shows that it is not only the ideology to which social actors are committed that accounts for their attitude to the value of human life; valuing human life is to a great degree a subjective choice. In the interwar period and in the shadow of the National Socialist mass murders, individual rights were accorded increasing importance in terms of political ethics. Yet it was also the growing significance of the individual during the rise of the post-modern era that served to alter the eugenic discourse after World War II, eventually leading to a reformulation of the questions raised by eugenicists.

8.8 *Under the Banner of Reformist Eugenics: The Greek Society of Eugenics*

In the post-war period eugenic associations gradually declined worldwide and had ceased to exist altogether by the end of the 1970s (Kühl 1997). But even up to their final abolition, the character of the eugenic agencies had changed once they had embraced a genetic discourse that corresponded to what is currently termed "genetic governmentality" (Lemke 2004). A central characteristic of the latter was the shift from the fetishization of community and the 'common good' (typical of the eugenic discourse) to the individualization of social risks, individual responsibility and self-technology. In any event, the decade of the 1950s still represented a phase of transition from a rather 'traditional' type of eugenic thinking to the burgeoning age of reform eugenics that was marked by progress in genetic technologies.

In these post-war developments, the first society for eugenics in Greece was founded in 1953 and its activities continued for at least two decades, until the end of the 1970s. Its founder was the gynaecologist Nikolaos Louros (1898–1986) who was also its president for twenty years, from 1953 to 1973. The transformation of the eugenic discourse and the increasing significance of genetics are reflected in the denomination of this agency. In its founding year it was named the Greek Society of Eugenics [Ελληνική Εταιρεία Ευγονικής] but in its very late phase it extended its name to the Greek Society of Eugenics and Human Genetics [Ελληνική Εταιρεία Ευγονικής και Γενετικής του Ανθρώπου].[126] The emergence of a eugenics society in Greece in the post-war period was facilitated first by the fact that eugenics did not have to be rehabilitated in this country since, despite all divergences of opinion and serious objections either to eugenics as a whole or to individual measures, it has never been completely discredited in Greece.[127]

This is scarcely surprising, since, as discussed above, prior to World War II there was a strong liberal current among the proponents of eugenic ideas in Greece. Generally, eugenics in Greece has been implicitly associated with a chapter of political and social history that is entrenched as a positive process of efforts to modernize society and state institutions. But the international environment was not completely negative to the establishment of a eugenic society in that period. Finally, eugenics was only discredited to any degree in the German fashion—indeed, the *Sonderweg* of racial hygiene. The reason

---

[126] In the third volume of the Minutes of the Public Debates it was called the Greek Society of Eugenics and Human Genetics. However, there is no doubt that it is the same agency.

[127] See for instance the private clinic in Athens that specialized in artificial insemination and is named 'Eugenics': http://www.eugonia.com.gr/content.php?cat=1&sub=0 (last accessed August 2010). See also the religious book on eugenics Mari (2004).

for this disapproval was the decay of the National Socialist regime in which racial hygiene flourished rather than the disapproval of eugenic control over the population generally and its misanthropic connotations. Illustrative of this fact is the example of the Danish geneticist and eugenicist Tage Kemp (1896–1964) who:

> continued to distinguish sharply between the "sober" eugenics and the "perverted" Rassenhygiene … In spite of the fact that the international eugenic movement had been a forum where German Nazi scientists had maintained friendly relations with scientists from the rest of the Western world, Kemp absolved the movement of all responsibility for the German development. Neither did he mention the close collegial relations that existed between Scandinavian and Nazi geneticists and anthropologists before the war broke out and which, in Denmark as in several other countries, were resumed shortly after the war [Koch 2002] (Koch 2009: 43).

As Koch remarks, for Kemp, as for many other geneticists, there was good eugenics and bad eugenics (Koch 2009: 43).

In several European countries eugenic agencies continued their working unabated after reforming their outlook.[128] Reformist eugenics societies embraced the newest achievements in human genetics, which sought to improve the chances of tracing genetic illnesses or deficiencies. Eugenic measures continued to be contentious, but they were also applied in several countries. Along with forced sterilization, the implementation of which was carried forward in several countries (including the United States, Norway, Finland, Switzerland, Slovakia, Hungary and even post-Nazi Germany),[129] prenatal genetic consultation surfaced to become a focus for eugenicists who continued their activities through networks operating internationally.[130] A choice of restrictions on childbearing formed the core demand of prenatal genetic examinations, as advanced by the reformist eugenicists. The focus of their arguments now addressed self-technology instead of enforcing measures imposed by the state;

---

128 On reformist eugenicists in the post-war period see Kühl (1997): 174f.

129 For the United States, see Bruinius (2006); for Denmark, Sweden, Norway and Finland, see Broberg and Roll-Hansen (1996); for the Federal Republic of Germany, see Sierck (1995): 45. For Switzerland, see Huonker (2002): *passim*. In Slovakia and Hungary, Roma women were systematically sterilized from 1970 to 1990. See http://www.errc.org/cms/upload/media/03/A9/m000003A9.pdf (last accessed November 2012).

130 Kühl (1997: 191–240) presents the new forms of networking among eugenicists in the aftermath of the Second World War. Although numerous eugenics societies continued their work, they were involved in activities initiated by institutions of human genetics or population sciences.

population politics faded from the spotlight in favour of family policy. The debate on a new eugenic technology has been fiercely contested since the 1980s (e.g. Lewontin, Rose and Kamin 1984; Rothman and Katz 1986; Kitcher 1996; Duster 1990; Lemke 2002). According to the historian Regina Wecker, prevention via prenatal examinations and impeding the birth of children who will probably have physical or mental disabilities is still thought of as being a decision between *life worth of living* versus *life unworthy of living* (Wecker 2009: 36).

Wecker is surely not alone in this conviction; however, opinions regarding eugenics differ remarkably. They range from radical rejection of prenatal controls, to implicit or conditional approval, to assertions that prenatal control can be both "pronatalist and simultaneously anti-eugenic", as Ruth Schwartz Cowan claims in her book on thalassaemia (Schwartz-Cowan 2008). The Danish historian Lene Koch labels the criticism addressed to eugenics as "the present's view of the past" (Koch 2009: 42), and claims that eugenics "is open to pejorative use because it is rarely, or only superficially, defined" (Koch 2009: 44). From a different point of view, however, in the aftermath of World War II:

> society does indeed experience new eugenics—or rather, the old eugenics in a new form, and it even uses the same term: "soft eugenics" is the expression used for techniques which promise the optimization of human nature, such as prenatal selection of characteristic traits or selective techniques to prolong life. It does not seem so very new if you consider the intention (Wecker 2009: 37).

The Greek Society of Eugenics stood close to the reformist eugenic current of the post-war period. This affinity radiates from the society's published material as well as from additional sources that provide an account of its manifold workings, which included research and the establishment of consulting agencies in hospitals. Consistent with prevailing trends in post-war eugenic movements, family planning was a major concern of the leading members of this society. This preference is revealed in the establishment of an experimental centre for premarital consultation that operated under the aegis of the eugenics society. This centre was integrated into the maternity hospital Alexandra [Αλεξάνδρα] in Athens, where Nikolaos Louros held a leading position from the 1950s onwards. An extensive report on the centre with respect to its founding, the ends it pursued and its activities was submitted by the theologian Alexandre M. Stavropoulos to the Belgian Catholic University of Louvain [Université Catholique de Louvain]. According to Stavropoulos the experimental centre opened on 22 February 1966 and closed on 5 July 1968.[131] Louros himself announced at a meeting of the Society of Eugenics that the centre functioned as long as he was director of the Alexandra Hospital but that it was abolished after his retirement.[132] On the basis of patient records, Stavropoulos outlined in his study the activities of the centre, which focused on counselling individuals or couples who desired but were unable to have children for medical or social reasons, including psychological and sexual factors.

The Greek Society of Eugenics also arranged a series of workshops that were closed to the public and attended exclusively by selected experts.[133] However, the activities were not restricted to the interior of the country; relations were maintained with sister-societies abroad. Specifically, the Greek Society of Eugenics had collaborated closely with the British Eugenics Society in a joint study, the conclusions of which, according to Nikolaos Louros and Ioannis Danezis at the society's meeting of 17 February 1965, were soon to be announced.[134] A series of public debates on topics related to eugenics also started to take place at the end of the 1950s that continued into the 1970s. The debates were accessible for the public, and the associated printed material provides the best available source for researching its profile.

8.8.1 *Post-War Eugenics in Public Debates: Old Issues—New Controversies*

The transition which the eugenic discourse went through in the post-war period was reflected in the topics discussed in the public debates that were arranged by the Greek Society of Eugenics. There is likely to have been a discrepancy (as far as available information allows for conclusions) between the reformist character of the society's activities, such as the experimental centre for premarital consultation, and the public debates. The latter were characterized by a rigid insistence on 'traditional' eugenic concerns and anxieties. Apart from questions whose attractiveness had very recently increased (such as, for instance, artificial insemination),

---

[131] The directorial commission of the centre included Vasileios Valaoras, Ioannis Danezis and the (female) doctor M. Karavariotou, according to Stavropoulos (1970): 15.

[132] Greek Society of Eugenics and Human Genetics (1978): 7.

[133] Details were revealed by Ioannis Danezis (the society's president at that time) in the introduction of the third volume: Greek Society of Eugenics and Human Genetics (1978): 5–6.

[134] According to the second volume of the public debates: Greek Eugenics Society (1976): 12.

issues that were well known from the eugenic discourse prior to World War II also reappeared on the agenda. These included, among other issues, the population question, heredity and criminality, premarital health certification, healthcare and restrictions on childbearing, physical education, sexual education of the youth, heredity and the limits of psychic health, environment and survival, drugs, criminality, abortion and the 'human and machine'. Most of the discussants involved in the eugenics society's public debates were outside experts invited by the organizers.[135] With respect to their expertise, they covered a wide spectrum of disciplines, ranging from medicine and the natural sciences to the social sciences, theology, law and even art and theatre.[136]

The discussions in many cases were controversial. It is likely that dissent was desired and even provoked by the organizers who invited prominent scholars whose views were well known to the public. Contentious points were often related to ethics and human rights. If, during the period prior to World War II, questions of the community's priorities towards individuals and their rights had been only partly problematized and usually tended to be answered for the benefit of the community, in the debates of the Greek Society of Eugenics individual rights were favoured by most of the external discussants. The topics to be discussed were specified by the organizers themselves. The population question, hereditary illnesses and restrictions on childbearing preoccupied Greek eugenicists just as much in the post-war period as they had done before the war. However, the perspective now changed to surpass the narrow frames of Greek society. The population question in particular was treated in the Greek Society of Eugenics not only in regard to Greece but also from a global perspective,[137] subjoining racist facets that had been absent in the Greek population discourse prior to World War II.

Overpopulation was now considered as a matter restricted to 'underdeveloped countries', whereas in Greece the more pressing issue was

135 There was also a remarkable level of audience participation, as reported in several passages of the edited volumes. The audience was not, however, actively involved in the discussions.

136 For example, prominent actors like Alexis Minotis (1898–1990) and Dimitris Chorn (1921–98) and the fine artist Ioannis Pappas (1913–2005) were discussants in the panel on heredity and talent.
"Heredity and Talent" ["Κληρονομικότης και ταλέντο"], meeting on 18 March 1968: Greek Eugenics Society (1976): 92–112.

137 The population question was discussed at the meeting on 15 March 1959: "The Population Question" ["Το πληθυσμιακό πρόβλημα"], Greek Eugenics Society (1965): 149–68.

thought to be the declining birth rate. The association of overpopulation with developing countries was congruent with the reorientation of post-war, reformist eugenics whose concerns included underdevelopment, overpopulation and the "backwardness of the Third Word" (Kühl 1997: 174ff.). Some participants in the public debates shared their preoccupation with the rapid proliferation of the 'yellow race', which was allegedly on its way to becoming a majority in the world's population. With respect to population development in Greece, the question was raised as to whether reproduction in the lower social classes and in agrarian areas should be controlled or not. The discussants unanimously rejected imposing any restrictions, although each did so for a different reason; some due to their conviction that the leading classes in a country were recruited also from the lower classes, while others insisted on the free choice of individuals to make decisions concerning their own families. The latter believed that family size was an individual decision regardless of social class, and that state and welfare institutions should enable the population, and hence the lower strata as well, to have a family of their own choosing.

Quite controversial was the session on "Blood and Heredity", where issues related to hereditarily transmitted illnesses such as Downs' Syndrome and thalassaemia were discussed.[138] Provocative questions posed by Nikolaos Louros, the society's president who also chaired the sessions, animated the debates. Arguing in an intriguingly similar way to Ioannis Koumaris when he promoted racist arguments in the 1920s, Louros formulated queries such as, "how is it possible that children [with Downs' Syndrome] acquire a racial complexion that, although normal for yellows, for us is a grievous degenerative anomaly?"; or, "why does this illness not render man a Black/Arab[139] but Chinese?"[140] Participants responded to the racism inherent to this provocation with a sense of repulsion. Faidon Fessas remarked that "the appearance of individuals with trisomy has no relation to the complexion of Chinese people.... I think we should not speculate on any racial correlation."[141] Generally, one can recognize that the achievements of genetics and medicine motivated the majority of

138 "Blood and Heredity" ["Αίμα και κληρονομικότητα"], Greek Society of Eugenics and Human Genetics (1978): 7–28.

139 Arab [*Αράπης* in original], pejorative (if not vulgar) expression meaning 'Black/Nigger'. This is obviously a play on words.

140 "Blood and Heredity", Greek Society of Eugenics and Human Genetics (1978): 11. Cf. similar arguments by Ioannis Koumaris in his 1938 paper on "Pseudo-Mongol Face and Mongol Eye in the Case of Anaemia", *PGAS* (14 May 1938): 2–12.

141 "Blood and Heredity", Greek Society of Eugenics and Human Genetics (1978): 12.

the discussants to voice their optimism for the successful treatment of illnesses of a hereditary character. The participants favoured convenient and intensive medical care during the period of pregnancy, minimizing the consequences of predisposition, and protecting any affected progeny rather than restricting childbearing. It is worth mentioning that prenatal examinations in order to diagnose genetic defects of the embryos and to terminate pregnancy were not at issue.

Restrictions on childbearing were an especially pointed issue in 1965 when premarital health examinations and the establishment of health certification for couples prior to marriage were being discussed.[142] The bone of contention was the limits of medical care and restrictions on reproduction in the case of a proven genetic predisposition to an illness. The examples of Mediterranean anaemia and other illnesses linked to procreation by the blood of the parents were the focus. The arguments of the discussants (who were jurists, hygienists, physicians and economists) differed significantly from those of their counterparts a generation earlier. First of all, the rhetoric of degeneration and social disease was completely absent. Instead, the focus had shifted to the consequences for the individual rather than for the national body. The main speaker in this discussion, the demographer Vasileios Valaoras, pleaded for simply promoting examinations prior to marriage, rather than enforcing health certification by law as a precondition for allowing marriage. The discussants shared his scepticism, claiming that even the introduction of examinations could have grievous social consequences for a young couple. The relationship could break up, or the couple could be stigmatized. In order to avoid discrimination, the economist and politician Virginia Tsouderou (*1924) (who opposed the introduction of prenuptial certification by law) suggested conducting examinations in schools in order to arrive at early diagnoses of possible hereditary illnesses. Pupils did not need to know about the results of the examinations at that point of time, but in the case of a positive diagnosis, the individuals concerned should be informed about their disease at a future date and in a discrete way. Tsouderou also suggested conducting medical examinations when individuals applied for health insurance or, for men, during their military service.

In spite of initial scepticism, the society's president, Nikolaos Louros, remarked in the course of the discussion that premarital examinations

142 The meeting on 17 February 1965 was dedicated to this issue: "Premarital Certification" ["Το Προγαμιαίον Πιστοποιητικόν"], Greek Eugenics Society (1976): 5–29.

should be established by law. But he continued to be sceptical, asking: "Who would dare to make such a suggestion? This has not yet occurred in any democratic regime."[143] This is a clear indication that the shadow of the National Socialist experience still hung over the eugenic discourse. A 'milder' option was proposed by the jurist Panagiotis Zepos, who advocated the establishment of premarital health certification in the 'French version'.[144] Decisive opposition came from the jurist Faidon Vegleris (1903–98) who rejected this measure as a violation of individual rights. Vegleris was a university professor for law and a defender of human rights. During the military dictatorship (1967–74), which started only two years after this public debate, he lived in exile in France where he was academically and politically active.[145] At that session of the Greek Society of Eugenics Vegleris expressed his conviction that the current progress in medicine could guarantee the treatment of poor health in favour of the affected individuals. Progress in medicine facilitated the defence of individual rights, notably the right to free choice of spouses. In a retrospective viewing of past legislation, Vegleris referred to the Greek Civil Law of 1940, remarking that it included too many impediments to marriage. One of these was the ban on marriage between Orthodox Christians and members of other religions. This had occurred in a period in which, he claimed, the influence of National Socialism and its eugenic ideas upon the Greek dictatorial regime of Ioannis Metaxas was growing.

Questions on individual rights and ethics, in particular bio-ethics, also arose in connection with several topics including euthanasia.[146] Serious objections to euthanasia were raised by a number of discussants, notable among them the theologian Konstantinos Mponis and the professor of philosophy Ioannis Theodorakopoulos. The prominent philosopher and politician Evangelos Papanoutsos (1900–82) was particularly pointed in his argument. He distinguished between two forms of euthanasia, one being voluntary and the other forced. The former was applied in the

---

143 "Premarital Certification" ["Το Προγαμιαίον Πιστοποιητικόν"], Greek Eugenics Society (1976): 17.

144 In France, the Vichy regime enforced prenuptial health certification by law, in 1942. Accordingly, a doctor needed to certify the ability or disability of the couple to get married without, however, needing to mention the diagnosed defect. See Sarasin (2001): 444.

145 Faidon Vegleris (1903–98) was professor for law. During the military dictatorship (1967–74) he lived and was politically active in France. As a professor associated with the university in Strasbourg, he taught human rights and international relations. Later he was a representative of Greece in the European commission of experts for human rights.

146 The issue of euthanasia was discussed on 7 March 1961: Greek Eugenics Society (1965): 89–113.

case of individuals "who suffer from incurable illnesses, are irrevocably doomed to die and beg for liberation".[147] The case of involuntary euthanasia applied to individuals who had a congenital or acquired disability (either mental or somatic) might be of interest to a eugenics society, but in Papanoutsos's opinion this was "euthanasia in an improper sense". Clearly rejecting, albeit indirectly, any euthanasia based on eugenic criteria, he decided to deal only with the dilemma of whether one should allow suicide in terminal cases.

The epitome of controversial opinions was the debate on criminality. Even the formulation of the panel's title was provocative: "The Ill Must Be Punished and the Criminals Cured".[148] In his introductory address, the chair of the meeting, Nikolaos Louros, admitted that this somewhat inflammatory title was aimed at attracting the audience's attention, giving as his motivation for choosing this title the famous dystopian novel by Samuel Butler (1835–1902) *Erewhon or Over the Range* (1981/1872). Using Butler's fiction as an exemplar, Louros outlined a totalitarian idea in which communality was prioritized over individuals and their rights. The idea he outlined was thoroughly in the spirit of the eugenics discourse that had flourished prior to World War II. He asserted that individuals who became ill in consequence of consuming drugs, tobacco or alcohol were alone responsible for their precarious situation. However, they burdened not only themselves but also the state and society at large. Hence, illness caused by individuals themselves was a crime against society, since an individual is simultaneously a member of society. This is why the state, along with the provision of healthcare, should also punish those who were at fault for their own illness. Punishment should occur, not least, for disciplinary reasons.

Louros's introduction communicated the message that victims must be blamed; individuality was abrogated in favour of the social totality. This was the main point of contention in the argument that followed. The controversy was carried out primarily between Louros on the one side and the hygienist Dimitris Trichopoulos (1938–?) and the psychiatrist Konstantinos Stefanis (1928–?) on the other. Trichopoulos insisted that the individual alone was responsible for their health and carried the consequences of the damages inflicted to their own health. He categorically

---

[147] Greek Eugenics Society (1965): 102.

[148] "The Ill Must Be Punished and Criminals Cured" ["Να τιμωρούνται οι ασθενείς και να θεραπεύονται οι εγκληματίες"], Greek Society of Eugenics and Human Genetics (1978): 88–116.

rejected the punishment of sick individuals, pointing to the spirit of social solidarity that, he argued, prevailed in contemporary society. Speaking as a hygienist and thus as a 'public health worker', as he argued, he would never approve of punishing poor health. He asserted that, in the final analysis, sick individuals had been punished enough in the gravest of ways, through their illness. Louros's ironic response was that Trichopoulos's statement on 'individual responsibility' was based on an "optimism of sin".[149] Louros's view did not meet with approval from the psychiatrist Konstantinos Stefanis either. Stefanis insisted emphatically that "there are no born criminals",[150] and rejected the association of criminality with disease and psychopathology. Instead, he underlined the complexity of factors that could lead to criminal acts. Simultaneously, he expounded on the problems deriving from subsuming all kinds of abuses under the collective term 'criminality'.

Taking into consideration all this disagreement, one might wonder why so many prominent and dissenting scholars agreed to participate in the public debates of the Greek Society of Eugenics. Certainly, this society was not a marginal institution, given its strong ties with health, medical and political institutions. It was integrated into institutional structures because of the important positions held by its leading members on the scientific and political scenes. Indicatively, the successor of Nikolaos Louros at the society's presidency was the paediatrician and minister of social care Spiros Doxiadis (1917–91). But the central figure was undoubtedly Nikolaos Louros, founder of the society and its president from 1953 to 1973.[151] When Louros created the society he was quite senior in age, with a wealth of professional experience. He was a contemporary witness of the rise of the golden age of the eugenic movement before World War II, although he was probably sparing with regard to his public statements on eugenic issues during that period.[152] An examination of his biographical details could thus serve an enlightening function as to the important junctions on his professional career path.

---

149 Greek Society of Eugenics and Human Genetics (1978): 101.

150 Greek Society of Eugenics and Human Genetics (1978): 103.

151 According to the introduction to the third volume by the then president, Ioannis Danezis. Greek Society of Eugenics and Human Genetics (1978): 5–6.

152 At least, I did not come across statements made by him in the sources considered in this study.

8.8.2 *Professional Networking and Political Opportunity: A Biographical Note on Nikolaos Louros*

Along with his posts as director of the Alexandra state maternity hospital, with which he was identified, as his biographer argues (Marketos 2001: 149), Nikolaos Louros was professor of medicine at the University in Athens and actively involved in the Humanitarian Association [Ανθρωπιστική Εταιρεία].[153] In the aftermath of the dictatorship (1967–74) he became minister of education for a brief period (1974) in the so-called 'Government of National Unity'. His biographer portrays him as the "most long-lived but also most creative statures of medicine" in 20th-century Greece (Marketos 2001: 1). In terms of political ideology, he is described as a royalist (he was also personal doctor to the royal family in the period 1947–64) and at the same time as a democrat; a conservative and simultaneously a revolutionary (Marketos 2001: 12).

The prominent gynaecologist studied medicine in Athens and Switzerland and earned his doctoral title in Berne (1919). He continued working in his specialization in Vienna, Munich and Berlin. In 1925–8 he was a docent at the state clinic in Dresden (Marketos 2001: 7). As German archival sources reveal, his ambition as a rising young scientist was to continue his career there, or at least to be institutionally affiliated with German establishments even though he was then living and working in Athens. However, his hopes in this regard seem to have been dashed by adverse calculations from the German side, and they were also undermined by his professional competitors in Greece. A landmark in Louros's professional trajectory was the turbulent period of the 1930s which saw the Nazis rise to power in Germany. One consequence of the changes in power relations in Germany was the rearrangement of coalitions between scholars abroad, including in Greece, who maintained their professional relations and affiliations with German academia. These conclusions are drawn on the basis of documents preserved in the archives of the Max-Planck-Society, the legal successor of the Kaiser-Wilhelm-Society [Kaiser-Wilhelm-Gesellschaft, hereafter: KWG]. Some files from this archive bring to light additional aspects of Louros's activities in the period 1930–2 and are summarized here as an ancillary note to his biography. Certainly, making a connection with the establishment and work of the Greek Society of Eugenics two decades later might be somewhat audacious. In any event, this information is interesting insofar as it can illuminate some additional

---

[153] See also Louros's autobiography: Louros (1980).

aspects of the question of how professional ambitions and political opportunity can influence networks and ideological affinities, and how individual and professional interests—and rivalries in particular—influence the trajectories of careers and coalitions, and vice versa.

The material in the KWG archive concerns a proposal submitted in 1930 by Nikolaos Louros on the founding of a German clinic for gynaecology in Athens, to be headed by Louros himself. It was proposed that his father's clinic in Athens (hitherto a private institution) would host this project. The proposal was first submitted in 1926 to the Section for Culture of the German Foreign Office [Kulturabteilung des Auswärtigen Amtes] and was supported by the German Ministry of Culture and the Education Foundation [Kulturministerium and Bildungsstiftung]. At that point in time, however, the project failed to materialize due to a shortage of funds.[154] Negotiations were reopened in 1930 and intensified between November 1930 and May 1931. The clinic was initially planned by Louros to be for women of the "second and third classes". This was the period in which the system of social welfare and health insurance was about to be established in Greece. Thus the middle and lower social classes were becoming a new clientele for doctors.

In the course of the negotiations for winning the KWG as a partner in his project, Louros's line of argument changed remarkably, shifting from a focus on its philanthropic character to emphasizing his intention to advance German influence in Greece. In the foreground his personal concern in facilitating his own research was clearly indicated. The young Louros confessed the fascination this project exerted on him in a letter from 1930 addressed to Dr Cranach, the executive director for the advancement of science of the KWG, who was the central figure in the negotiations on the German side. Louros wrote: "This plan appears to me like a bold dream. It captivates me in such a way that I would be ready to abdicate any other academic position in favour of this very plan."[155] He argued that this project would contribute to "founding a nucleus of German science" in Greece, gathering within it scholars who had studied in Germany.

---

154 Louros's letter (1 November 1930) addressed to Dr Cranach (Geschäftsführer der Kaiser-Wilhelm-Gesellschaft zur Förderung der Wissenschaften). *AMPG*, Abteilung 1, Repositur 1A, Best.-Nr. 317: 4–5.

155 Louros's letter (1 November 1930) addressed to Dr Cranach, (Geschäftsführer der Kaiser-Wilhelm-Gesellschaft zur Förderung der Wissenschaften). *AMPG*, Abteilung 1, Repositur 1A, Best.-Nr. 317: 4–5.

In the process of the negotiations, Louros became increasingly aware that the KWG supported research, rather than the founding of medical institutions. Hence in the months that followed he began to suggest that the planned clinic should operate in the fashion of a clinical institute. What he effectively needed was not financial support as such, but the status (and the prestige) of an institution affiliated to a German counterpart.[156] In realizing this project he felt that he would be able to advance German interests in Greece and, simultaneously, that he would be doing a service to his own fatherland.[157] In subsequent correspondence Nikolaos Louros disclosed that his motivation for arranging this German–Greek project was his preference for serving a German scientific institution rather than a Greek one. This was because Greek politics influenced scholarship in such a way that it was extremely difficult to engage in any serious scientific activity. In Germany he had no chance of being appointed to a leading position because of his status as a foreigner.[158] What he meant by government politics and its influence on scholarship were apparently the rivalries between doctors which, in the end, proved to be pivotal in the failure of his own project.

The KWG initially rejected his request on the grounds that the proposal was beyond its mandate. However, the suggestion itself was deemed worthy of consideration due to its outstanding significance for German interests in the "Near Orient" [*im nahen Orient*].[159] Greece, like the rest of the Balkans, belonged, for the National Socialists, to the Orient. In view of the planned German expansion into South-Eastern Europe (Beer and Seewann 2004) and the general German interests in Greece, the KWG did not close the file on Louros. Instead it asked for advice and information about him in order to decide the extent to which a possible collaboration would be beneficial for German interests. All informants (above all the German professors who had collaborated with Louros at the universities in Berlin and Dresden) praised his scientific expertise and professional competence. However, the position of the main informant (or rather, the

[156] In this same letter Louros provided the alternative option to treat patients at cost price instead of at no charge in order to cover a part of the expenses. In the following negotiations only treatment at cost price came into consideration. In 1932, when the private clinic of his father became affiliated with the University of Athens, he indicated the possibility of subventions from the Greek state for the joint project.

[157] He expressed his "commitment... to do my utmost in order to support, in any way, German scientific and cultural propaganda in my fatherland".

[158] Louros's letter (1 November 1930): *AMPG*, Abteilung 1, Repositur 1A, Best.-Nr. 317: 4–5.

[159] Lubarsch's letter to Dr Cranach (12 December 1930). *AMPG*, Abteilung 1, Repositur 1A, Best.-Nr. 317.

mediator in collecting information), namely the representative of the German Archaeological Institute in Athens, was negative.

This was not the only source of negative information. The legation counsellor and representatives in the German embassy in Athens provided detrimental assessments about Nikolaos Louros.[160] It appeared that the rivalries among Greek physicians had proved sufficient to play a decisive role in the final rejection of his project. Serious objections were raised by the German legate, asserting that Louros had a disturbed relationship with Greek institutions but also to those Greek physicians who were "German-educated" and "German-minded" [*Deutsch gesinnt*] and who had provided valuable services to German concerns for many years.[161] A controversy between Nikolaos Louros and Konstantinos Logothetopoulos seems in particular to have impeded the realization of Louros's project. The matter was not one of political significance but concerned an article in a German gynaecology journal, in which Louros criticized Logothetopoulos's approach as scientifically obsolete.[162] His criticism was clearly of medical relevance only, and thus had no political implications.

Louros's proposal was finally rejected in May 1931. The official reason was that the founding of a hospital was beyond the scope of the KWG.[163] This refusal notwithstanding, Louros did not abandon his dream project. Following the suggestion made by the executive director of the KWG, he looked at the German Red Cross as a possible collaborating institution, and to achieve this goal he modified the project from a "Clinic for Gynaecology" to a "Clinical Institute", indicating a research institution. He also suggested that the planned institution might be an affiliated company [*Tochtergesellschaft*] of the KWG in which the latter would have an intermediary role.[164] But in 1932 he submitted another 16-page confidential proposal to German institutions. The recipient of the proposal was not mentioned in the copy in the archive of the Max-Plank-Gesellschaft, but the context reveals that it was addressed to NSDAP.[165] The concern was

160 Letter dated 14 January 1931, by Georg Karo (Deutsches Archäologisches Institut Athen). *AMPG*, Abteilung 1, Repositur 1A, Best.-Nr. 317.

161 Deutsch Gesandtschaft [signed by Heberlein] to Cranach [Geschäftsführer der Kaiser Wilhelm-Gesellschaft zur Förderung der Wissenschaft], Athens, 14 Januar 1931. *AMPG*, Abt. I, Laufzeit 1931; Repositrur 1A, Best.-Nr. 317/2: 43–5.

162 Louros (1930), cf. Logothetopoulos (1926).

163 Extract from the Senat Protocole, 12 May 1931 [Auszug aus dem Senatsprotokoll vom 12. Mai 1931], AMPG, Abteilung I, Laufzeit 1931, Repositur 1A, Best.-Nr. 317/2: 70.

164 Louros's Letter, 22.V.31, *AMPG*, Abteilung I, Laufzeit 1931, Repositur 1A, Best.-Nr. 317/2: 73–4 and 76–8.

165 The addressee of the proposal was not mentioned in the copy available in the archive of the Max-Plank-Gesellschaft. "Denkschrift des Dr. N.C. Louros, a.o. Professors für

specified as "Suggestions on German cultural propaganda in Greece, and concrete means for the establishment of a German Hospital in Athens." His arguments now shifted to the advancement of German culture in Greece by "a German University professor of Greek nationality", namely himself. He reiterated his loyalty by closing the proposal "With German Regards".[166] In the text he ensured that several German referees were willing to endorse him. His references included, in addition to the executive director of the KWG, Dr Lucas von Cranach ("who provides Greek intellectuals with National Socialist literature", as Louros remarked), personalities from the German community and the NSDAP in Greece.[167]

Ultimately his project was not realized. Nevertheless, Louros had an outstanding career in Greece and belongs among the most prominent Greek gynaecologists of the 20th century. According to his biographer, his attitude during World War II was one of resistance to the Italian and German occupying forces (Marketos 2001: 10). He also declined the German offer to take office as minister of education during the occupation, in contrast to his rival, Konstantinos Logothetopoulos, who accepted the office and soon became prime minister in the second occupation government (1941–3). According to Marketos, Louros was even imprisoned in July 1944 and later transported to a concentration camp in Athens (Chaidari).

The case of Nikolaos Louros indicates that the commitment of intellectual and scholarly elites to a nation ('their' nation) has often been conditional. The one-dimensional model of national elites whose agency and attitude is revealed exclusively through their affiliation to a single national model has been shaken by post-colonial studies and the concepts of hybridity and 'in-between' spaces (Bhabha 1988 and 2003/1996; cf. Anderson 2005; Trubeta 2006). Enhancing this idea, I would suggest that the national loyalty of elites is interchangeable. It can be used as a symbolic resource which can quickly lose priority when professional or other interests come into play.

Geburtenhilfe und Gynäkologie an der Universität Arhen [*sic*], (bis 1932 a.o. Professor für Geburtshilfe und Gynäkologie an der Friedrich-Wilhelms Universität Berlin.) Betrifft: Vorschläge über eine deutsche kulturelle Propaganda in Griechenland namentlich eines Deutschen Krankenhauses in Athen": *AMPG*, Abteilung I, Laufzeit 1932, Repositur 1A, Best.-Nr. 317/4: 87–102.

166 "Denkschrift des Dr. N.C. Louros..." *AMPG*, Abteilung I, Laufzeit 1932, Repositur 1A, Best.-Nr. 317/4: 102.

167 He mentioned the German diplomat Dr Eisenlohr, personalities from the German community in Greece, such as the pastor Dr Kindermann, the head of the NSDAP in Greece and Mr Kuhdorfer, the external editor in chief of the telegraphic office 'Wolf' Mr Koch (Außerpolitischer Chefredakteur Wolfs' Telegraphischen Büros).

EXCURSUS: *FILI*, BIO-POWER AND AUTHORITATIVE BIOLOGISM

This section has reconstructed the concepts of Greek communality depicted in the notion of *fili* and traced its transformations from its religious to national, racial, bio-political and eugenic semantics. The demarcation of possible boundaries between each of the prevalent semantics is of analytical significance; the challenge, however, is to recognize the particular ways in which these semantics appear interwoven. The matter is actually a "coupling of notions" variable (Eßbach 2005), rather than a simple shifting, equalization or demarcation of these notions.[168] The interlinking occurs even when one of these notions prevailed.[169] In its racial and eugenic usages, *fili* refers to a large Greek community whose ties are intermediated by descent, sexuality and procreation.

Eugenic projects on communality were characterized by the contradiction between idealism and authoritarianism. The common cause for eugenic engagement was the conviction that intervention in human procreation was necessary in order to reach the desired betterment of society and humanity. But once the implementation of practical means had become the issue, the fascination with idealism abated and the eugenicists increasingly took their stand in a dispute about the priority given either to individual rights and rights to life or the well-being of the community. This dispute emerged out of the age of the Industrial Revolution and concerned universal issues, such as the responsibility of humans *vis-à-vis* the negative side effects of civilization; the limits of individual freedom and self-determination in the broader context of a community; and the boundaries between individual and collective rights. Eugenicists gave top priority to the common good. They claimed to defend solidarity and altruism against the increasing tendencies of individualism and atomization. In their rhetoric, not only were communality and the common good defended; they were literally sacralized, and the protagonists

---

[168] The sociologist Wolfgang Eßbach traces the transformations, actually the metonymy, of the three notions of race, class and mass in classical thought. Eßbach understands his contribution as a 'footnote' to Foucault's lecture, given on 17 March 1976; Foucault (2003).

[169] For instance, in the British case, eugenics primarily addressed class and the welfare of society; several eugenicists also emphasized race (Stone 2001).

of these ideas also sought to justify the sacrifices that were required from those who were deemed to hamper the 'common good'.

In what terms were 'common good' and 'communality' specified in eugenic projects? Who was eligible to belong to—and who was excluded from—such a community? Generally speaking, eugenicists worldwide usually claimed to have committed themselves to the welfare of the nation, the whole society, the race or to humanity. In their rhetoric, nation and national society were ranked centrally, rather than the more abstract ideal of humanity. Certainly, the accentuation of 'nation' was not fortuitous; 'nation' was the simplest form of communality that was able to produce consensus within each individual society in which eugenic measures were applied, despite all proclamations of universality. This was no accident, considering that the national ideal was not questioned by any of the elites in any country, regardless of their political preferences. The assumed sanctity of the nation was first challenged in the decades following World War II, when individualism was belatedly arising to become a counter-ideology (but certainly not a mainstream one) of and among grass-roots movements. For all the universal character of the eugenic idea and the international networking of eugenicists, the actual implementation of eugenic measures in fact occurred in individual societies, and the preferred reference to 'nation' acted as the vehicle for reaching a consensus within each individual society.

In the Greek case, the communality on to which mainstream eugenicists projected their proposals and visions was first and foremost the *fili*. Some eugenicists claimed to speak on behalf of the 'common good' without specifying the character of the communality to which they were referring. In some eugenic usages *fili* even transcended the national community and acquired an additional universal dimension insofar as it was considered part of a global whole (i.e. humanity and the human species). In this case, ensuring a sound Greek *fili* emerged was a contribution to a sound humanity. The reference to the Greek *fili* mobilized a consensus, allowing eugenics to appear as a project that addressed the advancement of the Greek community whose internal relations were represented as horizontal, since factual internal inequality was masked by a proclaimed 'common good'. This was all the more so since the members of the national community were called to contribute actively to the implementation of such projects even when this meant restrictions on their individual rights and even their lives.

The eugenic concept of *fili* differed significantly from the predominant Greek national idea. In their appeals to the welfare of the *fili*, mainstream

Greek eugenicists did not expand on any links to issues related to the prevalent national doctrine, such as the antiquity cult, or the imperative of national homogeneity and the threat of its violation by miscegenation with (national) aliens, typically minorities. Ioannis Koumaris and the high priority he gave to avoiding miscegenation was an exception. Paradoxically, the only related case was the 'refugees' from Asia Minor in the interwar period; they, however, did not make-up a minority in the classic sense since they were regarded as Greek co-nationals and their deviance was quickly repaired. Likewise, consideration of the Jews was strictly marginal in Greek eugenics, if it played any role at all; it occurred in the racial anthropological context rather than in racial hygiene (and certainly not at all in eugenics), and did not inform eugenic/racial hygienic propositions. Eugenic references to *fili* and its well-being did not communicate with the Hellenic past in seeking to render it topical, nor did they appeal to the continuity of *fili* over the centuries. Rather than a national ideal that aimed at conserving an ideal diachronic condition of the Hellenic communality, the eugenic notion of *fili* was conceptualized with a look towards the future and the vision of the betterment, the perfection of the community. In this eugenic usage, *fili* is relative to the Foucauldian concept of 'race' with which Michel Foucault describes a type of society that emerges in the age of bio-power, especially in the late 19th century, and is distinctive for the internal differentiation and inequality that take place along socio-biological lines and is transmitted by discourses of sexuality and reproduction; these discourses became the core issues of health, descent, race and class.[170] Foucault claims to have recognized the emergence of a "new concept of race [that] tended to obliterate the aristocratic particularities of blood, retaining only the controllable effects of sex" (Foucault 1998: 148).

By his concept of *race*, Foucault underlines the idea that traditional class divisions have been transcended and puts emphasis on an internal division into upper and lower races, rather than upper and lower classes. According to this scheme, the internal hierarchical division of society takes place in terms of a "race struggle", meaning a "polarity,... a binary rift within society" which is not, however, "a clash between two distinct races. It is the splitting of a single race into a super-race and a sub-race" (Foucault

170 Michel Foucault conceptualized his notion of race in the first volume of the *History of Sexuality* and further elaborated it in the series of lectures which are printed in the volume *Society Must be Defended* (Foucault 1998 and 2003).

2003: 60–1). The internal polarity of society reflects a global hierarchical order, since "the racist thematic is no longer a moment in the struggle between one social group and another; it will promote the global strategy of social conservatisms" (Foucault 2003: 62). What Foucault intends to describe here is a socio-biological discourse that is directed against 'abnormal' and 'degenerated' members of a society, rather than against the poor and the socially marginalized members. He designates this socio-biological discourse as "state racism" and specified it as a particular kind of racism that describes all politics of bio-power and the governmentality of human life in all its societal expressions. In his own words, this is:

> (racism in its modern, "biolozing," static form): it was then that a whole politics of settlement (*peuplement*), family, marriage, education, social hierarchization, and property, accompanied by a long series of permanent interventions at the level of the body, conduct, health, and everyday life, received their color and their justification from the mythical concern with protecting the purity of the blood and ensuring the triumph of the race (Foucault 1998: 149).

State racism is a novel type of state authority over society that is marked by a dual, yet ambivalent, quality: on the one hand, it targets the society's members as human beings, as members of humankind; but, at the same time, it professes to be a guardian of the whole community and to act as a visionary of its common welfare.

Foucault's concepts of race and racism have evoked criticism concerning various points, including, among others, his restricted European perspective and the underexposed issue of colonialism (Lemke 2007; Stoler 1995; Singelin 2003). Furthermore, he has been criticized for paying little attention to the interweaving concepts of nation and race, and the sexual components of colonial racial discourses (cf. Young 1995; Eßbach 2005). It seems likely, however, that Foucault was little interested in elaborating a comprehensive concept of race. Remaining faithful to his main concern to demarcate historical transitions, he instead utilized *race* for demarking the shift from antagonism between entities which define their differences to other antagonistic entities in terms of descent to internal antagonism within a single entity, a large community, which also occurs along the lines of descent (the shift from the plural of races, to the singular of race); but descent is now transcribed in terms of sexuality, reproduction, normality and abnormality.

> The discourse of race (in the singular) was a way of turning that weapon against those who had forged it, of using it to preserve the sovereignty

> of the State, a sovereignty whose luster and vigor were no longer guaranteed by magico-juridical rituals, but by medico-normalizing techniques (Foucault 2003: 81).

For all possible, and well-founded, reservations to the usages of the racial terminology by Foucault, his concept of 'state racism' is insightful and remains a valuable point of departure for exploring the transformed semantics of blood, heredity and reproduction in shaping social order in the late 19th and the early 20th century (i.e. the very period during which eugenics, one of the most aggressive forms of bio-power, was arising). Indeed, in the concept of *racism* there is codified a form of power that speaks in the name of the community and its common welfare, and promises to protect the whole. This type of power addresses a prophetic speech, viewing ahead into the future and claims that reaching the target demands an internal battle; according to Foucault, this is not "a battle in the sense that a warrior would understand the term, but a struggle in the biological sense: the differentiation of species, natural selection, and the survival of the fittest species" (Foucault 2003: 80). In order to reach the target, the power sets in motion selection within the community, which Foucault has called *race*. This is a form of selection beyond the social exclusion on the grounds of class hierarchy and social inequality in distributing resources and wealth. Rather, it is a matter of human economy that targets the very physical life of individuals, and its repressive, and murderous, function is justified on the grounds of promoting the common welfare: "racism justifies the death-function in the economy of biopower by appealing to the principle that the death of others makes one biologically stronger insofar as one is a member of a race or a population, insofar as one is an element in a unitary living plurality" (Foucault 2003: 258).

What Foucault describes as 'state racism' could be termed *authoritative biologism*, maintaining the core of his concept and avoiding the problematic usage of *race*. Indeed, the essence of Foucault's concept of racism is that it brings to light the biologistic discourse that has penetrated state power and which has become instrumental in the regulation of social functions as well as life itself. This occurred as soon as the scheme of evolutionism in the 19th century was used as a link between biological theory and the discourse of power.

> Basically, evolutionism, understood in the broad sense—or in other words, not so much Darwin's theory itself as a set, a bundle of notions... naturally became within a few years during the nineteenth century not simply a way of transcribing a political discourse into biological terms, and not simply a

> way of dressing up a political discourse in scientific clothing, but a real way of thinking about the relation between colonization, the necessity for wars, criminality, the phenomena of madness and mental illness, the history of societies with their different classes, and so on (Foucault 2003: 256–7).

The essence of this distinctive form of power is that, while society is perceived as a biological continuum, *authoritative biologism* (*racism* according to Foucault) allows for a distinction and social hierarchization along biological lines. This segmentation of the social whole makes possible the distinction between "what must live and what must die". At the same time—and this reveals the ambivalence of bio-power—it prescribes a positive relationship between the members of the entity and the protective state authority. This ambivalent relationship of the members of society to the protective, but simultaneously eliminating, state authority is described by Nikolas Rose and Carlos Novas (2004) as "biological citizenship", a term that includes all programmes of modern citizenship that have associated the concept of citizenship with views on the biological nature of human beings, as individuals, families, communities, populations, races and, finally, as humankind. 'Biological citizenship' supports the differentiation between diverse categories of citizens: the true, the disturbing citizens and so on, and some others who are deprived of the right to citizenship, and the right to belong to community, yet not only in political terms but moreover as physical persons. This ambivalent relationship is imprinted in the modern concept of population whose semantics are both universal and particular; population is a universal category, a 'global mass' pertaining to the human species, as Foucault himself specifies the term (2003: 242). Simultaneously, population accounts for the social body of particular societies on which the internal division is projected. Essentially, population is the plane of reference that unveils inequality and the interior borders of the (national) society, translated into terms of hereditary superiority versus inferiority.

In its eugenic usages, the Greek *fili* bore the semantics of a type of commonality defined in biologistic terms. When mainstream Greek eugenicists spoke on behalf, and for the welfare, of the Greek *fili*, biological determinism was implicit in their speaking. The source of this biological determinism was located in the concept of *population*, rather than in the racial connotations of *fili* that had indeed penetrated this notion since the late 19th century. Population is a convertible category, an economic and political problem to be governed once the threat to the community has been thought of as inherent to this. Other than nation and people, population is not thought of as an unchangeable imagined community, or as a political subject of history. Rather, it is an object to be governed by means of politics, using the instruments provided by modern sciences. It is thought of as a discrete biological entity, the body of the society that is defined by its immanent processes of birth and mortality rates, health condition, life span of individuals, and much more besides. In this, mainstream Greek eugenics was congruent with the currents of liberal eugenics internationally that handled population as a unity closely affiliated to projects of modernization and thus the main subject for human economy. (And, not least, population is a category created in the process of modernization by rising professional groups and technocratic elites that sought to advance their power over society by offering their expertise to the state.) Instead of presuming an ideal condition of the community, the eugenic notion of *fili* transmits the imperative to rescue the communality that is allegedly threatened by defective individuals.

Mainstream Greek eugenicists exclusively referred to social strata as the collective categories on which the betterment and the salvation of *fili* depended. Their elitist rhetoric reveals *classism*, but one that bore all the features of *authoritative biologism*. I will call it a *biologistic classism* in which 'class' becomes a category of social order that is perceived as a biological continuum as soon as sexuality and procreation become instrumental in governing social order and inequality within societies. The evaluation and selection of 'useful' and 'useless' parts of the population separate the bearers of 'good' from those of 'bad' genetic stock, who, again, are not thought of as mere individuals but were a priori localized in social strata. The lower strata were regarded as the main source of unfit individuals and useless members of the community. They hampered the *fili* in a dual way: with respect to the future vision of society due to their harmful biological stock, and in the present because most of the lower classes relied upon the state's welfare institutions to provide for the 'common good', which impinged on the state's finances. Coping with this situation required restricting the procreation of those who were deemed to be damaging the common good, and instead nurturing the potential members of the ideal community. The former belonged to the lower social strata, whereas the latter were mostly localized in the middle class. Both the *fili* and humanity whose Greek eugenicists dreamed of and asserted their commitment to were what was left after a selection had been made on the basis of the estimated value of each individual. What in fact was left were healthy (however 'health' was defined) members of the upper and middle classes; the elites and the upper classes were considered a priori the source of the best quality of human capital.

This concept of *fili* is clearly distinct from those elaborated in the Greek anthropological discourse, even if its echo certainly reached the anthropological institutions, including the debates of the Greek Anthropological Society. In the racial hygienic agenda of the only Greek anthropologist who raised claims to have arrived at a theory of race, namely Ioannis Koumaris, *fili* clearly had racial connotations, but his concept of race itself simply transcribed all social and political developments in racial terms in attempting to portray the Greek state as a racial state. The sharp emphasis given to avoiding emigration, or at least to avoiding marriages between racial Greeks and racial foreigners, presumed a horizontal model of society in which social stratification and the preoccupations of the mainstream eugenicists were abandoned from the landscape. In contrast to Koumaris's idiosyncratic racial hygienic project, the concept of the 'vital race', on which Stavros Zurukzoglu based his racial hygienic proposal, was a relatively instructive expression of the socio-biological discourse in which class and race were merged. Both Koumaris and Zurukzoglu perceived society as a racial entirety and their approaches were undoubtedly totalitarian. But the latter raised social stratification to a central point in his racial hygienic agenda and interpreted it in terms of race and reproduction, especially with respect to calculations regarding elite-building and class mobilization. Undoubtedly, one of the most important differences between the two was that Zurukzoglu contributed a great deal to implementing racial hygienic policies in practice.

The post-war developments in eugenics indicate that neither nation nor race is necessarily needed to govern a population. Progress in molecular biology in the post-war period and its usages inaugurated a new age of the social construction of the biological, which is described as a shift from biologism to geneticization (Lippman 1991). The latter refers to a discourse that explains human behaviour in genetic terms. The 'geneticization of society' describes the social control of genetic risks that involves a specific relationship of techniques of power and forms of knowledge (Lemke 2007: 133; Franklin 2003). Racial, national and class issues are redefined in accordance with changing forms of categorization in which individuality, rather than communality, gain an advanced rank. The relationship of equality–inequality is subject to transformation and new mechanisms of their production, masking and governing. Finally, in the age of the new genetic technologies, a condition which acted as an impetus to the rise of eugenics as an ideology of commitment to communality is gradually being abandoned; this is the welfare state and its ambivalent nature as protector of the community and eliminator of those members who threaten its well-being.

CHAPTER NINE

CONCLUDING REFLECTIONS: THE HIDDEN LEGACY OF RACIAL NATIONALISM

This study has traced the emergence and trajectory of anthropology in Greece over a period of almost 100 years. In the course of my research I became increasingly aware that this discipline, throughout the whole period under consideration, was closely interwoven with the national aspirations of its institutional representatives, Clon Stéphanos and Ioannis Koumaris, both of whom influenced its content and overall academic trajectory. Yet despite its largely national outlook, anthropology in Greece did not cease to be an integral part of a science with claims to universal knowledge in terms of epistemic content and methodology. The universality of scientific knowledge invests a discipline with objectivity and political neutrality and allows its representation as a regime of truth on which its institutional representatives draw a great deal of their scientific and symbolic capital. And it was exactly this universality that justifies the way in which Koumaris designed the Anthropological Museum to transcend a self-referential national system of ideas by inserting aspirations to absorb the colonial legacy into Greek science. On the other hand, the academic trajectory and the particular character of a scientific discipline in an individual country should be explained first and foremost within the specific framework on which this discipline and its representatives draw scientific capital and become embedded in regimes of truth.

This study traced the paths through which anthropological tenets became legitimate knowledge in the particular Greek conditions that advanced rational worldviews and, furthermore, the process through which anthropological science became integrated into Greek academia in a period where the specialization of scientific fields and the creation of separate academic departments was in progress. The downward trajectory of the anthropological institutions, which is illustrated in this book, inevitably leads to the question as to whether this outcome also means that the tenets of anthropology lacked credibility or whether anthropological assertions lost their legitimacy in Greece. But if the required legitimacy is a matter of what effects of power circulate in scientific assumptions and how the internal system of science is conditioned, as Michel Foucault suggests (1978: 26), then the legitimacy of anthropological science should also be looked for in sources external to science. In the Greek case, notably, such sources were the prevalent nationalist and biologistic tenets that were integrated into both society and academia.

Indeed, the Greek anthropological discipline acted as a forum in which biologistic, racial and racist models for approaching human life, history and society were represented and found fertile ground to disseminate, especially up to the end of World War II. The anthropological discourse was the main terrain in which racial anthropological theories were elaborated and scientific racism normalized; but it was certainly not the only vehicle of biologism in Greece. The internal debates of the Greek Anthropological Society echoed the biologistic programmes that were elaborated in other scientific disciplines and were regarded as secure knowledge. How profoundly biologism influenced Greek scholarly circles in Greece during the inter-war period is illustrated by the sociologist Alexandros Kirtsis (1996) in his study on sociological thinking and ideologies of modernization, and the mainstream eugenic movement also bears testimony to the rootedness of biological determinism in the modernization projects of a given society that lean on scientific knowledge. But how far did anthropology have a share in producing such knowledge? The internal debates of the GAS, and especially the analysis of the elaboration of concepts regarding Greek communality by racial, racial hygienic and eugenic means, might have revealed that the anthropological institutions failed to become an important partner for other disciplines that aspired to be involved in the modernization of several branches of Greek society. The loose collaboration with hygienists and eugenicists is an indication of this failure. Notwithstanding the broad scope of Greek anthropology and despite all the involvement of physicians and hygienists in the Greek Anthropological Society, very little cooperation seems to have taken place. The fact that the mainstream biologist discourse in Greece was embedded in liberal political currents, along with the incapability of Greek anthropology to adjust its methods and conceptualization to topical issues in science and society, especially during the long period of Koumaris's agency, played a crucial role in the dysfunctional nature of this collaboration with other disciplines. In this, Greek anthropology conformed to the paradigms of other Balkan states in which physical and racial anthropology played an insignificant role in scientific institutions, as well as in politics, up to the post-war period (Promitzer 2003; 2010a; 2010b; cf. Seisoro 1937; Škerlj 1937). The case of Romania was perhaps an exception in terms of its support for totalitarian thinking and, in the 1940s, of a totalitarian regime. Even though anthropology was established in Romanian academia later than in Greece, it was eventually more successful in terms of its innovative methods and networking (even coalescing) with other disciplines, in both the sciences and humanities. The conformity of leading anthropologists with totalitarian regimes facilitated their influence upon policy-making (Wedekind 2007). Unlike the Romanian case, leading Greek anthropologists and many of those involved in the anthropological discourse proclaimed the 'political neutrality of science', even though they enunciated an intention to place their knowledge at the service of the nation and society, or of engaging in politics.

However, despite the decaying trajectory of anthropological institutions and the failure of anthropology to reproduce its field by educating a generation of professional anthropologists, physical anthropology in Greece, as a field of knowledge, has not been discredited either in academia or in society. This remains the case despite the fact that the genetic-biological arguments for justifying Greek ancestry have been challenged by biologists and geneticists such as Krimpas (1993) and Tzavaras (1998). However, particular scholars and disciplines in Greek academia have not yet become a research subject in the history of science. For all his racist views, his sympathy for Nazi anthropology and contributions to the Italian fascist press, Ioannis Koumaris is still regarded by current anthropologists at the medical faculty of the University of Athens as an innovator in Greek anthropological science, whereas his racist views remain taboo. While silence rules in academia, the impact of the anthropological discourse over Greek society is latent but noticeable in the 'hidden legacy' of racial nationalism, the ideology which marked the Greek anthropological discourse in the period under consideration. Racial nationalism is a discourse that has been shaped in the interface of scientifically legitimate racial anthropological knowledge and its popular perception. As scientific racism, it is a form of populism 'from above'; scientific dispositions which are elaborated by drawing objective arguments from racial anthropological science and nationalism (the unchallenged national doctrine whose objectivity consists in it being a doctrine), but they also resort to irrational but secure means like the knowledge incorporated in human nature in the form of intuition or sense for the truth. The populism of racial nationalism is evident also in the intention of its representatives to reach and have an impact upon broader society.

For all of the irrationality inherent to them, racist tenets drew their validity from established regimes of truth that are based on rationality transmitted by anthropological knowledge, and this has not ceased to communicate a worldview that was liberated from fatality and coincidence. As a scientific regime of truth whose concrete contents have never been challenged in Greece, anthropological arguments were and are mobilized for buttressing the tenets of the national doctrine, which, finally, was at the centre of the anthropological conceptions of both of the main institutional representatives of Greek anthropology. The continuing effectiveness of racial nationalism is due less to the individual agency of certain anthropologists and more to the lack of any serious challenges to their agency and views. The only criticism addressed to Koumaris's statements and agency came during the Civil War from the communist wing, which was ultimately a loser in the political and social order that subsequently emerged. It was precisely this lack of criticism that contributed to the normalization of such arguments and their perpetuation in the fashion of a 'hidden legacy' of racial nationalism. The legacy of Koumaris's anthropological racism continued to be effective in the absence of its proponents in academia. The decaying trajectory of the anthropological institutions contributed to their marginalization in the historiography of sciences due first to their own incapability to found a school of thought in academia, but also to the embryonic stage of the research on the history of sciences and the social history of medicine in and on Greece.

The only protagonist of anthropology in the university contributed a great deal to the decay of the anthropological institutions by his subjective agency, but at the same time his extremely marginal position in academia might have facilitated the perpetuation of racial nationalism in the fashion of populist racism veiled in scientific clothes. This conclusion suggests a relative disconnection between the institutional production of scientific racism on the one hand and its legacy and effectiveness in populist discourses on the other. In view of this disconnection, the challenge for research is to ascertain how scientific anthropological knowledge has been transferred to non-scientific areas of knowledge; or, more specifically, whether, and if so how, scientific and popular knowledge have mutually interacted. A quite illustrative manifestation of this intersection is the perceptions of racial anthropological propositions (or even the mere terminology) in popular, scientific-like, literature and the public discourse in Greece. The voluminous literature on minorities, in particular the Muslim minority, which proliferated from the late 1950s onwards, illustrates the way in which popularized racial, anthropological-like, ideas were diffused in the fashion of 'objective knowledge' in seeking to corroborate Greek ancestry and to justify the exclusion of national deviants, notably minorities (Trubeta 1999; 2008: 93–125). One did not necessarily need to be an anthropologist by training to describe the 'anthropological type' of Roma, Pomaks or Turks. The close association of physical anthropological dispositions with national concerns facilitated the flow of the former into nationalist rhetoric.

But what is the relationship between knowledge provided by, let us say, 'practical researchers', and knowledge communicated by representatives of an institutionalized science? Is the former a legitimate form of knowledge and, if not, how can its proliferation and frequent citation be explained? In turn, one might also ask to what extent certain assumptions—such as the idea of 'racial intuition' as a criterion for ascertaining a race—should be considered scientifically legitimate? Current historians and sociologists of science suggest eliminating the rigid lines between scientific and non-scientific knowledge, given that modern societies are considered 'societies of knowledge' (Weingart 2003); the 'knowledge' category has acquired a new, even a central, significance in describing a society that transcends hitherto conventional modalities (Weingart 2003). A shift can thus be observed in the epistemic paradigm: scientific knowledge is no longer considered to be exclusive to scientific systems, but its role is also advanced in other societal systems in which diverse forms of knowledge are communicated. This view does not *contrast* with the concept of discourse (Foucault) or of 'field' (Bourdieu), but rather shifts the focus to the relationship between different forms of knowledge in a single space of their production and implementation. After all, Bourdieu himself suggested that the internal structure of the 'field' is open to any transformative challenge from within or from outside (Bourdieu 1993: 42).

However, in the recent paradigm of a 'society of knowledge', the matter is less the boundaries (either blurry or fixed) between 'fields', than the integration of scientific knowledge into a general concept of society in which all forms of knowledge are equally legitimate. This shift of paradigm pertains to the recent 'practical turn' (Lipphardt and Patel 2008) in studies of science, and is to a broad extent expressed in the concept of 'scientific culture' (Ass 2007; Labouvie 2007). The latter describes practical knowledge and the practical proficiency of the sciences in a process in which such knowledge is also transformed into professional knowledge. This indicates that the range of actors broadens as scientific elites lose their exclusivity in the production and diffusion of knowledge, while 'communities of practice' are structurally connected with 'scientific communities' (Willke 2002).

The 'practical turn' in studies of science has taken place in the foreground of the dramatically increasing significance of technologies in all sectors of society. The penetration of technologies to broader society may give the impression that knowledge has an egalitarian character. Nevertheless, questions arise as to whether, in the paradigm of a 'society of knowledge', issues of hierarchies of the diverse forms of knowledge and their legitimacy, as well as regimes of power that support the production and diffusion of knowledge, lose significance or are inclined to be transformed and thus require new theoretical models themselves? And what does this development mean for approaching physical anthropological knowledge in particular?

A challenge for future research would be to explore how anthropological knowledge is positioned in a 'society of knowledge', while at the same time taking into account the intensification of the production and transfer of knowledge in national society and in global networks. Moreover, the convergence of anthropology with genetics, which started in the early 20th century and has continued to intensify since World War II, has led the anthropological sciences to become embedded in a scientific terrain and discourse that to a great extent produces knowledge for implementation in practice. This kind of knowledge claims to dispense with justifications based on ideologies burdened with national, racial or sexist prejudices. In these frames, notions that are instrumental to anthropology, such as human diversity and race, could be rethought so as to take account of the changing semantics of 'human' and to produce new methods for dealing with human diversity. From the perspective of the history of ideas, research could explore the mutual interaction between concepts and ideas deriving not only from other scientific disciplines (foremost in the humanities), but also from religious and popular discourses.

Finally, are not the assumptions on the 'racial soul' and 'racial intuition' an appropriate example of the merger of scientific dispositions with popular and religious convictions? While these ideas appear nonsensical when they are expressed by the idiosyncratic Greek anthropologist, they proved influential in the German case as soon as their representatives began to play an instrumental role in shaping a totalitarian system based on racist determinism. How much 'common sense' might be inherent to scientific predications, especially with respect to racial ideas and programmes elaborated in the field of the anthropological sciences? And under which conditions can common sense emerge as legitimate knowledge?

APPENDIX

NEW MEMBERS WHO JOINED THE GAS, 1924–70
SOURCES: *PGAS* (1970)

Year	New Members
1924	68
1925	15
1926	61
1927	30
1928	19
1929	12
1930	12
1931	13
1932	3
1933	7
1934	4
1935	7
1936	10
1937	14
1938	10
1939	8
1940	6
1941	16
1942	19
1943	14
1944	9
1945	6
1946	7
1947	4
1948	3
1949–50	0
1951	1
1952–5	0
1956	3
1957–60	0
1961	4
1962	4
1963–4	0
1965	59
1966	5
1967	10
1968	21
1969	14
1970	5

SOURCES

Archival Sources

Historical Archives of the National University of Athens
Proceedings: *Meetings of the Medical Faculty* (1915–50)
Proceedings: *Meetings of the University Senate* (1915–50)
Portfolio: Anthropological Museum
General State Archives (Athens) [Γενικά Αρχεία του Κράτους]
Max-Planck-Gesellschaft (Berlin)
Humboldt Universität zu Berlin (Institute of Anthropology)
Bundesarchive (Berlin)

Published Sources

Private Archive: Nikos Hatzidimitrakos: Pamphlets and Press of the Physicians Involved in the Resistance during the Second World War.

Gazette of the Parliamentary Debates (Athens) [*Εφημερίδα Συζητήσεων της Βουλής*].

Official Governmental Gazette (Athens) [ΦΕΚ: *Εφημερίδα της Κυβερνήσεως*].

Gazette of the Debates on the New Civil Law [*Εφημερίς των Συζητήσεων του Νέου Αστικού Κώδικος*], 1931 (includes the minutes of the Revising Commission of the Civil Law in 1930).

Yearbooks of the Athenian University—Course Calendars (1925–70) [*Επετηρίδες του Εθνικού Πανεπιστημίου, Αθήνα*].

Proceedings of the Greek Anthropological Society (*PGAS*) (1924–70).

Bulletin de la Société Historique et Ethnique de la Grèce [*Δελτίον της Ιστορικής και Εθνολογικής Εταιρείας*] 1882–1950 (the edition was occasionally interrupted).

Greek Eugenics Society: Public Debates:

—— (1st Vol.). *Debates before the Public: Greek Eugenics Society* [*Συζητήσεις Ενώπιον του Κοινού: Ελληνική Εταιρία Ευγονικής*]. Athens: Parisianos [Εκδότης Γρηγόριος Παρισιάνος], 1965.

—— (2nd Vol.). *Debates before the Public: Greek Eugenics Society* [*Συζητήσεις Ενώπιον του Κοινού: Ελληνική Εταιρία Ευγονικής*]. Special Issue in *Κοινωνική Επιθεώρησις* (*Μηνιαίον Όργανον Ενημερώσεως του Υπουργείου Κοινωνικών Υπηρεσιών*) 4, no. 9–12 (Sept.–Dec. 1976).

—— (3rd Vol.). *Debates before the Public: The Greek Society of Eugenics and Human Genetics* [*Ελληνική Εταιρεία Ευγονικής και Γενετικής του Ανθρώπου*]. Athens: [no publisher], 1978.

Publications by Clon Stéphanos

Stéphanos, Clon (1875). "Unpublished Inscriptions of the Island Syros. Some Typographical and Historical Notes on Ancient Syros and Two Lithographic Tables" ["Επιγραφαί της νήσου Σύρου το πλείστον ανέκδοτοι. Μετά τυπογραφικών και ιστορικών παρατηρήσεων περί της αρχαίας Σύρου και δύο λιθογραφικών πινάκων"]. Athens: Varavarrigou [Τύποις Αδελφών Βαρβαρρήγου].

—— (1878). "Unpublished Documents in Cyclades from the Period of the Russian Occupation" ["Ανέκδοτα έγγραφα αποσταλέντα προς τους κατοίκους των Κυκλάδων κατά την υπό των Ρώσων κατοχήν αυτών"]. Athens: Ermou [Εκ του τυπογραφείου Ερμού].

—— (1879). "A Vocabulary from Syros" ["Γλωσσάριον Σύρου"]. *Bulletin de correspondance hellenique* 3: 20–9 [repr. Αθήνα: Καραβίας, 1973].

—— (1884). *La Grèce au Point de Vue Naturel, Ethnologique, Anthropologique, Démographique et Médical.* Paris: G. Masson, Libraire de L'Academie de Médecine.

—— (1911). *Contributions to the Physical Anthropology of Greece: Cephalic Indices* [*Συμβολαί εις την φυσικήν ανθρωπολογίαν της Ελλάδος: Εγκάρσιος κεφαλικός δείκτης*]. Athens: A. Raftani [Τυπογραφείον της Β. Αυλής, Α. Αραφτάνη].

Publications by Ioannis Koumaris
(Cited Texts)

Autobiography

Koumaris, Ioannis (1951). *50th Anniversary* [*50ετηρίς*]. Athens: [no publisher].

—— (1961). *50th Anniversary—Follow-up* [*50ετηρίδος συνέχεια*]. Athens: [no publisher].

—— (1970). *50th Anniversary—Follow-up* (2) *and Trilogy's Completion* [50ετηρίδος συνέχεια (Β'): Ολοκλήρωσις "Τριπτύχου"]. Athens: [no publisher].

Articles

Koumaris, Ioannis

—— (1903). [Kumaris, J.] and G. Sclavunos. "Über einige Varietäten der Muskeln, Gefässe und Nerven". *Anatomischer Anzeiger* 22: 142–52.

—— (1915). *Surgical Handbook for the Doctor on Campaign* [*Χειρουργικόν εγκόλπιον του ιατρού εν εκστρατεία*]. Athens: Paraskeva Leoni [Εκ του Τυπογραφείου Παρασκευά Λεώνη].

—— (1921). "One Museum and One Name" ["Ένα Μουσείο και ένα όνομα"]. *Εστία*, 19 March.

—— (1922). "From the Anthropological Museum of the National University: On the Occasion of the Broken Hill Cranium" ["Εκ του Ανθρωπολογικού Μουσείου του Εθνικού Πανεπιστημίου. Επί τη ευκαιρία του κρανίου του 'Broken Hil' "]. *Ιατρική. Ελληνική Ιατρική Επιθεώρησις* 1, no. 3 (September): 74–89.

—— (1923). "In Memoriam: Prince Albert I of Monaco" ["Νεκρολογία. Πρίγκηψ Αλβέρτος Α' του Μονακού"]. *Ιατρική. Ελληνική Ιατρική Επιθεώρησις* 2, no. 5–6: 142.

—— (1925). "Anthropology as a University Lesson" ["Η Ανθρωπολογία ως πανεπιστημιακόν μάθημα"]:
Ιατρική—Ελληνική Ιατρική Επιθεώρησις 3, no. 3 (1925a): 71–7.
Ιατρική—Ελληνική Ιατρική Επιθεώρησις 3, no. 4 (1925b): 99–105.

—— (1928) [Kumaris, J.]. "Die Blutgruppen bei den Griechen". *Zeitschrift für Rassenphysiologie* 1, no. 1: 16–20.

—— (1929a). "Ethnological or National?" ["Εθνολογικόν ή Εθνικόν;"]. *Εστία*, 21 January.

—— (1929b). "On the Intellectual Difference between Male and Female" ["Επί της διανοητικής διαφοράς μεταξύ άρρενος και θηλέος"]. *Ιατρικά Χρονικά* 2, no. 3: 126.

—— (1929c). "Intra Secretions and Intellectual Proficiency" ["Αι έσω εκκρίσεις και αι διανοητικαί λειτουργείαι"]. *Κλινική* 17 (27 April): 552.

—— (1931). [Koumaris, Jean]. "Comptes Rendus de la Société Hellénique d'Anthropologie (Office Hellénique de l'I.I.A.), Court résumé des principales communications, 1928–1930)", *Revue Anthropologique*, Quarante-et-unième annee, nos. 1–3 (January–March): 3–11.

—— (1931a). "The Legalization of Children: The Views of the Commission in Charge of the Civil Law" ["Η νομιμοποίησις των τέκνων—Αι απόψεις της επιτροπής του αστικού κώδικος"]. *Ελεύθερον Βήμα*, 14 January: 6.

—— (1931b). "A National Issue: Eugenics" ["Ένα εθνικόν ζήτημα. Δια την Ευγονίαν"]. *Εστία* 1 March: 1.

—— (1931c). "A National Issue: Eugenics" ["Ένα εθνικόν ζήτημα. Δια την Ευγονίαν"]. *Ιατρική Εφημερίς*, 5 April: 1, 4.

—— (1931d). "The Founding of an Ethnological Museum" ["Η Ίδρυσις Εθνολογικού Μουσείου"]. *Βραδυνή*, 24 November.
—— (1931e). "The Fair of Paris. On the Occasion of its Ending" ["Η Έκθεσις των Παρισίων. Επί τη λήξει της"]. *Πολιτεία*, 10 December: 3–4.
—— (1932). "Anthropology and Religion (Column: Popular University)" ["Το λαϊκόν Πανεπιστήμιον της Βραδυνής. Ανθρωπολογία και Θρησκεία"]. *Βραδυνή*, Δεκεμβρίου 7: 1, 3.
—— (1934). [Koumaris, Jean]. "Compte Rendu de la Société Hellénique d'Anthroplogie (Office Hellénique de l'I.I.A.), Court résumé des principales communications 1931–1933". *Revue Anthropologique*, Quarante-Quatrième Année, nos. 7–9 (July–September): 248–59.
—— (1935). "In Search of the Father" ["Η αναζήτησις του πατρός"]. Serialized:
Ιατρική Εφημερίς (8 December 1935): 1, 4.
Ιατρική Εφημερίς (15 December 1935): 4.
Ιατρική Εφημερίς (22 December 1935): 4.
Ιατρική Εφημερίς (29 December 1935): 4.
—— (1936). "The 'National Museum' " ["Το 'Εθνικόν Μουσείον"]. *Έθνος*, 7 January: 1.
—— (1937). "Mendelism and Humanity (Column: Popular University)" ["Λαϊκόν πανεπιστήμιον της 'Βραδυνής'. Μεντελισμός και ανθρωπότης"]:
Βραδυνή (10 March 1937): 1.
Βραδυνή (11 March 1937): 3.
Βραδυνή (12 March 1937): 1, 7.
Βραδυνή (13 March 1937): 1, 3.
Βραδυνή (16 March 1937): 1, 7.
—— (1937a) [Kumaris, J.]. "Die Anthropologie in Griechenland". *Zeitschrift für Rassenkunde und die gesamte Forschung am Menschen* 6, no. 2: 245–6.
—— (1938). "Race and Health" ["Φυλή και Υγεία"], serialized:
Βραδυνή (13 October 1938): 1, 5.
Βραδυνή (14 October 1938): 1, 5.
Βραδυνή (15 October 1938): 5.
Βραδυνή (16 October 1938): 1, 2.
—— (1938a). "Politica della razza in Grecia". *La difesa della razza* 2, no. 1 (November): 46.
—— (1939a). "Rasse und Gesundheit". *Ziel und Weg, Zeitschrift des Nationalsozialistischen Deutschen Ärzte-Bundes, e.V.* 9, no. 12: 386–8.
—— (1939b). [Koumaris, Giovani]. "La politica della Razza in Italia en el mondo". *La Difesa della razza* 3, no. 1: 38–43.
—— (1940). "The Emigrants and Our Race" ["Οι Ξενιτευμένοι και η φυλή μας"].
Βραδυνή (14 October 1940).
Βραδυνή (15 October 1940).
Βραδυνή (16 October 1940).
—— (1941). "Racial Politics in Greece and in the World" ["Η φυλετική πολιτική εν Ελλάδι και εν τω κόσμω"]. *Κουαδρίβιο* 1, no. 3 (15 November).
—— (1943). "Science and Propaganda" ["Επιστήμη και προπαγάνδα"]. *Ακτίνες* ('Οργανον της 'Χριστιανικής Ενώσεως Επιστημόνων') 6, no. 38: 245–7.
—— (1945a) "Greek Blood" [" 'Ελληνικόν αίμα' "], serialized:
Ελληνικόν αίμα (21 March 1945a): 1.
Ελληνικόν αίμα (22 March 1945a): 1.
—— (1945b). "For the Salvation of the Greek Race" ["Δια να σωθή η ελληνική φυλή"]. *Εστία* (10 March): 1.
—— (1945c). "The Nurture of the [Greek] Race" ["Ανατροφή της φυλής"]. *Εκκλησία* no. 5–6 (15 May): 37–8.
—— (1945d). "Racial Issues: 'Greek Blood'. Mixed Marriages are Prohibited" ["Φυλετικά Ζητήματα. 'Ελληνικόν αίμα'. Απαγορεύονται οι μικτοί γάμοι"].
Νέα Ιατρική Εφημερίς (15 May 1945d): 1.
Νέα Ιατρική Εφημερίς (3 June 1945d): 2.

—— (1945e). "Laboratories and Students" ["Εργαστήρια και φοιτητές"]. *Καθημερινή* (12 August).
—— (1945f). "The Predictions of Science: Man after 500 Thousand Years" ["Αι προβλέψεις της επιστήμης. Ο άνθρωπος μετά 500 χιλιάδες χρόνια. Ποιας μεταβολάς θα υποστεί"]. *Έθνος* (9 August): 3.
—— (1945g). "The Great Social Problems: Mixed Marriages" ["Τα μεγάλα κοινωνικά ζητήματα. Οι μικτοί γάμοι"]. *Έθνος* (19 September): 3.
—— (1945h). "The Scientist" ["Ο επιστήμων"]. *Ελληνικόν αίμα* (12 December): 3.
—— (1946a). "Science and Politics: The Consistency of the Races" ["Επιστήμη και πολιτική. Η αυτοτέλεια των φυλών"]. *Πολιτική Επιθεώρησις* 2, no. 1–2: 42–4.
—— (1946b). "On the Ethnological (?) Society" ["Νεοελληνισμοί. Δια την Εθνολογικήν (;) Εταιρείαν"]. *Έθνος* (5 June): 1.
—— (1946c). "The First Steps of Medicine (Stone Age)" ["Τα πρώτα βήματα της Ιατρικής (Λίθινη Εποχή)"]. Serialized:
Ήλιος (7 September 1946c): 633–4.
Ήλιος (14 September 1946c): 671–2.
—— (1946d). "What is 'Anthropology'?" ["Τι είναι 'Ανθρωπολογία' "]. Serialized:
Ήλιος (5 October 1946d): 70.
Ήλιος (12 October 1946d): 121.
—— (1946e). "The Anthropological Society. The Founding Principle of the 'Pre-historic Museum' " ["Η ανθρωπολογική εταιρεία. Αι πρώται βάσεις του 'προϊστορικού μουσείου' "]. *Καθημερινή* (8 October): 1.
—— (1947a). "The National Archaeological Museum: One of the Most Valuable Museums Worldwide" ["Αρχαιολογία. Το εθνικόν αρχαιολογικόν μουσείον. Ένα των πλέον αξιόλογων μουσείων του κόσμου"].
Ήλιος (18 January 1947a): 143.
Ήλιος (25 January 1947a): 195.
—— (1947b). "Between 'Human' and 'No-Human' " ["Μεταξύ 'ανθρώπου' και 'μη ανθρώπου' "].
Ήλιος (18 January 1947b): 134.
Ήλιος (25 January 1947b): 174.
Ήλιος (12 February 1947b): 225.
—— (1947c). "The Agonizing Dilemma of Humanity. Wars are Not Inevitable. What the Study of Human Evolution Teaches Us" ["Το αγωνιώδες δίλημμα της ανθρωπότητας. Οι πόλεμοι δεν είναι αναπόφευκτοι. Τι διδάσκει η μελέτη της ανθρώπινης εξελίξεως"].
Ήλιος 12, no. 187 (1947c): 593–4.
Ήλιος 12, no. 188 (1947c): 667–8.
—— (1948). [Koumaris, John]. "On the Morphological Variety of Modern Greeks". *Man* 48, no. 141: 126–7.
—— (1949). [Koumaris, J.G.]. "The Greek Race". *Man* 49, no. 183–7: 140.
—— (1949a). "The Negroes in Ancient Hellas. Observations by Hippocrates and Aristotle" ["Οι νέγροι εις την αρχαίαν Ελλάδα. Αι παρατηρήσεις του Ιπποκράτους και του Αριστοτέλους"]. *Ήλιος* (12 February): 165–6.
—— (1949b). "The Morphological Diversity of Greeks" ["Η μορφολογική ποικιλία των Ελλήνων"].
Ήλιος (5 March 1949b): 285–6.
Ήλιος (12 March 1949b): 336.
—— (1949c). "On the Occasion of a Donation. The Lack of an Ethnological Museum and the Urgent Need to Acquire One" ["Εξ αφορμής μιας δωρεάς. Η ανυπαρξία εθνολογικού μουσείου και η επιτακτική ανάγκη αποκτήσεως"].
"What the Head of a Savage from South America Teaches Us" ["Τι μας διδάσκει η κεφαλή ενός αγρίου της Ν. Αμερικής"]. *Ήλιος* (19 March 1949c): 365–6.
"An Appeal to Greeks" ["Μια έκκλησις προς όλους τους δυνάμενους Έλληνας"]. *Ήλιος* (26 March 1949c): 407–8.

—— (1949d). "The Function of the Brain and its Psychical Manifestations: The Existing Differences between Men and Women with Respect to Age" ["Η λειτουργία του εγκεφάλου και αι ψυχικαί εκδηλώσεις. Αι υπάρχουσαι διαφοραί μεταξύ ανδρών, γυναικών και ηλικίας"]. *Ήλιος* (30 April): 407–8.
—— (1949e). "The Function of the Brain and its Psychical Manifestations: Brain's Weight and Intellectual Proficiency" ["Η λειτουργία του εγκεφάλου και αι ψυχικαί εκδηλώσεις. Το βάρος του και αι διανοητικαί του ικανότητες"]. *Ήλιος* (7 May): 647–8.
—— (1951a). "Greek Race" ["Φυλή Ελληνική"]. In *Πανηγυρικός Τόμος, Επί τη Εβδημηκονταετηρίδι Μαθιού Μακκά*: 34–42. Athens: Elliniki Ekdotiki Etairia [Ελληνική Εκδοτική Εταιρία].
—— (1951b). "Anthropology and 'UNESCO'. The Mixing of Races" ["Η Ανθρωπολογία και η 'Ουνέσκο'. Η 'παμιξία' των φυλών"], serialized:
Ήλιος 19, no. 354 (10 February 1951b): 885–6.
Ήλιος 19, no. 355 (17 February 1951b): 929–30.
—— (1954). [Koumaris, Johannes]. "Die Homöomorphie in der Entwicklung der Menschen". *Zeitschrift für Morphologie und Anthropologie* (Special Issue: Eugen Fischer—Zur Vollendung des 80. Lebensjahres am 5. Juni 1954) 46, no. 2: 184–8.
—— (1955). [Koumaris, John G.]. "Greece: An Anthropological Review for 1952–1954". *Yearbook of Anthropology* 1: 471–80.
—— (1959). "The Autochtonous Character of the [Greek] Race" ["Το 'αυτόχθον' της φυλής"]. (Offprint from the volume dedicated to G. Ioakeimoglou): 129–37. Athens: Sxoli Monotypias [Τύποις Σχολή Μονοτυπίας].
—— (1961). [Koumaris, Jean-G.]. "Le caractère autochtone du peuple Grec". Extrait des 'Cahiers Ligures de Prehistoire et d'Archeologie', publies par les Sections Françaises de l' Institut International d' Etudes Ligures: 212–19.

Entries in the Encyclopaedia Helios

Ioannis Koumaris
—— [no year] "Race" [Φυλή] (Entry): *Νεώτερον Εγκυκλοπαιδικόν Λεξικόν, Ήλιος* 13: 284–7.
—— [no year]. "Blood. Blood Groups (Anthropology)" ["Αίμα. Αίματος ομάδες (Ανθρωπολογία)"] (Entry): *Εγκυκλοπαιδικόν Λεξικόν, Ήλιος* 1: 824–6.

Proceedings of the Greek Anthropological Society
(Cited Papers)

Ioannis Koumaris

Koumaris, Ioannis (11 May 1925). "Remarks on the Frontal Region of the Cranium" ["Κρανιομετρικαί παρατηρήσεις επί της μετωπικής χώρας"]: 4–11.
—— (22 February 1926). "On a Case of Deformation of the Cranium" ["Επίδειξις παραμορφώσεως κεφαλής"], *Proceedings of the Greek Anthropological Society*: 6–10.
—— (21 April 1927). "The Serological Distinction of Human Races" ["Η 'ορολογική' διάκρισις των ανθρωπίνων φυλών"]: 5–40.
—— (21 February 1928). "On the Capacity of the Crania" ["Επί της χωρητικότητας των κρανίων"]: 4–16.
—— (11 December 1928). "A visite in Glozel" ["Μια επίσκεψις στο 'Γκλοζέλ' "]: 82–94.
—— (3 December 1929). "A Case of Neanderthaloid Supraorbital Eyeshade" ["Περίπτωσις 'Νεαντερταλοειδούς' υπερκογχίου γείσου"]: 72–80.
—— (22 November 1930a). "On the Genetic Precodition in Ascertaining the Blood Group" ["Επί του ζητήματος της γεννητικής εξαρτήσεως κατά τον καθορισμόν των ομάδων αίματος]: 14–17.
—— (22 November 1930b). "On the Cephalic Index of the Crania in Greece" ["Ο εγκάρσιος κρανιακός δείκτης επί κρανίων της Ελλάδος"]: 18–38.

—— (2 May 1931). "Anthropological Observations in Crania Excavated in Agios Kosmas" ["Ανθρωπολογικαί παρατηρήσεις επί τινών κρανίων της εν Αγίω Κοσμά ανασκαφής"]: 45–53.
—— (22 December 1932). "The Facial Index of Greek Crania" ["Ο προσωπικός δείκτης επί ελληνικών κρανίων"]: 42–5.
—— (3 October 1933). "Preliminary Notes on the Caries" ["Πρόδρομον σημείωμα επί της τερηδόνος των οδόντων"]: 10.
—— (28 December 1933). "The Nasal Index of Greek Crania" ["Ο ρινικός δείκτης επί των ελληνικών κρανίων"]: 35–8.
—— (23 September 1934). "Osteometric Reports" ["Οστεομετρικαί Εκθέσεις"]: 2–8.
—— (7 December 1934). "The Orbital Index in Greece" ["Ο κογχικός δείκτης των κρανίων εν Ελλάδι"]: 31–5.
—— (11 April 1935). "On the Classification of the Primates" ["Διά την διαίρεσιν των Πρωτευόντων"]: 18–22.
—— (25 November 1935). "Anthropological Observations" ["Σημείωμα ανθρωπολογικών παρατηρήσεων"]: 26–7.
—— (23 May 1936). "Anthropological Observations on Anaemia" ["Ανθρωπολογικαί παρατηρήσεις επί αναιμιών"]: 3–19.
—— (7 April 1937). "Anthropological Notes" ["Σημείωμα ανθρωπολογικόν"]: 19–21.
—— (20 November 1937). "Anthropological Observations" ["Ανθρωπολογικαί παρατηρήσεις"]: 36–8.
—— (14 May 1938). "Pseudo-Mongolian Face and Mongolian Eye in the Case of Anemia" ["Ψευδομογγολοειδές πρόσωπον και μογγολικός οφθαλμός επί αναιμιών"]: 2–12.
—— (14 May 1939). "The Problem of Race" ["Το πρόβλημα της φυλής"]: 10–26.
—— (1 June 1940). "The Greece Race Outside of Greece" ["Η ελληνική φυλή εκτός της Ελλάδος"]: 3–7.
—— (1 May 1942). "The Diversity of Greeks" ["Η ποικιλομορφία των Ελλήνων"]: 7–28.
—— (15 April 1943). "Albanians and Greeks" ["Αλβανοί και Έλληνες"]: 30–41.
—— (21 February 1945). "The Unavoidable Wars" ["Το μοιραίον των Πολέμων"]: 31–44.
—— (7 January 1970a). "Valediction"[1] ["Χαιρετισμός"]: 14–15.
—— (7 January 1970b). "Neanderthal and Homo sapiens" ["Ο Νεαντερτάλιος και ο Έμφρων Άνθρωπος"]: 9–13.

Other Papers Included in the PGAS *(Cited Presentations)*

Amantos, Konstantinos (22 February 1926). "Slavs and Slavophones in the Hellenic Lands" ["Σλάβοι και Σλαβόφωνοι εις τας ελληνικάς χώρας"]: 10–31.
—— (27 December 1945). "Biology and History" ["Βιολογία και Ιστορία"]: 48–55.
Anagnostou, Ioannis (21 February 1945). "Are the Armenians of Greek Origins?" ["Είναι οι Αρμένιοι Ελληνικής καταγωγής;"]: 22–3.
Andriotis, Nikolaos (12 March 1929). "Prehistoric Tombs in Samothrace" ["Προϊστορικοί τάφοι εν Σαμοθράκη"]: 54–63.
Apostolakis, Georgios, Petros Petridis and Orestis Louridis (18 April 1951). "Biometrical Studies on the Face and the Ears of Adult Greeks" ["Βιομετρικαί έρευναι επί του προσώπου και των ώτων των ενηλίκων Ελλήνων"]: 14–32.

[1] The Greek title could also be translated to "Welcoming Speech"; however, "Valediction" seems to be the most appropriate translation since it was the last statement by the president of the Greek Anthropological Society and was made the year of his death.

Bees, Nikos (8 October 1943). "On the History of the Jewish Community of Tripoli in the Peloponnese" ["Εκ της Ιστορίας της Ιουδαϊκής Κοινότητος της Τριπόλεως της Πελοποννήσου"]: 50–9.
Charitakis, Kostis (12 October 1944). "The Order of the Letters in the Contemporary Greek Phonetic Alphabet" ["Η φυσιολογική σειρά των γραμμάτων του νεοελληνικού φωνητικού αλφάβητου"]: 62–7.
Chatzivasileiou, Grigorios P. (27 December 1945): 57–9.
Dendias, Michail (1 May 1948). "The Theory of 'Pure Race' and the Racial Mixtures in Greece" ["Η θεωρία της 'καθαράς φυλής' και αι φυλετικαί αναμίξεις εις την Ελλάδα"]: 10–27.
Exarchopoulos, Nikolaos I. (22 April 1941). "Conclusions on the Physical Growth of Greek Children (On the Basis of Laboratory Research)" ["Πορίσματα περί της σωματικής εξελίξεως των Ελληνοπαίδων (επί τη βάσει πειραματικών ερευνών)"]: 5–10.
Fronimopoulos, Ioannis (12 October 1944). "Contribution to Study of 'Epitarsus' " ["Συμβολή εις την μελέτην του 'επιτάρσου' "]: 67–71.
Gardikas, Konstantinos G. (28 December 1933a). "The Influence of Time of the Year and the Climate upon Criminality in Greece" ["Η επίδρασις της ώρας του έτους και του κλίματος επί της εν Ελλάδι εγκληματικότητος"]: 28–33.
—— (28 December 1933b). "The Frequency in which the Diverse Types of Finger Prints Appear in Greece" ["Η συχνότης των διαφόρων τύπων των δακτυλικών αποτυπωμάτων εν Ελλάδι"]: 33–5.
—— (7 December 1934). "Profession and Criminality" ["Επάγγελμα και εγκληματικότης"]: 36–47.
—— (23 May 1936). "Criminality in Greece by Region" ["Η εγκληματικότης εν Ελλάδι κατά τόπους"]: 21–41.
Katsas, Grigorios and Vasileios Konstantelos (8 October 1943). "A case of situs viscerum inversus totalis or heterotaxy" ["Επί της καθολικής αναστροφής των σπλάχνων, (Situs viscerum inversus totalis ή Heterotaxie)"]: 68–70 (discussion pp. 70–1).
Kodounis, A., G. Loukatos and E. Loutsidis (8 October 1946). "A New Hereditary Blood Illness, Cyanosis—Congenital Methemoglobinaemia" ["Η νέα κληρονομική νόσος του αίματος. Κληρονομική μεθαιμοσφαιρινική κυάνωσις"]: 30–9.
Konstantinidis, Konstantinos (12 March 1965). "Brain and Civilization" ["Εγκέφαλος και πολιτισμός"]: 53–8.
Kosmetatos, Georgios (18 December 1943). "On the Competence of the Genetic Theory of Heredity" ["Περί της επάρκειας ή μη της γονυλιακής θεωρίας της κληρονομικότητας"]: 72–80.
—— (30 December 1946). "The Frequency in which Daltonism Appears in the Greek *phili*/ Race" ["Η συχνότητα του Δαλτωνισμού εις την ελληνικήν φυλήν"]: 49–54.
Koukoules, Faidon (12 April 1927). "Mediaeval Publications" ["Από την μεσαιωνικήν διαπόμπευσιν"]: 59–68.
Koumari-Stathaki, Rodoula (12 November 1966). "On Folk Art from Crete" ["Από την Λαϊκήν τέχνην της Κρήτης"]: 15–21.
—— (12 November 1967). "The Skin and the Towel in Crete" ["Η πέτσα και η πετσέτα στην Κρήτη"]: 30–40.
Kouretas, Dimitrios (9 January 1944). "The Hypothalamus as a Formative Organ" ["Ο υποθάλαμος ως μορφοπλαστικόν όργανον"]: 12–28 (incl. discussion).
Krikos, Alexandros (11 April 1935). "The Development of Caries in Greece from Antiquity up to Present" ["Η εξέλιξις της τερηδόνος των οδόντων εν Ελλάδι από των αρχαιοτάτων μέχρι των καθ' ημάς χρόνων"]: 3–18.
—— (28 December 1936). "The Development of Caries in the Greek *fili*" ["Εξέλιξις της Τερηδόνος των οδόντων παρά την ελληνική φυλή"]: 50–4.
—— (29 December 1940). "Observations on the Physical Growth of Greek Children" ["Συμβολή εις την μελέτην της σωματικής αναπτύξεως του ελληνόπαιδος"]: 47–53.
—— (27 December 1941). "The Alimentary Problem and the Future of the fili" ["Το επισιτιστικόν πρόβλημα και το μέλλον της φυλής"]: 51–6.

Kritikos, Nikolaos N. (28 October 1941). "Volcanic Eruptions and Earthquake in the Aegean and Man in the Southeastern Mediterranean Sea" ["Ηφαίστειοι εκρήξεις και σεισμοί εν τω Αιγαίω και ο Άνθρωπος της Ν.Α. Μεσογείου"]: 22–5 (discussion: 26–7).

Krontiris, Athanasios and Pan. Petropoulos (30 December 1946). "Supernumerary Neck Vertebrae (*Klippel-Feil syndrome*)" ["Υπεράριθμος αυχενός σπόνδυλος (Σύνδρομον Klippel-Feil)"]: 54–61.

Lampadarios, Emmanuil (21 February 1928). "The Physical Growth of Greek Schoolchildren (Anthropological Auxology)" ["Η σωματική εξέλιξις του Έλληνος μαθητού (ανθρωπολογική αυξησιολογία)"]: 19–37.

—— (27 December 1941). "The Types of Body Constitution at School Age" ["Οι τύποι της ιδιοσυστασίας του σώματος εις την μαθητικήν ηλικίαν"]: 40–2.

Livas, Xerxis (29 November 1956). "The Aegean—Cradle of the Aryans and Hellenism: Cultural Antecedence and Radiation—Racial Unity—the Racially Negligible Intrusion of Slavs and Turks" ["Η Αιγηΐς, Κοιτίς των Αρίων και του Ελληνισμού. Πολιτιστική προήγησις και ακτινοβολία—Φυλετική ενότης—Αμελητέα φυλετικώς η παρείσδυσις των Σλάβων και των Τούρκων"]: 26–43.

—— (12 March 1965). "The Macedonian Question—A Non-Existing Question" ["Το 'Μακεδονικόν'' ζήτημα Ανύπακτον [*sic*] ζήτημα"]: 21–7.

Marinatos, Spiridon (21 February 1945). "Steatopygia in Pre-Historic Greece" ["Η Στεατοπυγία εν τη Προϊστορική Ελλάδι"]: 13–20.

—— (12 March 1965). "Could There be a 'Mental Anthropology?' " ["Δύναται να υπάρξη και 'πνευματική ανθρωπολογία';"]: 59–63.

Markovits, Adalbert (24 April 1928). "From the Dark Caves" ["Από τα σκότη των σπηλαίων"]: 45–61.

—— (3 December 1929). "Hitherto Studies on the Stone Age in Greece" ["Περί των μέχρι σήμερα ερευνών επί της λιθικής περιόδου της Ελλάδος"], *Proceedings of the Greek Anthropological Society* (3 December 1929): 114–34.

—— (29 November 1930). "On the Flint Industry in the Last 500 Years" ["Περί της βιομηχανίας του πυρίτου κατά τα τελευταία 500 έτη"]: 44–50.

—— (30 December 1931). "Microliths with Acanthi" ["Μικρόλιθοι μετ' ακάνθης"]: 115–24.

Mazaraki, Despoina (1 June 1940). "Crimes of Filicide in Greece" ["Το έγκλημα της παιδοκτονίας εν Ελλάδι"]: 8–23.

Megas, Georgios (12 November 1966). "Folklore, Ethnography, Ethnology" ["Λαογραφία, Εθνογραφία, Εθνολογία"]: 22–7.

Milonas, Georgios (2 May 1931). "The Prehistoric Cemetery and the Settlement of Agios Kosmas" ["Το προϊστορικόν νεκροταφείον και ο συνοικισμός του Αγίου Κοσμά"]: 31–44.

Mpelezos, Nikon (18 December 1943). "Heredity and Psoriasis" ["Κληρονομικότης και ψωρίασις"]: 85–9.

Mpezos, Dimitrios (8 October 1946). "Anthropological Observations on the Schoolchildren during the Occupation (1941–1944)" ["Ανθρωπολογικαί παρατηρήσεις επί των μαθητών (Κατοχή 1941–44)"]: 39–41.

Mpezos, Dimitrios (12 November 1967). "Anthropological Observations on the School Achievements" ["Ανθρωπολογικά από τας μαθητικάς επιδόσεις"]: 13.

Panagiotou, Panagiotis (18 December 1943). "Heredity and Deaf-Muteness" ["Κληρονομικότης και κωφαλαλία"]: 89–90.

—— (30 December 1946). "Natural Selection among Humans" ["Η φυσική επιλογή επί του ανθρώπου"]: 69–73.

—— (12 November 1967). "Contemporary Views on Genetic Equipment" ["Η γενετική αποσκευή κατά τας σύγχρονας αντιλήψεις"]: 8–12.

Papamarkou, Ploutarchos and P. Sklepa (28 October 1941). "The Frequency in which Blood Groups Appear in Greece" ["Η συχνότης των ομάδων αίματος εν Ελλάδι"]: 28–9.

Pezopoulos, Emmanuil (15 April 1943). "On the Continuity of the Greek Nation from Ancient to Present Time" ["Περί της συνεχείας του ελληνικού έθνους από των αρχαίων μέχρι των καθ' ημάς χρόνων"]: 12–17.

Piniatoglou, Lazaros (25 September 1942). "Pythagorean Taboo" ["Τα Πυθαγορικά Ταμπού"]: 36–9.
—— (15 April 1943). "Social Anthropology (Art) and Psychoanalysis" ["Κοινωνική Ανθρωπολογία (Τέχνη) και Ψυχανάλυσις"]: 46–9.
Poulianos, Aris (2 April 1962). "The Origins of the Greeks. Ethno-genetic Research" [Η προέλευση των Ελλήνων. Εθνογενετική Έρευνα]: 20–31.
—— (12 March 1965). "The Origins of the Greeks" ["Η προέλευσις των Ελλήνων"]: 27–38.
—— (12 November 1966). "Anthropology and Population Genetics (On the Origins of the Cretans)" ["Ανθρωπολογία και πληθυσμιακή γενετική (Περί της καταγωγής των Κρητών)"]: 43–52.
Pournaropoulos, Georgios (3 November 1938). "A Mineralogical Study in Bones Conserved in the Anthropological Museum in Athens" ["Επί εφαρμογής μεθόδων ορυκτολογικής ερεύνης εις παλαιοπαθολογικάς ερεύνας επί οστών του Ανθρωπολογικού Μουσείου Αθηνών"]: 27–41.
—— (29 December 1940a). "A Donation of Bones from the Tomb of the So-Called 'Leonidas' 300-Men' to Anthropological Museum" ["Προσφορά οστών τινών των άκρων και σπονδύλων εκ των τάφων των λεγομένων των τριακοσίων του Λεωνίδου εις το Ανθρωπολογικό Μουσείον του Εθνικού Πανεπιστημίου"]: 54.
—— (29 December 1940b). "Collective Cases of Syndactylism" ["Ομαδικαί περιπτώσεις συνδακτυλίας"]: 53–4.
Prince Peter, His Majesty [*sic*] (1 February 1961). "Introduction into the Anthropological Sciences" ["Εισαγωγή εις τη Επιστήμην της Ανθρωπολογίας"]: 24–38.
—— (3 February 1961). "A Lecture on Ethnology" ["Εισήγησις εις την Εθνολογίαν"]: 24–38.
—— (6 February 1961). "On My Anthropological Studies. The Practical Significance of Anthropology for Greece" ["Από τας Μελέτας μου επί της Ανθρωπολογίας. Η πρακτική σημασία της επιστήμης αυτής δια την Ελλάδα"]: 42–57.
—— (14 March 1962). "The Historical Development of Ethnology" ["Η ιστορική εξέλιξις της εθνολογίας"]: 12–22.
—— (19 March 1962). "Progress in Ethnology" ["Οι Πρόοδοι της Εθνολογίας"]: 24–36.
—— (21 March 1962). "Contemporary Ethnology" ["Η σύγχρονος Εθνολογία"]: 37–51.
—— (2 May 1963). "A Scientific Theory of Culture" ["Μια επιστημονική θεωρία του πολιτισμού"]: 12–23.
Rivouar, Ioannis (1 June 1944). "Redeeming Craniology" ["Κρανιολογία η σωτήριος"]: 34–42.
Romaios, Konstantinos (12 November 1966). "The Borderlines of Folklore and Anthropology" ["Στα σύνορα της Λαογραφίας και της Ανθρωπολογίας"]: 29–35.
Sakellariou, Georgios (27 January 1932). "On Crimes from a Psychological Perspective" ["Περί του εγκλήματος εν Ελλάδι από ψυχολογικής απόψεως"]: 3–24.
—— (1 May 1942). "Contribution to Clinical Psychology—School Retardation" ["Μια συμβολή εις την Κλινική Ψυχολογίαν Σχολική Καθυστέρησις"]: 2–6.
—— (12 December 1943). "Woman and Scientific Research" ["Γυνή και επιστημονική έρευνα"]: 80–5.
Stavropoulos, I. (28 October 1941). "Blood Groups in Infancy" ["Αι ομάδες αίματος εν τη παιδική ηλικία"]: 30–1.
Triantafyllidou, Anna (27 June 1945). "Ethnological Notes from Macedonia" ["Εθνολογικά Σημειώματα από την Μακεδονίαν"]: 25–31.
Xirotiris, Nikolaos (12 November 1967). "Anthropological Issues: 1) The Negroes of Thrace; 2) Blood Groups of Women in Northern Greece" ["Ανθρωπολογικά, 1) Οι Νέγροι της Θράκης, 2) Αι ομάδες αίματος εις τας γυναίκας Β. Ελλάδος"]: 24–6.
Zurukzoglu, Stavros (11 May 1925). "On Eugenics" ["Περί ευγονίας"]: 14–40.
—— (3 December 1929). "On a Congenital Deformation of the Small Toe (Clinodactylism)" ["Περί τινός κληρονομικής παραμορφώσεως του μικρού δακτύλου (κλινοδακτυλία)"]: 80–4.

REFERENCES

[Anonymous] (1874). "Epidemics and Population" ["Επιδημίαι και πληθυσμός"]. *Οικονομική Επιθεώρησις* 2, no. 13: 8–17.

—— (1925). "The Medical Examination Prior to Marriage" ["Η προ του γάμου ιατρική εξέτασις"]. *Υγεία* 2, no. 22: 435–7.

—— (1930). "The Hygenic Programme (Social Hygiene)" ["Το υγειονομικόν πρόγραμμα (Κοινωνική υγεινή)"]. *Εργασία* 1, no. 27: 19.

—— (1933). "Forced Sterilisation" ["Η αναγκαστική στείρωσις"]. *Εκκλησία* 47, 25 November: 371.

—— (1934). "From Horror-Germany. The Barbaric Implementation of Forced Sterilisation. German Scientists Condemn the Measure" ["Από την Γερμανίαν του τρόμου. Η βάρβαρος εφαρμογή της αναγκαστικής στειρώσεως. Οι γερμανοί επιστήμονες αποδοκιμάζουν το μέτρον"]. *Ιατρική Εφημερίς*, 14 January: 3.

—— (1936). "Instead of a Record. The Primitives" ["Αντί δελτίου. Οι πρωτόγονοι"]. *Ιατρική Εφημερίς*, 22 November: 1 [reprinted from the newspaper *Τύπος* (22 October 1936)].

—— (1945). "The National Committment of Greek Physicians During the Occupation. G. Sklavounos, Speech at the Academy of Athens in 1942" ["Η εθνική δράσις των Ελλήνων ιατρών επί Κατοχής. Γ Σκλαβούνος, Ομιλία εν τη Ακαδημία Αθηνών τω 1942"]. *Νέα Ιατρική Εφημερίς*, 18 April: 2.

—— (1962). "Speech at the Academy in Winter 1942: The Academy Declined" ["Λόγος στην Ακαδημία το Χειμώνα 1942: Η Ακαδημία αρνήτω"]. *Επιθεώρηση Τέχνης*, March–April 8 (15): 298–9.

[Correspondence] (1938). "Rassenpflege in Griechenland". *Kölnische Volkszeitung*, 27 November.

[K.B.K. (K.B. Kalogeris)] (1945). "The Massacred Doctors" ["Οι σφαγιασθέντες ιατροί"]. *Νέα Ιατρική Εφημερίς*, 14 March: 1.

Agapitidis, Nikitas I. [Αγαπητίδης, Νικήτα I.] (1996). *A Captivity Chronicle (1949–1956)* [*Το χρονικό μιας αιχμαλωσίας (1949–1956)*]. Athens: [no publisher].

Agelakis, Anagnostis (1997). "Ioannis Koumaris". In *History of Physical Anthropology, an Encyclopedia*, vol. 1, edited by Frank Spencer, 455. New York and London: Garland Publishing.

Agelopoulos, Georgios (2010). "Contested Territories and the Quest of Ethnology: Peoples and Places in Iżmir 1919–1922". In *Spatial Conceptions of the Nation: Modernizing Geographies in Greece and Turkey*, edited by Nikiforos Diamantouros, Thalia Dragona and Çağlar Keyder, 181–299. London and New York: I.B. Tauris Academic Studies.

Aggelidis, V. (1945). "The Pseudo-Science of Racism and its Representatives in Greece" ["Η ψευτοεπιστήμη του ρατσισμού και οι αντιπρόσωποί της στην Ελλάδα"]. *ΚΟΜΕΠ (Communist Review)* (July) 39: 36–40.

Agriantoni, Christina (1986). *The Beginning of Industrialisation in Greece in the 19th Century* [*Οι απαρχές της εκβιομηχάνισης στην Ελλάδα του 19ου αιώνα*]. Athens: Historical Archives of the Commercial Bank [Ιστορικό Αρχείο Εμπορική Τράπεζα της Ελλάδος].

Alexiou, Athanassios (1994). *Zur Frage der Entstehung und Formierung der griechischen Arbeiterbewegung—Eine sozialhistorische Untersuchung*. Frankfurt am Main [et al.]: Peter Lang.

Almgren, Birgitta, Jan Hecker-Stampehl and Ernst Piper (2008). "Alfred Rosenberg und die Nordische Gesellschaft. Der 'nordische Gedanke' in Theorie und Praxis". *Nordeuropa Forum* 2: 7–51.

Amantos, Konstantinos (1920). *Macedonian Issues: A Contribution to Mediaeval History and Ethnology of Macedonia* [*Μακεδονικά: Συμβολή εις την μεσαιωνική ιστορίαν και εθνολογίαν της Μακεδονίας*]. Athens: P. Petrakos [Τυπ. Π. Πετράκος].

—— (1923). *The Nordic Neighbours of Greece: Bulgarians, Albanians, South-Slavs* [*Οι βόρειοι γείτονες της Ελλάδος: (Βούλγαροι—Αλβανοί—Νοτιοσλάβοι)*]. Athens: Eleftheroudakis [Ελευθερουδάκη].

—— (1934). *The Ethnological Make-Up of the Greek Nation* [*Η εθνολογική διαμόρφωσις του ελληνικού έθνους*]. Athens: Pirsou [Τυπ. Πυρσού].

—— (1952). *The Macedonian Hellenes during the Late Middle Ages and Turkish Rule to the 18th Century* [*Ο μακεδονικός ελληνισμός κατά το τέλος του μεσαίωνος και την παλαιοτέραν τουρκοκρατίαν μέχρι του δεκάτου ογδόου αιώνος. (Διάλεξις γενομένη εν τη αιθούση της Εταιρείας Μακεδονικών Σπουδών)*]. Thessaloniki: Society of Macedonian Studies [Εταιρεία Μακεδονικών Σπουδών].

Anastasiadis, Giorgos (1998). "The University of Thessaloniki and the Resistance (1941–1944)" ["Το Πανεπιστήμιο της Θεσσαλονίκης και η Αντίσταση (1941–1944)"]. In *Macedonia and Thrace 1941–1944: Occupation—Resistance—Liberation* [*Μακεδονία και Θράκη, 1941–1944. Κατοχή Αντίσταση Απελευθέρωση*], edited by the Institute for Balkan Studies (IMXA), 262–77. Thessaloniki: Institute for Balkan Studies (IMXA).

Anderson, Benedict (2005). *Under Three Flags: Anarchism and the Anti-Colonial Imagination*. London and New York: Verso.

Apostolidès N.X. (Communiqué par M. Deniker) (1883a). "Quelques mesures sur le vivant prises en Grèce". *Bulletins de la Société d'Anthropologie de Paris* 6 (3d série): 614–16.

—— (1883b). "Contribution to Greek Anthropology: Cephalometric Notes B" ["Συμβολαί εις την ελληνικήν Ανθρωπολογίαν, Κεφαλομετρικαί σημειώσεις B"]. *Δελτίον της Ιστορικής και Εθνολογικής Εταιρίας της Ελλάδος* 1: 365–7.

Apostolidis, Petros (1981). *All I Remember 1900–1969* (vol. 1: 1922–1923) [*Όσα θυμάμαι, 1900–1969, Γκαρνιζόν Ουσιάκ, 1922–1923* (vol. 1)]. Athens: Kedros [Κέδρος].

—— (1983). *All I Remember 1900–1969* (vol. 2: 1923–1969) [*Όσα θυμάμαι 1900–1969, Η συνέχεια 1900–1923 και 1923–1969* (vol. 2)]. Athens: Kedros [Κέδρος].

Ardavanis-Limperatos, Spiridon (1934). "From the Forensic Institute of the National University. The Founding of the Association for Criminal Biology. Its Great Mission" ["Εκ του ιατροδικαστικού Ινστιτούτου του Εθνικού Πανεπιστημίου. Σύστασις Ελλ. Εταιρείας Εγκληματολογικής Βιολογίας. Ο μέγας αυτής προορισμός"], serialized:
Ιατρική Εφημερίς (21 October 1934): 1.
Ιατρική Εφημερίς (28 October 1934): 1.
Ιατρική Εφημερίς (4 November 1934): 1.

Ardavanis-Limperatos, Spiridon (1935a). "Eugenics and Social Hygiene. Prophylaxis and Therapy of Criminality. The Prisons of the Civilized States" ["Η ευγονία και η κοινωνική υγιεινή. Προφύλαξις και θεραπεία εγκληματικότητος. Αι φυλακαί των πολιτισμένων κρατών"]. *Ιατρικη Εφημερίς*, 20 January: 1.

—— (1935b). "Eugenics and Social Hygiene. The Obligation of the State to Observe the Criminals after their Discharge. How to Protect Society" ["Η ευγονία και η κοινωνική υγιεινή. Το καθήκον του κράτος να παρακολουθή τους εγκληματίας μετά την αποφυλάκισίν των. Πώς θα προφυλαχθή η κοινωνία"]. *Ιατρική Εφημερίς* (3 March 1935b): 2.
Ιατρική Εφημερίς (10 March 1935b): 2.

—— (1935c). "New Methods of State Thinking in Italy" ["Αι *νέαι μέθοδοι της* κρατικής αντιλήψεως εις την Ιταλίαν"].
Ιατρική Εφημερίς (26 May 1935c): 1.
Ιατρική Εφημερίς (2 June 1935c): 3.
Ιατρική Εφημερίς (9 June 1935c): 3.
Ιατρική Εφημερίς (16 June 1935c): 2.
Ιατρική Εφημερίς (23 June 1935c): 2.
Ιατρική Εφημερίς (30 June 1935c): 2.
Ιατρική Εφημερίς (7 July 1935c): 2.
Ιατρική Εφημερίς (14 July 1935c): 2.
Ιατρική Εφημερίς (21 July 1935c): 2.
Ιατρική Εφημερίς (4 August 1935c): 2.

Ιατρική Εφημερίς (11 August 1935c): 2.
Ιατρική Εφημερίς (18 August 1935c): 2, 3.
Ιατρική Εφημερίς (25 August 1935c): 2.
Ιατρική Εφημερίς (1 September 1935c): 2.
Ass, Mitchell (2007). "Von Vielschichtigkeiten und Verschrägungen. Kulturen der Wissenschaft—Wissenschaften in der Kultur". *Berichte der Wissenschaftsgeschichte* 30: 91–105.
Assimacopoulou, Fotini (1999). *Gobineau et la Grèce*. Frankfurt am Main [et al.]: Peter Lang.
Baader, Gerhard (2008). "Eugenische Programme in der sozialistischen Parteienlandschaft in Deutschland und Österreich in Vergleich". In *Eugenik in Österreich. Biopolitische Strukturen von 1900 bis 1945*, edited by Gerhard Baader, Veronika Hofer and Thomas Mayer, 66–139. Vienna: Czernin Verlag.
Baader, Gerhard, Veronika Hofer and Thomas Mayereds, eds (2008). *Eugenik in Österreich. Biopolitische Strukturen von 1900 bis 1945*. Vienna: Czernin Verlag.
Balibar, Étienne (2004). *We, the People of Europe? Reflections on Transnational Citizenship*. Princeton, NJ: Princeton University Press.
Barth, Frederik (2005). *One Discipline, Four Ways: British, German, French and American Anthropology*. Chicago, IL: Chicago University Press.
Barzun, Jacques (1965). *Race, A Study in Superstition* (Revised with a New Preface "Racism Today"). New York, Evanston and London: Harper Torchbooks.
Bauman, Zygmunt (1991). *Modernity and Ambivalence*. Cambridge: Polity Press.
Beer, Mathias and Gerhard Seewann, eds (2004). *Südostforschung im Schatten des Dritten Reiches. Institutionen—Inhalte—Personen*. Munich: Oldenbourg Wissenschaftsverlag.
Belliger, Andréa and David J. Krieger, eds (2006). *Anthology: Ein einführendes Handbuch zur Akteur-Netzwerk-Theorie*. Bielefeld: Transcript.
Bennett, Tony (1995). *The Birth of the Museum: History, Theory, Politics*. London: Routledge.
Berner, Margit (2007). "From 'Prisoners of War Studies' to Proof of Paternity: Racial Anthropologists and the Measuring of 'Others' in Austria". In *"Blood and Homeland": Eugenics and Racial Nationalism in Central and Southeast Europe, (1900–1940)*, edited by Marius Turda and Paul Weindling, 41–54. Budapest and New York: CEU Press.
Bhabha, Homi K. (1988). "The Commitment to Theory". *New Formations* 5: 5–23.
—— (1994). *The Location of Culture*. London and New York: Routledge.
—— (2003). "Culture's In-Between". In *Questions of Cultural Identity*, edited by Stuart Hall and Paul du Gay, 53–60. London, New York and Thousand Oaks, CA: Sage Publications [first edition 1996].
Black, Edwin (2003). *War Against the Weak: Eugenics and America's Campaign to Create a Master Race*. New York and London: Four Walls Eight Windows.
Black, Lydia T. (1977). "The Concept of Race in Soviet Anthropology". *Studies in Soviet Thought* 17, no. 1: 1–27.
Bourdieu, Pierre (1988). *Homo Academicus*. Stanford California: Stanford University Press.
—— (1992). *Die verborgenen Mechanismen der Macht*. Hamburg: VAS-Verlag.
—— (1993). *The Field of Cultural Production: Essays on Art and Literature*, edited by Randal Johnson. London: Polity Press.
—— (1998). *Vom Gebrauch der Wissenschaft. Für eine klinische Soziologie des wissenschaftlichen Feldes*. Konstanz: Universitätsverlag Konstanz.
Brabrook, Edward William (1898). "Presidential Address to Section H-Anthropology". *Report of the BAAS* (British Association for the Advancement of Science): 999–1010 [cf. *Nature* 58/1509 (1898): 527–32].
Brinton, Daniel Garrison (1894). "The Nation as an Element in Anthropology". In *Memoirs of the International Congress of Anthropology*, edited by C. Staniland Wake, 19–36. Chicago: The Schulte Publishing Company.
Broberg, Gunnar and Nils Roll-Hansen (1996). *Eugenics and the Welfare State: Sterilisation Policy in Danemark, Sweden, Norway, and Finland*. East Lansing, MI: Michigan State University Press.

Brubaker, Rogers (2004). *Ethnicity Without Groups*. Cambridge and Massachusetts: Harvard University Press.

Bruinius, Harry (2006). *Better for all the World: The Secret History of Forced Sterilisation and America's Quest for Racial Purity*. New York: Alfred A. Knopf.

Bukharin, Nicolai I., ed. (1971). *Science at the Crossroads*. New York: Frank Cass & Co. [repr. from 1931 edition].

Burleigh, Michael and Wolfgang Wipperman (1991). *The Racial State: Germany 1933–1945*. Cambridge: Cambridge University Press.

Burrow, John W. (1966). *Evolution and Society: A Study in Victorian Social Theory*. Cambridge: Cambridge University Press.

Butler, Judith (1991). *Das Unbehagen der Geschlechter*. Frankfurt am Main: Suhrkamp [orig. *Gender Trouble: Feminism and the Subversion of Identity: Thinking Gender*. New York and London: Routledge, 1990].

Butler, Samuel (1981/1872). *Erewhon or Over the Range*, edited by Hans-Peter Breuer. Newark: University of Delaware Press.

Calcagno, James M., ed. (2003). "Biological Anthropology: Historical Perspectives on Current Issues, Disciplinary Connections, and Future Directions". Special Issue in *American Anthropologist* 15, no. 1.

Callon, Michael (1986). "Some Elements of a Sociology of Translation: Domestication of the Scallops and the Fishermen of St. Brieuc Bay". In *Power, Action and Belief: A New Sociology of Knowledge*? edited by John Law, 196–233. London: Routledge and Kegan Paul.

Caminopetros, J. (1938a). "Recherches sur l'anémia erythroblastique infantile, des peoples de la Médirerranée Orientale". *Annales de medicine* 43 (January): 27–43.

—— (1938b). "Recherches sur l'anémia erythroblastique infantile, des peoples de la Médirerranée Orientale". *Annales de medicine* 43 (February): 104–25.

Canguilhem, Georges (1979). *Wissenschaftsgeschichte und Epistemologie. Gesammelte Aufsätze*. Frankfurt am Main: Suhrkamp.

—— (2007). *The Normal and the Pathological*. New York: Zone Books.

Carabott, Philip (2003). "The Politics of Constructing the Ethnic 'Other': The Greek State and its Slav-Speaking Citizens, ca. 1912–ca. 1949". In *Minorities in Greece—Historical Issues and New Perspectives*, edited by Sevasti Trubeta and Christian Voss (Special Issue of *History and Culture of South Eastern Europe: An Annual Journal*), 5: 141–59.

Carson, John (2003). "The culture of intelligence". In *The Modern Social Sciences*, edited by Theodore M. Porter and Dorothy Ross (The Cambridge History of Sciences, vol. 7), 635–638. Cambridge: Cambridge University Press.

Caspari, Rachel (2003). "From Types to Populations: A Century of Race, Physical Anthropology, and the American Anthropological Association". *American Anthropologist* 105, no. 1: 65–76.

Chalaby, Jean K. (1996). "Beyond the Prison-House of Language: Discourse as a Sociological Concept". *British Journal of Sociology* 47, no. 4: 685–95.

Chapoutot, Johann (2008). *Le national-socialisme et l'Antiquité*. Paris: Presses universitaires de France.

Charamis, Spilios I. (1926). "Trachoma and Crime" ["Τράχωμα και έγκλημα"]. *Ιατρική, Ελληνική Ιατρική Επιθεώρησις* 4 (1 July): 25–6.

Charitakis, Kostis (1928a). "Social Hygiene" ["Κοινωνική Υγεινή"]. *Ελληνίς* 8 (1 January): 10–12

—— (1928b). "Social Hygiene" ["Κοινωνική Υγεινή"]. *Ελληνίς* 8 (2 February): 63–5.

—— (1929). *Updated State of Affairs on Social Hygiene: Principles and Criteria for the Organisation of the Public Hygiene* [*Τα νεώτατα δεδομένα επί της κοινωνικής υγιεινής: αρχαί και κριτήρια οργανώσεως της δημόσιας υγιεινής*]. Athens: Ethniko Tipografeio [Εκ του Εθνικού Τυπογραφείου].

Chatzivasileiou, Grigorios P. (1925). "The Population Problem in Greece" ["Το πρόβλημα του πληθυσμού εν Ελλάδι"]. *Αρχείον Οικονομικών και Κοινωνικών Επιστημών* 5, no. 3: 257–66.

—— (1929a). "Coefficient of Mortality as Hygiene Indicator: Idiostasia and Anthropometrics" ["Ο συντελεστής θνησιμότητος ως δείκτης υγιεινής. Ιδιοστασία και ανθρωπομετρία"]. *Ιατρικά χρονικά* 2, no. 9: 531–46.

—— (1929b). "The Social Legislation in Greece as Factor of Social Hygiene" ["Η κοινωνική νομοθεσία παρ' ημίν ως παράγων της κοινωνικής υγιεινής. Η ενδεικνυόμενη κατεύθυνσις αυτής"]. *Ιατρικά Χρονικά* 2, no. 12: 784–91.

Chouzouris, Stefanos (1988). *Doctor in Three Wars (Autobiography)* [*Γιατρός σε τρεις πολέμους (αυτοβιογραφία)*]. Thessaloniki: [no publisher].

Christian Union of Scientists [Χριστιανική Ένωσις Επιστημόνων] (1957). *Ioannis Skaltsounis (1821–1905) on the Occasion of the 50th Anniversary of his Death* [*Ιωάννης Σκαλτσούνης (1821–1905) επί τη συμπληρώσει πεντηκονταετίας από το θανάτό του*]. Athens: Edition "Damaskus" [Εκδ. "Η Δαμασκός"].

Clogg, Richard, ed. (2008). *Bearing Gifts to Greeks: Humanitarian Aid to Greece in the 1940s*. Basingstoke [et al.]: Palgrave Macmillan.

Communist Review [ΚΟΜΕΠ] (1945). "Professor Koumaris and Racism" ["Ο καθηγητής Κούμαρης και ο ρατσισμός"] 43 (November): 42–4.

Conrad, Peter (2007). *The Medicalization of Society: On the Transformation of Human Conditions into Treatable Disorders*. Baltimore: The Johns Hopkins University Press.

Coon, Carleton Stevens (1939). *The Races of Europe*. New York: The Macmillan Company.

Coray, [Adamantius] (1800/1801). *Traité d'Hippocrate des airs, des eaux et des lieux*. Paris: Baudelot et Eberhart, 1st Volume 1800; 2nd Volume 1801.

—— (1803). *Mémoire sur l'etat actuel de la civilisation dans la Grèce: lu à la Société des observateurs de l'homme*, le 16 nivôse an XI (6 January): Paris [repr. Hors commerce, Institute Néo-hellénique de la Sorbonne, Université Paris IV, 1979].

Count, Earl W., ed. (1950). *This is Race: An Anthology Selected from the International Literature on the Races of Man*. New York: Henry Schuman.

Dalkavoukis, Vasilis and Filimon Karamitsos (2008). "Ioannina in the 1940s—On the Basis of Memoirs of Petros Apostolidis, EAM-Mayor" ["Τα Γιάννενα στη δεκαετία του 1940 μέσα από τα απομνημονεύματα του Πέτρου Αποστολίδη, Δημάρχου του ΕΑΜ"]. *Ηπειρωτικά Χρονικά* 42: 189–213.

Damaskos, Dimitris and Dimitris Plantzos, eds (2008). *A Singular Antiquity—Archaeology and Hellenic Identity in Twentieth-Century Greece*. Athens: Mouseio Benaki.

Deimling, Wilhelm (1862). *Die Leleger. Eine ethnographische Abhandlung*. Leipzig: Druck und Verlag von B.G. Teubner.

Depaepe, Marc (1992). "Experimental Research in Education 1890–1940. Historical processes behind the development of a discipline in Western Europe and the United States. Aspects of Education". *Journal of the Institute of Education—The University of Hull* 47: 67–93.

Diamantopoulos, G. (1880). *Anthropology. For Use in Gymnasium, Schools, and for Self-Education* [*Ανθρωπολογία. Εις χρήσιν των Γυμνασίων και Διδασκαλείων και εις αυτοδιδασκαλίαν μεμορφωμένων. Μετά 95 εικόνων και 1 ανατομικού πίνακος*]. Izmir: Edition "New Izmir" [Τύποις "Νέας Σμύρνης"].

Dimakopoulou, Charikleia (1979). "The Saint-Simonist François Graillard on the Greek Political Question (Notes and Suggestions)" ["Ο Σανσιμονιστής François Graillard περί των ελληνικών πολιτικών πραγμάτων (Παρατηρήσεις και Προτάσεις"]. *Bulletin De la Société Historique et Ethnologique de la Grèce* 22: 368–448.

Dimaras, Konstantinos (1989). *New-Greek Enlightenment* [*Νεοελληνικός Διαφωτισμός*] (5th edition). Athens: Ermis [Ερμής].

Dordanas, Stratos (2004). "The German Attitude towards the Jewish 'Problem' in Thessaloniki (April 1941–January 1943)" ["Η γερμανική στάση απέναντι στο εβραϊκό 'πρόβλημα' της Θεσσαλονίκης (Απρίλιος 1941–Ιανουάριος 1943)"], *Ελληνική Ιστορική Εταιρεία* [Proceedings of the 14th Panhellenic Historical Conference, 30–31 May–1 June 2003]: 607–18. Thessaloniki: Greek Historical Society [Ελληνική Ιστορική Εταιρεία].

Doxiadis, Apostolos (1928a). "Social Biology: Biological Politics" ["Κοινωνική βιολογία—Βιολογική Πολιτική"]. *Ελληνικά Γράμματα* 3, no. 2: 49–51.

—— (1928b). "Biological Politics based on the Growth of the Country's Population" ["Βιολογική πολιτική με βάση την αύξηση του πληθυσμού της χώρας"]. *Ελληνικά Γράμματα* 3, no. 3: 95–9.

—— (1930a). "(Problems of Life) Biology and Church" ["(Προβλήματα της ζωής) Βιολογία και Εκκλησία". *Ελεύθερον Βήμα* (7 September): 1.

—— (1930b). "(Problems of Life) Biology and Church" ["(Προβλήματα της ζωής) Βιολογία και Εκκλησία"]. *Ελεύθερον Βήμα* (8 September): 1.

—— (1930c). "(Problems of Life) The American Apostle and the Abundance of Children" ["(Προβλήματα της ζωής) Η Αμερικανίς απόστολος κατά της πολυτεκνίας"]. *Ελεύθερον Βήμα* (9 September): 1.

—— (1930d). "(Problems of Life) Reduction or Regulation of Births" ["(Προβλήματα της ζωής) Περιορισμός ή διαρρύθμισις των γεννήσεων"]. *Ελεύθερον Βήμα* (10 September): 1.

—— (1930e). "(Problems of Life) Reduction or Regulation of Births" ["(Προβλήματα της ζωής) Περιορισμός ή διαρρύθμισις των γεννήσεων"], *Ελεύθερον Βήμα* (11 September): 1.

—— (1930f). "(Problems of Life) The Role of Women" ["(Προβλήματα της ζωής) Η θέσις της γυναικός"]. *Ελεύθερον Βήμα* (12 September): 1, 2.

—— (1930g). "(Problems of Life) The Overpopulation in Greece" ["(Προβλήματα της ζωής) Ο αναπληθυσμός της Ελλάδος"]. *Ελεύθερον Βήμα* (13 September): 1.

—— (1930h). "(Problems of Life) The Two Prevalent Instincts" ["(Προβλήματα της ζωής) Τα δύο κυρίαρχα ένστικτα"]. *Ελεύθερον Βήμα* (14 September): 1.

—— (1930i). "(Biological Problems:) Heredity-Environment" ["(Βιολογικά προβλήματα) Κληρονομικότης—Περιβάλλον"]. *Ελεύθερον Βήμα* (1 November): 1.

—— (1931a). "(Biological Problems) Mendel—The Father of the Science of Heredity" ["(Βιολογικά προβλήματα) Ο πατήρ της κληρονομικής επιστήμης—Ο Mendel"]. *Ελεύθερον Βήμα* (13 January): 1.

—— (1931b). "(Biological Problems) Heredity—Answers to Questions" ["(Βιολογικά Προβλήματα) Η κληρονομικότης—Λύσεις Αποριών"]. *Ελεύθερον Βήμα* (14 January): 1, 5.

—— (1931c). "(Biological Problems) The Mystery of Life" ["(Βιολογικά προβλήματα) Το μυστήριον της ζωής"]. *Ελεύθερον Βήμα* (15 January): 1, 5.

—— (1931d). "(Biological Problems) The Determination of the Sex of the Child" ["(Βιολογικά προβλήματα) Ο καθορισμός του γένους του παιδιού"]. *Ελεύθερον Βήμα* (16 January): 1.

—— (1931e). "(Biological Problems) The Human Evolution" ["(Βιολογικά προβλήματα) Η εξέλιξις του ανθρώπου"]. *Ελεύθερον Βήμα* (17 January): 1.

—— (1931f). "(Biological Problems) The Weakening of Human Kind" ["(Βιολογικά προβλήματα) Η εξασθένησις του ανθρώπινου είδους"]. *Ελεύθερον Βήμα* (18 January): 5.

—— (1931g). "(Biological Problems) More Children" ["(Βιολογικά προβλήματα) Περισσότερα Παιδιά"]. *Ελεύθερον Βήμα* (19 January): 3.

—— (1931h). "(Biological Problems) Which Tactic Should the State and Society Pursue? ["(Βιολογικά Προβλήματα) Ποιάν τακτικήν οφείλει να ακολουθήση η κοινωνία και η πολιτεία;"]. *Ελεύθερον Βήμα* (15 March): 3.

—— (1933). "Eugenics (Library of Social Hygiene)" ["Ευγονική"], *Βιβλιοθήκη Κοινωνικής Υγιεινής*], (Special Edition on "Hygiene of Marriage") 2: 40–7.

Drakoulidis, Nikolaos (1932a). "Some Words on the 'Library of Social Hygiene'" ["Δύο λόγια για την 'Βιβλιοθήκη Κοινωνικής Υγιεινής'"]. *Library of Social Hygiene* [*Βιβλιοθήκη Κοινωνικής Υγιεινής*], (Special Issue on "Social Diseases") 1: 5–13.

—— (1932b). "Propaganda for Social Hygiene" ["Η προπαγάνδα υπέρ της κοινωνικής υγιεινής"]. *Ελληνίς* 12, no. 10: 201–2.

—— (1932c). "The Prenuptial Medical Examination" ["Η προ του γάμου ιατρική εξέτασις"]. *Ελεύθερος Άνθρωπος* (17 February).

—— (1933). "The Prenuptial Medical Examination" ["Η προ του γάμου ιατρική εξέτασις"], Library of Social Hygiene [*Βιβλιοθήκη Κοινωνικής Υγιεινής*] (Special Edition on "Hygiene of Marriage") 2: 50–80.

—— s.v. [no year]. "Marriage" ["Γάμος"]. In the *Great Greek Encyclopaedia* [*Μεγάλη Ελληνική Εγκυκλοπαίδεια*], vol. 8: 126–7.

Drakoulis, Platon (1927). *Criminality and Humanism: A Study on the Revision of the Penal Law* [*Εγκληματικότης και Ανθρωπισμός. Μελέτη περί αναμορφώσεως του ποινικού συστήματος*]. Athens: Avdi [Εκδοτικός οίκος Αύδη].

Droysen, Johann Gustav (1833). *Geschichte Alexanders des Großen*. Berlin: Finke.
—— (1836). *Geschichte des Hellenismus*. Theil 1: *Geschichte der Nachfolger Alexanders*. Hamburg: Perthes.
—— (1843). *Geschichte des Hellenismus*. Theil 2: *Geschichte der Bildung des hellenistischen Staatensystems*. Hamburg: Perthes.
Duster, Troy (1990). *Backdoor to Eugenics*. New York: Routledge.
Eickstedt, Egon Freiherr von (1934). *Rassenkunde und Rassengeschichte der Menschheit*. Stuttgart: Enke.
—— (1936). *Grundlagen der Rassenpsychologie*. Stuttgart: Enke.
Eleftheriadis, Dimosthenis (1933). "Heredity in the Family" ["Η κληρονομικότης εις την οικογένεια"], *Library of Social Hygiene* [*Βιβλιοθήκη Κοινωνικής Υγιεινής*] (Special Edition on "Hygiene of Marriage") 2: 81–103.
—— (1948). *Lessons of Social Biology* [*Μαθήματα Κοινωνικής Βιολογίας*]. Athens: [no publisher].
Elias, Nobert (1993). *Über den Prozess der Zivilisation* (vol. 1). Frankfurt am Main: Suhrkamp.
Eßbach, Wolfgang (2005). "Elemente ideologischer Mengenlehre: Rasse, Klasse, Masse". In *Grenzen des Menschseins—Probleme einer Definition des Menschlichen*, edited by Justin Stagl and Wolfgang Reinhard, 727–55. Vienna, Cologne and Weimar: Böhlau.
Exarchopoulos, Nikolaos (1928). *Science of the Body of the Children* [*Σωματολογία του παιδός*] (2nd edition). Athens: Dimitrakos [Δ. *και* Π. Δημητράκος].
Fallaize, E.N. (1921). "Suggestions for the Classification of the Subject-Matter of Anthropology". *The Museums Journal* 20: 149–56 and 180–3.
Fallmerayer, Jakob Philipp (1830–6). *Geschichte der Halbinsel Morea während des Mittelalters: ein historischer Versuch* (2 volumes). Stuttgart [et al.]: Cotta.
—— (1857–60). *Das Albanische Element in Griechenland* (2 volumes). Munich: [no publisher].
Ferdinand, Ursula (2007). "Geburtenrückgangstheorien und 'Geburtenrückgangs-Gespenster' 1900–1930". In *Herausforderung Bevölkerung. Zu Entwicklungen des modernen Denken über die Bevölkerung vor, im und nach dem "Dritten Reich"*, edited by Ursula Ferdinand and Josef Ehmer, 77–98. Wiesbaden: VS-Verlag für Sozialwissenschaften.
—— (2009). "Der 'faustische Pakt' in der Sozialhygiene Alfred Grotjahns (1869–1931). Sozialhygiene und die Beziehungen zur Eugenik und Demographie". In *Wie Nationalsozialistisch ist die Eugenik? / What is National Socialist about Eugenics?* edited by Renate Wecker et al., 173–85. Vienna, Cologne and Weimar: Böhlau.
Fischer, Isidor, ed. (1933). *Biographisches Lexikon der hervorragenden Ärzte der letzten fünfzig Jahre* (vol. 2). Berlin: Urban & Schwarzenberg.
Fischer, Peter (2007). *Mitte, Maß und Mäßigkeit. Zur Idee und Relevanz eines gesellschaftlichen Mittebezuges*. Hamburg: Verlag Dr Kovač.
Fleischer, Hagen (2008). *The Wars of Memory: The Second World War in the Public History* [*Οι πόλεμοι της μνήμης. Ο Β' Παγκόσμιος Πόλεμος στη Δημόσια Ιστορία*]. Athens: Nefeli [Νεφέλη].
Folkmar, Daniel (1893–9). "Sociology as Based upon Anthropology". *Tracts on Social and Industrial Questions*: 1–4 [repr. from *Journal of Social Science* (1898)].
Foucault, Michel (1972). *Archaeology of Knowledge*. London and New York: Routledge Classics.
—— (1978). *Dispositive der Macht. Über Sexualität, Wissen und Wahrheit*. Berlin: Merve Verlag.
—— (1986). "Of Other Spaces". *Diacritics* 16, no. 1: 22–7.
—— (1988; 1990; 1990). *The History of Sexuality* (3 volumes). New York: Vintage Books.
—— (1992). *Die Ordnung des Diskurses*. Frankfurt am Main: Fischer-Wissenschaft [*L'Ordre du discours*. Paris: Gallimard, 1972].
—— (1994). *Power. Essential works of Foucault, 1954–1984: Volume 3*. Rabinow, P. & Faubion, J. D. (eds). New York: New York Press.

—— (1998). *History of Sexuality 1: the Will to Knowledge*. London: Penguin Books.
—— (2003). *Lectures at the Collège De France 1975–1976*, edited by Mauro Bertani and Alessandro Fontana. New York: Picador.
—— (2006). *The Birth of the Clinic: An Archaeology of Medical Perception*. London: Routledge.
Franklin, Sarah (2003). "Re-Thinking Nature-Culture: Anthropology and New Genetics". *Anthropological Theory* 3, no. 1: 65–85.
Frazer, James George (1890). *The Golden Bough: A Study in Comparative Religion* (2 vols). London: Macmillan and Co. (expanded to 12 volumes in 1900).
Freudenthal, Gideon (2005). "The Hessen-Grossman Thesis: An Attempt at Rehabilitation". *Perspectives on Science* 13, no. 2: 166–93.
Freudenthal, Gideon and Peter McLaughlin (2009). *The Social and Economic Roots of the Scientific Revolution: Texts by Boris Hessen and Henryk Grossmann* [Boston Studies in the Philosophy of Science, vol. 278]. Heidelberg and New York: Springer.
Fuchs, Brigitte (2003). *"Rasse", "Volk", Geschlecht. Anthropologische Diskurse in Österreich 1850–1960*. Frankfurt am Main and New York: Campus Verlag.
Galton, Francis (1904). "Eugenics: Its Definition, Scope, and Aims". *The American Journal of Sociology* 10, no. 1: 1–6.
—— (2004). *Essays in Eugenics*. Honolulu, Hawaii: University Press of the Pacific [reprint from the 1909 edition].
Gardikas, Konstantinos D. (2002). *The Chronicle of an Incurable Doctor* [*Το χρονικό ενός αγιάτρευτου… γιατρού*]. Athens: Parisianou [Επιστημονικές Εκδόσεις Παρισιάνου] (1st edition 1985).
Gardikas, Konstantinos G. (1934a). "Über den Einfluß der Jahreszeiten auf die Kriminalität in Griechenland". *Monatsschrift für Kriminalpsychologie und Strafrechtsreform* 25, no. 1: 24–9.
—— (1934b). "Beruf und Kriminalität in Griechenland". *Monatsschrift für Kriminalpsychologie und Strafrechtsreform* 25, no. 10–12: 549–57.
—— (1934c). "Eugenik im alten Griechenland". *Monatschrift für Kriminalpsychologie und Strafrechtsreform* 25, no. 6: 340–6.
—— (1934d). "Ein griechischer Fortbildungskurs der Strafrechtswissenschaft". *Monatsschrift für Kriminalpsychologie und Strafreform* 25, no. 1: 47–8.
Garoufalias, Petros (1930). "Our Demographic Problem: The Overpopulation of the Cities" ["Το δημογραφικόν *μας* πρόβλημα. Ο υπερπληθυσμός *των πόλεων*"]. *Εργασία* 1, no. 8: 17.
Gedeon, Sofia (1930). *Paedometrical Researches in Greece* (vol. 2) [*Παιδομετρικαί έρευναι εν Ελλάδι*]. Athens: Laboratory of Experimental Paedagogics of the University [Δημοσ. Εργαστ. Πειραματικής Παιδαγωγικής Πανεπιστημίου].
Gedeon, Sophie (1935). "Jugendkunde und pädagogische Psychologie in Griechenland. Aus der Tätigkeit des Pädagogischen Instituts der Universität zu Athen. I. Wissenschaftliche Arbeiten des Laboratoriums für experimentale Pädagogik der Universität zu Athen. II. Versuch und Forschung in der Versuchsschule". *Zeitschrift für Jugendkunde* 5: 129–32.
Gefou-Madianou, Dimitra (2000). "Disciples, Discipline and Reflection: Anthropological Encounters and Trajectories". In *Audit Cultures, Anthropological Studies in Accountability, Ethics and the Academy*, edited by Marilyn Strathern, 256–78. London: Routledge.
Georgiadou, Maria (2004). *Constantin Carathéodory, Mathematics and Politics in Turbulent Times*. Berlin: Springer.
Giannios, Nikolaos (1914). *State Socialism* [*Ο σοσιαλισμός του κράτους*]. Athens: Edition of the Socialist Centre [Εκδόσεις Σοσιαλιστικού Κέντρου].
Gkarmati-Theodoropoulou, Anna (1952). *Intellectual Work and its Relationship to Bodily Work and the Psychic Functions* [*Ο διανοητικός κάματος και η σχέσις αυτού προς τον σωματικόν κάματον και τας ψυχικάς λειτουργίας*]. Athens: [no publisher].
Glinos, Dimitris (1976). *War Trilogy* [*Η Τριλογία του Πολέμου. Οι μονόλογοι του ερημίτη της Σαντορίνης*]. Athens: Athina [Εκδόσεις Αθηνά] (1st edition 1945).
Gobineau, Comte Arthur de (1853a; 1853b; 1855a; 1855b). *Essai sur l'inégalité des races humaines* (4 volumes). Paris: Firmin-Didot.

Goldstein, Jan (2003). "Bringing the Psyche into Scientific Focus". In *The Modern Social Sciences*, edited by Theodore M Porter and Dorothy Ross (The Cambridge History of Sciences, vol. 7), 131–53. Cambridge: Cambridge University Press.

Goschler, Constantin (2003). *Rudolf Virchow: Mediziner, Anthropologe, Politiker*. Cologne, Weimar and Vienna: Böhlau.

—— (2004). "'Wahrheit' zwischen Seziersaal und Parlament. Rudolf Virchow und der kulturelle Deutungsanspruch der Naturwissenschaften". *Geschichte und Gesellschaft* 30, no. 2: 217–49.

Gould, Stephen Jay (1981). *The Mismeasure of Man*. New York: Norton.

Graikos, Panos (1930). "The Problem of Urbanisation" ["Το πρόβλημα της αστυφιλίας"], *Εργασία* 1, no. 14: 10–11, and 1, no. 15: 15–16.

Grosse, Ernst (1900). *Kunstwissenschaftliche Studien*. Tübingen: Mohr.

Hamilakis, Yannis (2007). *The Nation and its Ruins: Antiquity, Archaeology, and the National Imagination in Greece*. Oxford: Oxford University Press.

Hampe, Roland (1941). "Griechenland, das Land der Gegensätze". *Volk und Rasse* 16, no. 7–8: 117–21.

Hann, Chris, ed. (2005). *Studying Peoples in the People's Democracies: Socialist Era Anthropology in East-Central Europe*. Münster: LIT Verlag.

Hartfiel, Günter (1976). *Wörterbuch der Soziologie*. Stuttgart: Alfred Kröner Verlag.

Hauss, Gisela and Béatrice Ziegler (2009). "Sterilisation bei Armen und Unmündigen. Eine Untersuchung der Vormundschaftspraxis in den Städten St. Gallen und Bern". In *Wie Nationalsozialistisch ist die Eugenik? What is National Socialist about Eugenics*? edited by Regina Wecker et al., 75–91. Vienna, Cologne and Weimar: Böhlau.

Heald, Suzette and Ariane Deluz, eds (1994). *Anthropology and Psychoanalysis: an encounter through culture*. London and New York: Routledge.

Heberer, Gerhard, Gottfried Kurth and Ilse Schwidetzky-Roesing (1959). *Anthropologie*. Frankfurt am Main: Fischer Verlag.

Hering, Gunnar (1992). *Die politischen Parteien in Griechenland 1821–1936* (2 volumes). Munich: Oldenbourg.

Herzfeld, Michael (1982). *Ours Once More: Folklore, Ideology, and the Making of Modern Greece*. Austin, TX: Univerity of Texas Press.

Hessen, Boris (1971). "The Social and Economic Roots of Newton's Principia". In *Science at the Crossroads*, edited by Nicolai I. Bukharin, 151–212. New York: Frank Cass & Co. [repr. from 1931 edition].

Hillesheim, Jürgen and Elisabeth Michael (1993). *Lexikon nationalsozialistischer Dichter: Biographien, Analysen, Bibliographien*. Würzburg: Königshausen & Neumann.

Hionidou, Violetta (2006). *Famine and Death in Occupied Greece, 1941–1944*. Cambridge: Cambridge University Press.

Hirsch, August (1860/1862–4). *Handbuch der historisch-geographischen Pathologie*. Erlangen: Verlag von Ferdinand Enke, vol. 1: 1860; vol. 2: 1862–4.

Hohmann, Joachim S. (1991). *Robert Ritter und die Erben der Kriminalbiologie. Zigeunerforschung im Nationalsozialismus und in Westdeutschland im Zeichen des Rassismus*. Frankfurt am Main [et al.]: Peter Lang.

Hoßfeld, Uwe (2005). *Geschichte der biologischen Anthropologie in Deutschland: von den Anfängen bis zur Nachkriegszeit*. Stuttgart: Steiner.

Hunter, Richard A. and Ida Macalpine (1960). "Harvey's 'Boy abowt Holborn Bridg'". *Journal of the History of Medicine and Allied Sciences* (Column: Notes and Events) 15, no. 3: 295–6.

Huonker, Thomas (2002). *Anstaltseinweisungen, Kindswegnahmen, Eheverbote, Sterilisationen, Kastrationen. Fürsorge, Zwangsmassnahmen, "Eugenik" und Psychiatrie in Zürich zwischen 1890 und 1970*. Zurich: [Sozialdepartement der Stadt Zürich, Edition Sozialpolitik Nr. 7, Sozialberichterstattung].

Iosifidis, Simeon (1925). "Turkey, Islamism and Civilization" ["Η Τουρκία, ο Ισλαμισμός και ο Πολιτισμός"]. *Η Παμπροσφυγική* (8 February): 1.

—— (1936). "Filetic/Racial Affinity Between Greeks and Turks" ["Αι φυλετικαί συγγένειαι μεταξύ Ελλήνων και Τούρκων"], serialized:
Προσοχή (8 April 1936): 1.
Προσοχή (29 March 1936): 1.
Προσοχή (15 April 1936): 1.

Jeffreys, M.D.W. (1949). "The Greek Race?" *Man* 49: 36.

Jensen, Arthur R. (1969). "How Much Can We Boost I.Q. and Scholastic Achievement? *Harvard Educational Review* 33: 1–123.

—— (1982). "The Debunking of Scientific Fossils and Straw Persons". *Contemporary Education Review* 1, no. 2 (Summer): 121–35.

Kahl, Thede (2003). "Aromanians in Greece: Minority or Vlach-Speaking Greeks?" In *Minorities in Greece—Historical Issues and new Perspectives*, Special Issue of *History and Culture of South Eastern Europe: An Annual Journal* 5: 205–19.

Kairis, Michail N. (1917). *Eugenics* [*Η Ευγονία*]. Athens: Printing House Frantzeskaki [Εκ του Τυπογραφείου Αδελφών Β Φραντζεσκάκη].

Kakouri, Katerina (1946). *Pre-Aesthetic Forms of Theatre* [*Προαισθητικές μορφές Θεάτρου*]. Athens: Friends of the Book [Οι φίλοι του βιβλίου].

Kalfoglous, H.I. (2002). *Historical Geography of the Peninsula of Asia Minor* [*Ιστορική Γεωγραφία της Μικρασιατικής Χερσονήσου*]. Athens: Centre of Asia Minor Studies [Κέντρο Μικρασιατικών Σπουδών] [first edition 1899].

Kannelis A. and A. Savvas (1964). "Craniological Study on the Homo neanderthalensis of Petralona" ["Κρανιολογική μελέτη του Homo neanderthalensis των Πετραλώνων"]. *Επιστημονική επετηρίς Φυσικομαθηματικής Σχολής Πανεπιστημίου Θεσσαλονίκης* 9: 65–82.

Karakasidou, Anastasia N. (1997). *Fields of Wheat, Hills of Blood: Passages to Nationhood in Greek Macedonia, 1870–1990*. Chicago, IL [et al.]: University of Chicago Press.

Karanikas, Dimitris (1934). "Berichte aus Griechenland". *Monatsschrift für Kriminalpsychologie und Strafrechtsreform* 25, no. 2–3: 153–4.

Kardamatis, Ioannis (1934). *Racial Land Minde Entorpeumiento de razar* [*Νάρκη φυλετική: Entorpeumiento de razar*]. Athens: Framma [Φλάμμα].

Karolides, Paul K. (1913). *Bemerkungen zu den alten kleinasiatischen Sprachen und Mythen*. Straßburg i.E.: Schmidt.

—— (1922–9). *Modern History of Greeks and Other Peoples of the Orient from 1821 to 1921* [*Σύγχρονος Ιστορία των Ελλήνων και των λοιπών λαών της Ανατολής από το 1821 μέχρι του 1921*], 7 volumes. Athens: Paraskeva Leoni [Τύποις Παρασκευά Λεώνη].

—— (1992). *Notes on the Arian Filetic Affinity in Asia Minor* [Σημειώσεις τινές περί της Μικρασιανής Αρίας ομοφυλίας/*Μικρασιανή Άρια Ομοφυλία*]. Thessaloniki: Kiromanos [Εκδ. Οίκος Κυρομάνος] (first edition 1886).

Katopodis, Georgios Ap. (1934). "The Greek Correctional Issue and Hitler's Law on Sterilisation" ["Το ελληνικόν σωφρονιστικόν πρόβλημα και ο χιτλερικός νόμος της στειρώσεως"], serialized:
Καθημερινή (22 February 1934).
Καθημερινή (25 February 1934).
Καθημερινή (26 February 1934).

Kazolea-Kavolari, Nagia (2002). "Psychology and Eugenics in Greece" ["Ψυχολογία και Ευγονική στην Ελλάδα"]. *Τετράδια Ψυχιατρικής* 78 (April, May, June): 52–64.

Kazolea-Tavoulari, Panagiota (2002). *History of Psychology in Greece (1880–1987)* [*Η ιστορία της Ψυχολογίας στην Ελλάδα (1880–1987)*]. Athens: Ellinika Grammata [Ελληνικά Γράμματα].

Keim, Wiebke (2008). *Vermessene Disziplin: zum konterhegemonialen Potential afrikanischer und lateinamerikanischer Soziologien*. Bielefeld: Transcript.

Keiter, Friedrich (1941). "Nordisch: Vorderasiatisch as rassenpsychologische Polarität". *Zeitschrift für Rassenkunde und die gesamte Forschung am Menschen* 12, no. 1: 60–5.

Keller, Reiner (2001). "Wissenssoziologische Diskursanalyse". In *Handbuch Sozialwissenschaftliche Diskursanalyse*, vol. 1: *Theorie und Methoden*, edited by Reiner Keller et al., 113–43. Opladen: Leske and Budrich.

Keller, Reiner et al., eds (2001). *Handbuch Sozialwissenschaftliche Diskursanalyse*, vol. 1: *Theorie und Methoden*. Opladen: Leske and Budrich.

Kesper-Biermann, Sylvia (2007). "'Ehegesundheit' als bevölkerungspolitisches Problem. Internationale Dimensionen von Diskussion und Gesetzgebung in der Weimarer Republik". In *Herausforderung Bevölkerung. Zu Entwicklungen des modernen Denken über die Bevölkerung vor, im und nach dem "Dritten Reich"*, edited by Ursula Ferdinand and Josef Ehmer, 123–32. Wiesbaden: VS-Verlag für Sozialwissenschaften.

Kiriakidou-Nestoros, Alki (1993). *Laographic Essays* [*Λαογραφικά μελετήματα*]. Athens: Poreia [Πορεία].

Kirtsis, Alexandros-Andreas (1996). *Sociological Thought and Modernization Ideologies in the Greek Inter-War Period* [*Κοινωνιολογική σκέψη και εκσυγχρονιστικές ιδεολογίες στον ελληνικό μεσοπόλεμο*]. Athens: Nisos.

Kitcher, Philip (1996). *The Lives to Come: The Genetic Revolution and Human Possibilities*. New York: Touchstone.

Klee, Ernst (2003). *Das Personenlexikon zum Dritten Reich. Wer war was vor und nach 1945*. Frankfurt am Main: Fischer Verlag.

Koch, Lene (2002). "The Ethos of Science: Relations between Danish and German Scientists around World War II". *Scandinavian Journal of History* 27: 167–73.

—— (2009). "How Eugenic was Eugenics? Reproductive Politics in the Past and the Present". In *Wie Nationalsozialistisch ist die Eugenik?* edited by Renate Wecker et al., 39–63. Vienna, Cologne and Weimar: Böhlau.

Koger, Friedrich (2008). *Die Anfänge der Ethnologie in Wien. Ein Beitrag zur Wissenschaftsgeschichte*. Vienna: LIT Verlag.

Kontogiannis P.M. (1995). *Geography of Asia Minor* [*Γεωγραφία της Μικράς Ασίας*]. Athens: Club for Advancing Useful Books [Σύλλογος προς Διάδοσιν Ωφέλιμων Βιβλίων] [first edition 1921].

Korasidou, Maria (2002). *When Disease Threatens: Surveillance and Health Control of the Population in Greece in the 19th Century* [*Όταν η αρρώστια απειλεί. Επιτήρηση και Έλεγχος της Υγείας του Πληθυσμού στην Ελλάδα του 19ου αιώνα*]. Athens: Tipothito [Τυπωθήτω].

—— (2004). *The Wretched of Athens and their Healers: Poverty and Philanthropy in the Greek Capital in the 19th Century* [*Οι Άθλιοι των Αθηνών και οι Θεραπευτές τους. Φτώχεια και Φιλανθρωπία στην Ελληνική Πρωτεύουσα τον 19ον αιώνα*]. Athens: Tipothito [Τυπωθήτω] [first edition 1995].

Kostis, Kostas (1988). "Agrarian Reform and Economic Development in Greece, 1917–1940" ["Αγροτική μεταρύθμιση και οικονομική ανάπτυξη στην Ελλάδα, 1917–1940"]. In [*Βενιζελισμός και Αστικός εκσυγχρονισμός*], edited by Georgios Mavrogordatos and Christos Chatziiosif, 149–57. Irakleion: University Press of Crete.

Kostopoulos, Tasos (2003). "Counting the 'Other': Official Census and Classified Statistics in Greece (1830–2001)". In *Minorities in Greece—Historical Issues and New Perspectives*. Special Issue of *History and Culture of South Eastern Europe: An Annual Journal*, edited by Sevasti Trubeta and Christian Voss, 5: 55–78.

—— (2005). *The self-censored memory. Security Battalions and the post-war nationalism* [Η αυτολογοκριμένη μνήμη Τα τάγματα ασφαλείας και η μεταπολεμική εθνικοφροσύνη]. Athens: Filistor.

—— (2008). *The Forbidden Language* [*Η απαγορευμένη γλώσσα*]. Athens: Vivliorama [Βιβλιόραμα] (4th edition).

Kozintsev, Alexander G. (1997). "Debetz, Georgy Frantsevich (1905–1969)". In *History of Physical Anthropology, An Encyclopedia*, vol. 1, edited by Frank Spencer, 322–3. New York and London: Garland Publishing, 1997.

Kretsi, Georgia (2002a). "Austauschbar-nicht-austauschbar. Albanophone Muslime (Çamen) und andere Grenzbevölkerungen des Epirus der Zwischenkriegszeit in Kräftefeld zwischen ethnischer Identitätskonstruktion und Entmischungspolitik". *History and Culture of South Eastern Europe: An Annual Journal* 4: 205–31.

—— (2002b). "The 'Secret' Past of the Greek–Albanian Borderlands. Cham Muslim Albanians: Perspectives on a Conflict over Historical Accountability and Current Rights". *Ethnologia Balkanica* 6: 171–95.

—— (2003). "From Landholding to Landlessness: The Relationship between the Property and Legal Status of the Cham Muslim Albanians". *Minorities in Greece—Historical Issues and New Perspectives*. Special Issue of *History and Culture of South Eastern Europe: An Annual Journal*, edited by Sevasti Trubeta and Christian Voss, 5: 125–38.

Krimpas, Kostas (1993). *Splinters of Mirror—Essays* [*Θραύσματα Κατόπτρου, Δοκίμια*]. Athens: Themelio [Θεμέλιο].

Krumbacher, Karl (1886). *Griechische Reise: Blätter aus dem Tagebuch einer Reise in Griechenland und in die Türkei*. Berlin: Hettler.

Kühl, Stefan (1997). *Die Internationale der Rassisten. Aufstieg und Niedergang der internationalen Bewegung für Eugenik und Rassenhygiene im 20. Jahrhundert*. Frankfurt am Main: Campus.

Labouvie, Eva (2007). "Alltagswissen—Körperwissen—Praxiswissen—Fachwissen: Zur Aneignung, Bewertungs- und Orientierungslogik von Wissenskulturen". *Berichte der Wissenschaftsgeschichte* 30: 119–34.

Lada, Sasa, ed. (2007). *The Image of Schientists in Greece 1900–1980* [*Η Εικόνα του/της Επιστήμονα στην Ελλάδα 1900–1980. [Εν]τοπισμένες ταυτότητες—ταυτότητες τόπων*]. Thessaloniki: Museum of Photography [ΑΠΘ, Μουσείο Φωτογραφίας Θες/κης].

Lagos, Katerina (2005). "The Metaxas Dictatorship and Greek Jewry, 1936–1941". Unpublished DPhil thesis, St Antony's College, Oxford University, UK.

Lampadarios, Emmanouil (1940). *School Hygiene with Remarks on Paedology* [*Σχολική Υγιεινή μετά στοιχεία παιδολογίας*]. Athens: [no publisher].

Latour, Bruno (1983). "Give Me a Laboratory and I Will Raise the World". In *Science Observed: Perspectives on the Social Study of Science*, edited by Karin Knorr-Cetina and Michael Mulkay, 141–70. London: Sage.

—— (1988). *The Pasteurization of France*. Cambridge: Harvard University Press.

—— (1993). *We Have Never been Modern*. New York: Harvester Wheatsheaf.

—— (1999). *Pandora's Hope: Essays on the Reality of Science Studies*. Cambridge, MA: Harvard University Press.

Leaf, Murray J. (1974). "The Development of Modern Anthropology". In *Frontiers of Anthropology: An Introduction to Anthropological Thinking*, edited by Murray J. Leaf, 1–47. New York: D. Van Nostrand Company.

Lebzelter, Victor (1929). "Rasse und Volk in Südosteuropa (Eine paläo-ethnologische Studie)". *Mitteilungen der Anthropologischen Gesellschaft in Wien* 59, 2–3: 61–126.

Leininger-Miller, Theresa A. (2001). *New Negro Artists in Paris: African American Painters and Sculptors in the City of Light, 1922–1934*. Piscataway, NJ: Rutgers University Press.

Lemke, Thomas (2002). "Genetic Testing, Eugenics and Risk". *Critical Public Health* 12, no. 3: 283–90.

—— (2003). "Rechtssubjekt oder Biomasse? Reflexionen zum Verhältnis von Rassismus und Exklusion". In *Biopolitik und Rassismus, Biopolitk und Rassismus*, edited by Martin Stingelin, 160–83. Frankfurt a.M: Suhrkamp.

—— (2004). "Disposition and determinism—genetic diagnostics in risk society". *The Sociological Review* 52, no 4: 550–566.

—— (2007). *Biopolitik zur Einführung*. Hamburg: Junius Verlag.

Lentin, Alana (2000). "Race, Racism, Anti-Racism: Challenging Contemporary Classifications". *Social Identities* 6, no. 1 (March): 91–106.

Leontis, Artemis (1995). *Topographies of Hellenism: Mapping the Homeland*. Ithaca, NY, and London: Cornell University Press.

Lester, Paul and Jacques Millot (1947). *Grundriß der Anthropologie*. Lahr: Schauenburg.

Lewin, Gerson (1934). *Ochrona zdrowia i eugenika w Biblii i Talmudzie*. Warsaw: Brothers Wójcikiewicz Publishing House.

Lewontin, R.C., Steven Rose and Leon J. Kamin (1984). *Not in Our Genes: Biology, Ideology and Human Nature*. New York: Pantheon Books.

Liakos, Antonis (1993). *Labour and Politics in the Interior Greece: The International Labor Office and the Rise of Welfare Institutions* [*Εργασία και Πολιτική στην Ελλάδα του*

Μεσοπολέμου. Το Διεθνές Γραφείο Εργασίας και η ανάδυση των κοινωνικών θεσμών]. Athens: Foundation of the Commercial Bank [Ίδρυμα Έρευνας και Παιδείας της Εμπορικής Τράπεζας της Ελλάδος].
Linke, Uli (1999). *Blood and Nation: The European Aesthetics of Race*. Philadelphia, PA: University of Pennsylvania Press.
Lipphardt, Veronika and Kiran Klaus Patel (2008). "Neuverzauberung im Gestus der Wissenschaftlichkeit. Wissenschaftspraktiken im 20. Jahrhundert am Beispiel menschlicher Diversität". *Geschichte und Gesellschaft* 33, no. 4: 425–54.
Lippman, Abby (1991). "Prenatal Genetic Testing and Screening: Constructing Needs and Reinforcing Inequalities". *American Journal of Law and Medicine* 17: 15–50.
Livas, Xerxis (1963/1956). *Aegean—Cradle of the Arians and Hellenism: Cultural Antecedence and Radiation, Filetic/Racial Unity, Racially Negligible Penetration of the Slavs and Turks* [*Η Αιγηΐς κοιτίς των αρίων και του ελληνισμού: Πολιτιστική προήγησις και ακτινοβολία, φυλετική ενότης, αμελητέα φυλετικώς η παρείσδυσις των σλάβων και των τούρκων*]. Athens: Sinodinow [Τύποις Χ. Συνοδινός] (second edition Athens: [no publisher], 1963).
Logothetopulos, K[onstantinos] (1926). "Eine absolut sichere Blutstillungsmethode bei vaginalen und abdominalen gynäkologischen Operationen". *Zentralblatt für Gynäkologie* 50, no. 50: 3202–4.
Lösch, Niels C. (1997). *Rasse als Konstrukt: Leben und Werk Eugen Fischers*. Frankfurt am Main [et al.]: Peter Lang.
Louros, N[ikolaos] (1930). "Bemerkungen zu einer sogenannten 'absolut sicheren Blutstillungsmethode' nach Logothetopoulos". *Zentralblatt für Gynälogie* 54, no. 44: 2781–3.
Louros, Nikolaos (1980). *Yesterday: Memoirs, Impressions and Reflections in Seven Acts* [*Χτές: Αναμνήσεις, εντυπώσεις καί στοχασμοί σε επτά πράξεις*]. Athens: Elliniki Paideia [Εκδ. Ελληνική Παιδεία].
Lüddecke, Andreas (2000). *Rassen, Schädel und Gelehrte. Zur politischen Funktionalität der anthropologischen Forschung und Lehre in der Tradition Egon von Eickstedts*. Frankfurt am Main: Peter Lang.
Luschan, Felix Ritter von and Eugen Petersen (1889). *Reisen in Lykien, Milyas und Kibyratis*. Vienna: Gerold.
Mackensen, Rainer (2002). "Ursprünge, Arten und Folgen des Konstrukts 'Bevölkerung' vor dem 'Dritten Reich'". In *Bevölkerungslehre und Bevölkerungspolitik vor 1933*, edited by Rainer Mackensen, 27–38. Opladen: Leske und Budrich.
Marchand, Suzanne L. (1996). *Down from Olympus: Archaeology and Philhellenism in Germany, 1750–1970*. Princeton, NJ: Princeton University Press.
Macherey, Pierre (2009). *La force des normes: de Canguilhem à Foucault*. Paris: La Fabrique.
Margaritis, Giorgos (1993). *From the Defeat to Riot: Greece, Spring 1941–Autumn 1942* [*Από την ήττα στην εξέγερση. Ελλάδα: άνοιξη 1941-φθινόπωρο 1942*]. Athens: Politis [Πολίτης].
—— (2005). *Undesired Compatriots: Facts on the Persecution of Minorities in Greece: Jews, Chams* [*Ανεπιθύμητοι συμπατριώτες. Στοιχεία για την καταστροφή των μειονοτήτων της Ελλάδας. Εβραίοι Τσάμηδες*]. Athens: Vivliorama [Βιβλιόραμα].
Mari, Ioanna (2004). *Eugenics. Prenatal Care. Care of Children begins from Conception* [*Ευγονία. Προγεννητική Αγωγή. Η αγωγή του παιδιού αρχίζει από τη σύλληψη*]. Athens: Pirinos Kosmos [Πύρινος Κόσμος].
Marketos, Spyros Ger (2001). *History of Medicine of the 20th Century. I Greek Pioneers: 5: Nikolaos Louros (1898–1986)* [*Ιστορία της Ιατρικής του 20ού Αιώνα. I. Οι Έλληνες Πρωτοπόροι 5: Νικόλαος Κ. Λούρος (1898–1986)*]. Athens: Iatrikes Ekdoseis Zita [Ιατρικές Εκδόσεις Ζήτα].
Marten, Heinz-Georg (1983). *Sozialbiologismus: Biologische Grundpositionen der politischen Ideengeschichte*. Frankfurt am Main and New York: Campus Verlag.
Martin, Rudolf (1914). *Lehrbuch der Anthropologie in systematischer Darstellung*. Jena: Verlag von Gustav Fischer.
Mason, O.T. (1887). "A Hairy Human Family". *Science* 9, no. 205: 16–17.

Massin, Benoit (1996). "From Virchow to Fischer: Physical Anthropology and 'Modern Race Theories' in Wilhelmine Germany". In *Volksgeist as Method and Ethic: Essays on Boasian Ethnography and the German Anthropological Tradition*, edited by George W. Stocking, 79–154. Madison, Wisconsin: Univerity of Wisconsin Press.

Matalas, Paraskevas (2002). *Nation and Orthodoxy: Aventures of a Relationship* [*Έθνος και ορθοδοξία. Οι περιπέτειες μιας σχέσης. Από το 'Ελλαδικό' στο Βουλγαρικό σχίσμα*]. Irakleion: University Press of Crete [Πανεπιστημιακές Εκδόσεις Κρήτης].

Mavrogordatos, Georgios and Christos Chatziiosif, eds (1988). *Venizelism and Modernization* [*Βενιζελισμός και Αστικός εκσυγχρονισμός*]. Irakleion: University Press of Crete [Πανεπιστημιακές Εκδόσεις Κρήτης].

Mazower, Mark, ed. (2000). *After the War was Over: Reconstructing the Family, Nation, and the State in Greece, 1943–1960*. Princeton and Oxford: Princeton University Press.

Michalopoulos, Ioannis (1931). "The Decrease of Births" ["Η ελλάτωσις των γεννήσεων"]. *Ελεύθερον Βήμα*, 22, 23, 24 and 26 October.

Mihok, Brigitte (2008). "Verdrängte Erinnerung. Die Deportation der rumänischen Roma nach Transnistrien 1942–1944". In *Der nationalsozialistische Genozid an den Roma Osteuropas. Geschichte und künstlerische Verarbeitung*, edited by Felicitas Fischer von Weikersthal et al., 29–42. Cologne: Böhlau Verlag.

Miliarakis, Spiridon (1880). *Carl Darwin* (translated from German by William T. Preyer) [*Κάρολος Δάρβιν*, (Μετάφρασις *εκ του* γερμανικού Preyer, William T.)]. Athens: Ell. Anexartisia [Εκ *του* Τυπογραφείου *της* ,Ελλ. Ανεξαρτησίας'].

—— (1884). "Verkieselung lebender Elementarorgane bei den Pflanzen". Phil. Diss., Würzburg University.

—— (1889). *Handbook of Zoology Including Complete Anthropology—For the Use at the Gymnasiums* [*Εγχειρίδιον ζωολογίας περιέχον και πλήρη ανθρωπολογίαν—Προς χρήσιν των Γυμνασίων*]. Athens: Kasdonis, Perri [Εκδότης Γεώργιος Κασδόνης, Τύποις Αδελφών Περρή].

Miller, Geneviene (1962). "'Airs, Waters, and Places' in History". *Journal of the History of Medicine and Allied Sciences* 17, no. 1: 129–40.

Mocek, Reinhard (2002). *Biologie und soziale Befreiung: Zur Geschichte des Biologismus*. Frankfurt am Main: Peter Lang.

Moiseidis, Moisis (1912). *Eugenics Erstwhile and Today* [*Η καλλιπαιδία άλλοτε και σήμερον*]. Constantinople: Aristovoulou [Τύποις Αριστοβούλου, Αναστασιάδου *και* Σας].

—— (1922). *Eugenics and Marriage* [*Ευγονισμός και γάμος*]. Constantinople: [no publisher].

—— (1925). *Eugenics and Childbearing in Ancient Hellas* [*Ευγονική και Παιδοκομία παρά τοις αρχαίοις Έλλησιν*]. Athens: Makridou and Alevropoulou [Εκ του Τυπογραφείου ΚΓ Μακρίδου και ΙΑ Αλευροπούλου].

—— (1930). "The Medical Examination Prior to Marriage" ["Η προ του γάμου ιατρική εξέτασις (Εισήγησις εις το Α' Πανελλήνιον Συνέδριον της Υγιεινής, 29 Απριλίου 1930)"], *Εργασία* 1, no. 19: 17–19.

—— (1932). *Malthusianism Then and Now: Birth Control and Sterilisation, A Medical, Eugenic, Social Study* [*Ο μαλθουσιανισμός άλλοτε και νυν: έλεγχος γεννήσεων και αποστείρωσις, Μελέτη Ιατρική, Ευγονική, Κοινωνική*]. Athens: Gerardon [Τυπογραφικά καταστήματα Αδελφών Γεράρδων].

—— (1933). "Marriage Hygiene" ["Υγιεινή του γάμου"], Library of Social Hygiene [*Βιβλιοθήκη Κοινωνικής Υγιεινής*] (Special Edition on "Hygiene of Marriage"), 2: 7–39.

—— (1934). *Eugenic Sterilisation: Principles-Methods-Implementation* [*Ευγονική αποστείρωσις. Αρχαι- Μέθοδοι- Εφαρμογή*]. Athens and Alexandria: A. Kasigoni [Εκδοτικός οίκος Α. Κασιγόνη].

Montagu, Ashley (1947). *Man's Most Dangerous Myth: The Fallacy of Race* (2nd ed., rev. and enl., 3. print). New York: Columbia Univ. Press.

—— (1951). *Statement on Race: An Extended Discussion in Plain Language of the UNESCO Statement by Experts on Race Problems*. New York: Schuman.

—— (1972). *Statement on Race: An Annotated Elaboration and Exposition of the Four Statements on Race Issued by the United Nations Educational, Scientific, and Cultural Organization* [3rd edition]. London [et al.]: Oxford University Press.

Moskof, Kostis (1988). *Introduction to the History of the Movement of the Labour Class: The Formation of National and Social Consciousness in Greece* [*Εισαγωγικά στην Ιστορία του κινήματος της εργατικής τάξης. Η διαμόρφωση της εθνικής και κοινωνικής συνείδησης στην Ελλάδα*]. Athens: Kastaniotis [Καστανιώτης] [first edition 1979].

Moutousis, Konstantinos (1933). "Hygiene and Eugenics in the Current Societies: Inaugural Lecture" ["Η Υγιεινή και η Ευγονία εις τας συγχρόνους κοινωνίας (Εναρκτήριον μάθημα εις το Πανεπιστήμιον τη 9 Νοεμβρίου 1933)"], *Κλινική* 9, no. 47: 863–74.

Mpampiniotis, Georgios (1998). *Lexicon of the Modern Greek Language* [*Λεξικό της νεοελληνικής γλώσσας*]. Athens: Kentro Lexikologias [Κέντρο Λεξικολογίας].

Mprougkas, Apostolos S. (1945). "The Activity of Doctors during the Occupation" ["Η δράσις των ιατρών επί Κατοχής"]. *Νέα Ιατρική Εφημερίς*, 15 May: 1.

Mühlmann, Wilhelm E. (1986). *Geschichte der Anthropologie*. Wiesbaden: Sammlung Aula [first edition: Bonn: Universitätsverlag 1948].

Müller, Dietmar (2005). *Staatsbürger auf Widerruf: Juden und Muslime als Alteritätspartner im rumänischen und serbischen Nationscode; Ethnonationale Staatsbürgerschaftskonzepte 1878–1941*. Wiesbaden: Harrassowitz.

Niemann-Findeisen, Sören (2004). *Weeding the Garden: Die Eugenik-Rezeption der frühen Fabian Society*. Münster: Westfälisches Dampfboot.

Nikoklis, Nikolaos G. (1855). *On the Indigeneity of the Albanians* [*Περί της αυτοχθονίας των Αλβανών ήτοι Σκιπιτάρ: Πραγματεία ιστορικοφιλολογική ην επ' αισίοις οιωνοίς ψήφω και δοκιμασία της των φιλοσόφων υπερτίμου τάξεως προς έννομον επίτευξιν των υψίστων εν φιλοσόφοις τιμών παρά της Βασιλικής εν Γοττίγγη Ακαδημίας Γεωργίας Αυγούστης* [De Albanensium sive Schkipitar origine et prosapia / Dissertatio inauguralis quam ... scripsit], Gotingae Dietrich [repr. Athens: Dionisiou Noti Karavia (Βιβλιοπωλείο Διονυσίου Νότη Καραβία), 2000].

Ntoulas, K. (1936). *The Health of the Couple about to Get Married and the Medical Certificate of Marriage* [*Η Υγεία των Μελλονύμφων και το Ιατρικό Πιστοποιητικό του Γάμου*]. Thessaloniki: [no publisher].

Oikonomopoulos, Nikolaos (1922). *Social Hygiene: Social Welfare* [*Κοινωνική υγιεινή: Κοινωνική πρόνοια—κρατική μέριμνα*]. Athens: Petrakis [Πετράκης].

—— (1934). "The Health Politics in Our Country" ["Η υγιειονολογική πολιτική εις την χώραν μας"]. *Ιατρική Εφημερίς* 14, no. 1: 2.

Oldenburg, Volker (1996). *Der Mensch und das Monströse: Zu Vorstellungsbildern in Anthropologie und Medizin in Darwins Umfeld*. Essen: Verlag die Blaue Eule.

Ornstein, Bernhard (1881). "Über die physischen Verhältnisse Griechenlands und seiner Bewohner, mit besonderer Berücksichtigung der Langlebigkeit der letzteren und deren Ursachen". *Zeitschrift für Ethnologie (Organ der Berliner Gesellschaft für Anthropologie, Ethnologie und Urgeschichte)* 13: 11–95.

Osborne, Michael A. (2000). "The Geographical Imperative in Nineteenth-Century French Medicine". In *Medical Geography in Historical Perspective* (Medical History, Supplement no. 20), edited by Nicolaas A. Rupke, 31–50. London: The Wellcome Trust Centre for the History of Medicine at UCL.

Özbudun Demirer, Sibel (2011). "Anthropology as a Nation-Building Rhetoric: The Shaping of Turkish Anthropology (From 1850s to 1940s)". *Dialectical Anthropology* 35, no. 1: 111–29.

Panagiotakou, Panag. G. (1928). "Social Medicine or Social Hygiene" ["Κοινωνική Ιατρική ή Κοινωνική Υγιεινή"]. *Ιατρικά Χρονικά* 2: 95.

Panopoulos, Panayotis (2003). "Between Self and Others: The Academic Establishment of Greek Anthropology". In *Educational Histories of European Social Anthropology*, edited by Dorle Drackle, Iain Edgar and Thomas K. Schippers, 193–207. New York, Oxford: Berghahn Books.

Papadiamantopoulos, A. (1877a). "The Human is Born and dies in the Lap of the Christian Religion" ["Ο άνθρωπος γενώμενος εις τας αγκάλας της χριστιανικής θρησκείας και εις τους κόλπους αυτής αποκοιμώμενος"]. *Σωτήρ* (October).
—— (1877b). "Theories on Materialism in Greece" ["Αι θεωρίαι του υλισμού εν Ελλάδι"]. *Σωτήρ* (December).
—— (1878). "Dialogue between Materialists and Spiritualists" ["Διάλογος μεταξύ υλιστού και πνευματιστού"]. *Σωτήρ* (November).
Papanastasiou, Nikos (2006). "Collaborators against 'Metaxists Collaborators and Impostor'" ["Δωσίλογοι εναντίον μεταξικών 'δωσιλόγων και καταχραστών'. Η θεσμική ασυνέχεια της κατοχικής κυβέρνησης Τσολάκογλου"]. In *"Enemy Inside the Wall": Aspects of Collaboration in Occupated Greece* [*'Εχθρός' εντός των τειχών. Όψεις του Δωσολογισμού στην Ελλάδα της Κατοχής*], edited by Iakovos Michailidis, Ilias Nikolakopoulos and Hagen Fleischer, 107–45. Athens: Ellinika Grammata.
Paparrigopoulos, Konstantinos (1848). *On the settlemet of Slav philes in Peloponesse* [*Περί της εποικήσεως σλαβικών τινών φυλών εις την Πελοπόννησον*]. Athens: Em. Antoniadou [Εκ του Τυπογραφείου Εμ. Αντωνιάδου] [repr. Athens: Karavias (Καραβίας), 1986].
—— (1850/1851a). "Introduction to the History of the Enlightenment of the Greek Nation" ["Εισαγωγή εις την ιστορίαν της αναγεννήσεως του ελληνικού έθνους"]. *Πανδώρα* 1, no. 9: 199–203.
—— (1850/1851b). "Introduction to the History of the Enlightenment of the Greek Nation" ["Εισαγωγή *εις την* ιστορίαν *της* αναγεννήσεως *του* ελληνικού έθνους"]. *Πανδώρα* 1, no. 10: 230–3.
—— (1970). *Prolegomena* [*Προλεγόμενα*], edited by K. Th. Dimaras. Athens: Ερμής.
Papastefanaki, Leda (2011). "Politics, Modernization, and Public Health in Greece: The Case of Occupational Health, 1900–1940". In *Health, Hygiene and Eugenics in Southeastern Europe to 1945*, edited by Christian Promitzer, Sevasti Trubeta and Marius Turda, 165–91. New York and Budapest: CEU Press.
Papastratigakis, Konstantinos (1915). *The Deception of Mental Diseases* [*Η υπόκρισις των φρενικών νοσημάτων*]. Athens: V. Fratzeskaki [Τύποις Αδελφών Β. Φρατζεσκάκη].
—— (1929). *General Therapeutics of Neurological and Psychic Diseases* [*Γενική θεραπευτική των νευρικών και ψυχικών παθήσεων*]. Athens: [no publisher].
Papataxiarchis, Efthimios (2003). "Social Anthropology in Post-War Greece: The First Steps" ["Η κοινωνική ανθρωπολογία στη μεταπολεμική Ελλάδα: τα πρώτα βήματα"]. In *Social Sciences and Avant-Garde in Greece 1950–1967* [*Κοινωνικές επιστήμες και πρωτοπορία στην Ελλάδα 1950–1967*], edited by Ioanna Lampiri-Dimaki. Athens: EKKE/ Gutenberg.
Papoulia, Fanny Marina (2007). "Griechenlands Weg in die Moderne? Die Reformen der Regierung Trikoupis and Venizelos 1875–1935". In *Schnittstellen. Gesellschaft, Nation, Konflikt und Errinerung in Südosteuropa. Festschrift für Holm Sundhaussen zum 65. Geburtstag*, edited by Ulf Brunnbauer, Andreas Helmedach and Stefan Troebst, 81–93. Munich: R. Oldenbourg Verlag.
Pappalouka Eftaksias, Ioan (1874). "Paupers" ["Πτωχολογιά"]. *Οικονομική Επιθεώρησις* 2, no. 19: 298–309.
Paraskeva, Sofia (1954). *The Occupational Education in Theory and Practice* [*Η επαγγελματική κατεύθυνσις εν θεωρία και πράξει*] (Library of the Athenian Society of Education, no. 5) [Βιβλιοθήκη της εν Αθήναις Φιλεκπαιδευτικής Εταιρείας, αριθ. 5]. Athens: The Athenian Society of Education.
Pentedekas, Kostas (2007). *The Assistant Medical Department in the Democratic Army of Roumelis 1946–1949* [*Το Επιμελητειακό καί Υγειονομικό στο Δημοκρατικό Στρατό Ρούμελης 1946–1949*]. Athens: Entos [Εντός].
Person, Jutta (2005). *Der pathographische Blick, Physiognomik, Atavismustheorien und Kulturkritik, 1870–1930*. Würzburg: Königshausen & Neumann.
Petersen, Eugen and Felix von Luschan (1889). *Reisen in Lykien, Milyas und Kibyratis* (2nd vol. of *Reisen im südwestlichen Kleinasien*). Vienna: Gerold.

Pezold, Hans von (1932). "Moses als Eugeniker". *Deutsche Medizinische Wochenschrift* 58, no. 35: 1370.

Picon, Antoine (2003). "Utopian Socialism and Social Science". In *The Modern Social Sciences* [The Cambridge History of Science, vol. 7], edited by Theodore M. Porter and Dorothy Ross, 71–82. Cambridge: Cambridge University Press.

Piniatoglou, Lazaros (1938). *The Archaic Form of the Royalism: An Essay on Social Anthropology* [*Η αρχαϊκή μορφή του βασιλικού θεσμού: δοκίμιον κοινωνικής ανθρωπολογίας*]. Athens: Pirsou [Τύποις Πυρσού].

Pitsios, Theodoros (1993). "Educational and Investigatory Content of Physical Anthropology (Human Biology)" ["Εκπαιδευτικό και ερευνητικό περιεχόμενο της φυσικής ανθρωπολογίας (Ανθρώπινη Βιολογία)"]. In *Anthropology-Bulletin of the Anthropological Society of the Friends of the Anthropological Museum* [*Ανθρωπολογία -Δελτίο της Ανθρωπολογικής Εταιρείας των Φίλων του Ανθρωπολογικής Μουσείου*], 1 (March): 33–41.

Pizanias, Petros (1993). *The Poor in the Cities: Know-How of Survival in Interwar-Greece* [*Οι φτωχοί των πόλεων. Η τεχνογνωσία της επιβίωσης στην Ελλάδα το μεσοπόλεμο*]. Athens: Themelio [Θεμέλιο].

Plantzos, Dimitris (2008). "Archaeology and Hellenic Identity, 1896–2004: The Frustrated Vision". In *A Singular Antiquity—Archaeology and Hellenic Identity in Twentieth-Century Greece*, edited by Dimitris Damaskos and Dimitris Plantzos, 11–30. Athens: Mouseio Benaki.

Ploetz, Alfred (1895). *Die Tüchtigkeit unserer Rasse and der Schutz der Schwachen. Ein Versuch über Rassenhygiene und ihr Verhältnis zu den humanen Idealen, besonders zum Socialismus*. Berlin: S. Fischer.

Poliadis, Nik. K. (1936). "Social Welfare" ["Κοινωνική Πρόνοια. Περί κρατικής κοινωνικής αντιλήψεως"]. *Ιατρική Εφημερίς*, 20 September: 1.

Porter, Theodore (1986). *The Rise of Statistical Thinking: 1820–1900*. Princeton, NJ: Princeton University Press.

—— (2003). "Statistics and Statistical Methods". In *The Modern Social Sciences* (The Cambridge History of Sciences, vol. 7), edited by Theodore Porter and Dorothy Ross, 238–50. Cambridge: Cambridge University Press.

Porter, Theodore M. and Dorothy Ross, eds (2003). *The Modern Social Sciences* (The Cambridge History of Sciences, vol. 7). Cambridge: Cambridge University Press.

Poulianos, Aris [N.] (1960). *The Origins of the Greeks: An Ethnogenetical Study* [*Η προέλευση των Ελλήνων. Εθνογενετική Έρευνα*]. Athens: Morfosi [Εκδοτικός Οίκος Μόρφωση].

—— (1974). "In Petralona the Oldest (Hitherto Known) Signs of Fire used by Humans" ["Στα Πετράλωνα τα αρχαιότερα (γνωστά μέχρι σήμερα) ίχνη φωτιάς που χρησιμοποίησε ο άνθρωπος"]. *Άνθρωπος* 1: 144–6.

—— (1976). "Greek Ape Erectus. (Is Africa the only Continent of Origins of the Humankind?)" ["Ελλαδοπίθηκος ο ημιόρθιος. (Είναι η Αφρική η μόνη ήπειρος καταγωγής των Ανθρωπιδών;)"]. *Άνθρωπος* 3: 3–29.

—— (1978). "The Oldest Artefacts in Petralona until Today" ["Τα αρχαιότερα μέχρι σήμερα εργαλεία στα Πετράλωνα"]. *Άνθρωπος* 5: 74–80.

—— (1982). *The Cave of Archanthropos in Petralona* [*Το Σπήλαιο του Αρχανθρώπου των Πετραλώνων*] (Βιβλιοθήκη Ανθρωπολογικής Εταιρείας Ελλάδος, αρ. 3). Athens: Library of the Anthropological Society of Greece.

—— (2004). *The Origins of the Greeks: An Ethnogenetical Study—Mathematical Evidence for the Descent of the Greeks* [*Η προέλευση των Ελλήνων. Εθνογενετική Έρευνα. Η μαθηματική απόδειξη για την καταγωγή των Ελλήνων*]. Athens: Library of the Anthropological Society of Greece [Βιβλιοθήκη της Ανθρωπολογικής Εταιρείας της Ελλάδας] [13th edition].

Poulianou, Dafni (2006a). *Aris Poulianos: His Life and Work. Anthropology's repatriation* [*Άρης Πουλιανός: Η ζωή και το έργο του: Ανθρωπολογίας επαναπατρισμός*]. Athens: Copy Express.

—— (2006b). *Aris Poulianos: "Reversals" from his Life and Work* [*Άρης Πουλιανός: "Ανατροπές" από τη ζωή και το έργο του*]. Athens: In tempo.

Pouliopoulos, Pantelis (1934). *Democratic or Socialist Revolution in Greece?* [*Δημοκρατική ή Σοσιαλιστική Επανάσταση στην Ελλάδα*]. Athens: Gkovosti [Γκοβοστή].
Presner, Todd Samuel (2007). *Muscular Judaism: The Jewish Body and the Politics of Regeneration*. London and New York: Routledge Jewish Studies Series.
Prestwich, Patricia E. (1997). "Paul-Maurice Legrain (1860–1939)". *Addiction* 92, no. 10: 1255–64.
Preyer, William T. (1896). *Darwin: Sein Leben und Wirken*. Berlin: Hofmann.
Prince Peter of Greece and Denmark (1963). *A Study of Polyandry*. The Hague: Mouton.
Proctor, Robert (1988). "From *Anthropologie* to *Rassenkunde* in the German Anthropological Tradition". In *Bones, Bodies, Behavior: Essays on Biological Anthropology*, edited by George W. Stocking Jr, 138–79. Madison and London: The University of Wisconsin Press.
Promitzer, Christian (2003). "The Body of the Other: 'Racial Science' and Ethnic Minorities in the Balkans". In *Minorities in Greece—Historical Issues and New Perspectives*, edited by Sevasti Trubeta and Christian Voss (Special Issue of *History and Culture of South Eastern Europe: An Annual Journal*) 5, no. 40: 27–40.
—— (2010a). "Physical Anthropology and Ethnogenesis in Bulgaria (1878–1944)". *Focaal* 58: 47–62.
—— (2010b). "'Betwixt and Between': Physical Anthropology in Bulgaria and Serbia until the End of the First World War". In *Doing Anthropology in Wartime and War Zones: World War I and the Cultural Sciences in Europe*, edited by Reinhard Johler, Christian Marchetti and Monique Scheer, 141–65. Bielefeld: Transcript.
Promitzer, Christian, Sevasti Trubeta and Marius Turda (2011). "Framing Issues of Health, Hygiene and Eugenics in Southeastern Europe", Introduction to *Health, Hygiene and Eugenics in Southeastern Europe to 1945*, edited by Christian Promitzer, Sevasti Trubeta and Marius Turda, 1–24. New York, Budapest: CEU Press.
Pusman, Karl (2008). *"Die Wissenschaften vom Menschen" auf Wiener Boden (1870–1959). Die Anthropologische Gesellschaft in Wien und die anthropologischen Disziplinen im Fokus von Wissenschaftsgeschichte, Wissenschafts- und Verdrängungspolitik*. Vienna: LIT Verlag.
Ripley, William Zebina (1899). *The Races of Europe: A Sociological Study*. New York: Appleton.
Ritter, Hans Jakob (2009). *Psychiatrie und Eugenik: Zur Ausprägung eugenischer Denk- und Handlungsmister in der schweizerischen Psychiatrie, 1850–1950*. Zürich: Chronos.
Rohlfs, Gerhard (1879). "Bu-Dabus: Ein Beitrag zur medicinischen Geographie". *Deutsches Archiv für Geschichte der Medicin und medicinische Geographie* 2: 236–8 [repr.: Hildesheim and New York: Georg Olms Verlag, 1971].
—— (1880). "Beiträge zur Geschichte der Medicin und medicinischen Geographie Merokko's". *Deutsches Archiv für Geschichte der Medicin und medicinische Geographie* 3: 65–73 [repr.: Hildesheim and New York: Georg Olms Verlag, 1971].
Rohlfs, Heinrich and Gerhard Rohlfs (1878). "Was wir wollen". *Deutsches Archiv für Geschichte der Medicin und medicinische Geographie* 1: 1–6 [repr.: Hildesheim and New York: Georg Olms Verlag, 1971].
Roper, A.G. (1992). "Ancient Eugenics". *Mankind Quarterly* 32, no. 4: 383–419.
Rose, Nikolas and Carlos Novas (2004). *Biological citizenship*. In *Global assemblages: technology, politics, and ethics as anthropological problems*, edited by Ong, Aihwa and Stephen J Collier, 439–463. Oxford: Blackwell Publishing.
Rosen, Christine (2004). *Preaching Eugenics: Religious Leaders and the American Eugenics Movement*. Oxford: Oxford University Press.
Rothman, Barbara Katz (1986). *The Tentative Pregnancy: Prenatal Diagnosis and the Future of Motherhood*. New York: Viking Press.
Rozou, Vasileiou P. (1976). *The First Care on Public Hygiene, and Welfare in Greece (1821–1862)* [*Αι Πρώται Εν Ελλάδι Μέριμναι περί της Δημόσιας Υγιεινής, της Ιατρικής Αντιλήψεως και της Κοινωνικής Πρόνοιας (1821–1862)*]. Habilitation, University of Athens.
Rübel, H. (1942). "Rassenkräfte in der hellenischen Geschichte". *Volk und Rasse* 17, no. 10: 169–75.

Ruisinger, Marion Maria (1997). *Das griechische Gesundheitswesen unter König Otto (1833–1862)*. Frankfurt am Main and Berlin: Peter Lang.
Ryan, William (1976). *Blaming the Victim*. New York: Vintage Books.
Sakelariou, Nontas [no year]. *Organisation of the Hygiene Service in the Democratic Army* [*Η οργάνωση της υγειονομικής υπηρεσίας του Δημοκρατικού Στρατού*]. Athens: Tolidi [Αφοί Τολίδη].
Sakellariou, Georgios (1932). *Crime in Greece: A Psychological and Social Study* [*Το έγκλημα εν Ελλάδι: Μελέτη ψυχολογική και κοινωνική*]. Athens: [no publisher].
—— (1939). "He emphanisis tes ebes eis tous neous en Elladi (The Appearance of Puberty Among Boys in Greece)". *Psychological Abstracts* 13, no. 2: 110.
—— (1940a). "Psychologia tou ephevou (Psychology of the Adolescence)". *Psychological Abstracts* 14, no. 1: 56.
—— (1940b). "Hai anysychiai kai ta pnevmatika symplegmata ton Ellinon ephevon (Anxieties and Complexes of Greek Youth)". *Psychological Abstracts* 14, no. 1: 56.
—— (1941). "Documents of the Psychological Laboratory during the First Three Years of its Operation (1937–1940)" ["Τα πεπραγμένα του ψυχολογικού εργαστηρίου κατά την πρώτην τριετίαν της λειτουργίας αυτού 1937–1940"], *Annex of the Scientific Yearbook of the Philosophical Faculty* [*Παράρτημα Επιστημονικής Επετηρίδος Φιλοσοφικής Σχολής*]. Thessaloniki: Theodoridou [Τύποις Κορ. Θεοδωρίδου].
—— (1946). "Documents of the Psychological Laboratory during the Last Five Years (1941–1946)" ["Τα πεπραγμένα του ψυχολογικού εργαστηρίου κατά την τελευταίαν πενταετίαν (1941–1946)"]. Athens: [no publisher].
Salaman, Redcliffe N. (1911). "Heredity and the Jew". *Journal of Genetics* 1, no. 3: 273–92.
Salimpa, Zizi (2004). *Working Women in Greek Industry and Manufacturing (1870–1922)* [*Γυναίκες Εργάτριες στην Ελληνική Βιομηχανία και στη Βιοτεχνία (1870–1922)*]. Athens: EIE [Κέντρο Νεοελληνικών Ερευνών E.I.E].
Salvatore, Ricardo D. (2005). "Positivist criminology and state formation in modern Argentina, 1890–1940". In *Criminals and their Scientists, The History of Criminology in International Perspective*, edited by Peter Becker and Richard F. Wetzell, 253–280. Cambridge: Cambridge University Press.
Sarasin, Philipp (2001). *Reizbare Maschinen. Eine Geschichte des Körpers 1765–1914*. Frankfurt am Main: Suhrkamp.
—— (2003). *Geschichtswissenschaft und Diskursanalyse*. Frankfurt am Main: Suhrkamp.
—— (2009). *Darwin und Foucault: Genealogie und Geschichte im Zeitalter der Biologie*. Frankfurt am Main: Suhrkamp.
Saratsis, Dimitris (1931). "Is Humanity Going to Become Degenerate?" ["Η Ανθρωπότης εκφυλίζεται;"]. *Ελεύθερον Βήμα* (3 February): 3.
Satzinger, Helga (2004). *Rasse, Gene und Geschlecht. Zur Konstituierung zentraler biologischer Begriffe bei Richard Goldschmidt und Fritz Lenz 1916–1936*. Berlin: Max-Planck-Gesellschaft zur Förderung der Wissenschaften e.V.
Savvakis, Manos (2008). *The Lepers of Spinalinga. (1903–1957)* [*Οι Λεπροί της Σπιναλόγκας. Ιατρική, Εγκλεισμός, Βιωμένες Εμπειρίες (1903–1957)*]. Athens: Plethron [Πλέθρον].
Scheid, Walter (1939). "Die europäischen Rassen, ihre Zuchträume und ihre Kulturleistungen". *Zeitschrift für Rassenkunde und die gesamte Forschung am Menschen* 10, no. 2–3: 198–205.
Schmuhl, Hans-Walter (2003). "Rassen, Rassenforschung, Rassenpolitik. Annährungs an das Thema". In *Rassenforschung an Kaiser-Wilhelm-Instituten vor und nach 1933*, edited by Hans-Walter Schmuhl, 7–37. Göttingen: Wallstein.
—— (2005). *Grenzüberschreitungen. Das Kaiser-Wilhelm-Institut für Anthropologie, menschliche Erblehre und Eugenik 1927–1945*. Göttingen: Wallstein.
Schwartz Cowan, Ruth (2008). *Heredity and Hope: The Case of Genetic Screening*. Cambridge, MA, and London: Harvard University Press.
Seidler, Horst (2005). "Die biologi(sti)schen Grundlagen des Rassismus". In *Grenzen des Menschseins. Probleme einer Definition des Menschlichen*, edited by Justin Stagl and Wolfgang Reinhard, 705–26. Vienna, Cologne and Weimar: Böhlau.

Seisoro, Chr. (1937). "Rassenforschung in Bulgarien". *Zeitschrift für Rassenkunde und die gesamte Forschung am Menschen* 6, no. 2: 245.

Sierck, Udo (1995). "Prävention und Normalität. Zum Verhältnis der Behindertenbewegung zur Expertokratie". In *Doppelcharakter der Prävention*, edited by Heidrun Kaupen-Haas and Christiane Rothmaler, 43–8. Frankfurt am Main: Mabuse-Verlag.

Sigalas, Nikos (2000). "Hellénistes, hellénisme et idéologie nationale. De la formation du concept d' 'hellénisme' en grec moderne". In *L'antiquité grècque au XIXème siècle. Un exemplum contesté*? edited by Chryssanthi Avlami, 239–91. Paris: L'Harmattan.

—— (2001). "The Emergence of the Modern Greek Concept of Hellenism" ["Ο σχηματισμός *της νεοελληνικής έννοιας ελληνισμός*"]. *Τα Ιστορικά* 18, no. 34: 3–70.

Singelin, Martin, ed. (2003). *Biopolitik und Rassismus*. Frankfurt am Main: Suhrkamp.

Skalieris, Georgios K. (1922). *Peoples and Files of Asia Minor. With Carts and Tables* [*Λαοί και φυλαί της Μικράς Ασίας. Μετά πινάκων και χαρτών*]. Athens: Risos [Ρήσος] [reprint 1990].

—— [no year]. *The Empire of Trapezounta: On the Ethnographic, Demographic and Social Constitution of Pontos and Pontic Paphlagonia* [*Η Αυτοκρατορία της Τραπεζούντος: περί της εθνογραφικής, δημογραφικής και κοινωνικής συγκρότησης του Πόντου και της ποντιακής παφλαγονίας*]. Athens: Eleftheri Skepsis [Ελεύθερη Σκέψις] [reprinted 1995].

Skalnik, Peter, ed. (2002). *A Post-Communist Millennium: The Struggles for Sociocultural Anthropology in Central and Eastern Europe*. Prague: Set Out.

Skaltsounis, Ioannis (1884). *Religion and Science: Essays* [*Θρησκεία και Επιστήμη. Μελέται*]. Τεργέστη: Tipois Afstroouggrikou Loud [Τύποις του Αυστροουγγρικού Λόυδ].

—— (1888a). "'Anthropology': On the Theory of the Zoological Origins of Humans" ["'Ανθρωπολογία': Περί γενέσεως *του* ανθρώπου *δια* ζωολογικών μεταμορφώσεων θεωρία"]. *Ανάπλασις* 15.

—— (1888b). "Man is in His Own Kingdom on the Earth" ["Ο Άνθρωπος αποτελεί ίδιον επί *της* γης βασίλειον"]. *Ανάπλασις* 16.

—— (1889a). "Conclusions from Scientific Exercises and Questions" ["Εξαγόμενα επιστημονικών πειραμάτων και εκζητήσεων"]. *Ανάπλασις* 17.

—— (1889b). "Theories Dismissed from the Contemporary Sciences and their Teaching in Greece as Scientific Truths" ["Θεωρίαι καταδικασθέσαι υπό της νεωτέρας επιστήμης, και εν Ελλάδι διδασκόμεναι ως επιστημονικαί αλήθειαι"]. *Ανάπλασις* 19.

—— (1893). *On the Origins of Man: Harmony between Cristianity and Science* [*Περί γενέσεως ανθρώπου: Αρμονίαι χριστιανισμού και επιστήμης*]. Athens: Anesti Konstantinidou [Ανέστη Κωνσταντινίδου].

—— (1894a). "The Demolition of the Greek University. PhD submitted by Theodoros Tsikopoulos to University in Leipzig" ["Η κατεδάφισις *του* Ελληνικού Πανεπιστημίου. Διατριβή του Θεοδώρου Τσικόπουλου, εν Λειψία 1894"]. *Ανάπλασις* 157.

—— (1894b). "The Demolition of the Greek University. PhD submitted by Theodoros Tsikopoulos to University in Leipzig" ["Η κατεδάφισις *του* Ελληνικού Πανεπιστημίου. Διατριβή του Θεοδώρου Τσικόπουλου, εν Λειψία 1894"]. *Ανάπλασις* 158.

—— (1898). *Folk Christian Apologetism* [*Δημώδης του Χριστιανισμού Απολογητική*]. Athens: Tip. Konstantinidis Anestis [Τυπ. Κωνσταντινίδης, Ανέστης Π.].

Škerlj, B. (1937). "Anthropologie in Jugoslawien". *Zeitschrift für Rassenkunde und die gesamte Forschung am Menschen* 6, no. 2: 246–7.

Skliros, Georgios (1907). *Our Social Question* [*Το κοινωνικο μας ζήτημα*]. Athens: Anesti Konstantinidou [Εκ του Τυπογραφείου των Καταστημάτων Ανέστη Κωνστατινίδου].

—— (1919). *Contemporary Problems of Hellenism* [*Τα σύγχρονα προβλήματα του Ελληνισμού*]. Alexandria: Grammata [Γράμματα] [repr. Athens: Mati (Μάτι), 2000].

Skopetea, Elli (1999). *Fallmerayer: Wonderment of a Rival Awe* [*Φαλμεράυερ. Τεχνάσματα του αντιπάλου δέους*]. Athens: Themelio [Θεμέλιο].

Sotiriadou, Anthi (1990). "The Beginning of the Theory of Evolution: Facts from Greece" ["Η εμφάνιση *της* θεωρίας *της* εξέλιξης *των* ειδών. Δεδομένα *από τον* ελληνικό χώρο"]. PhD Thessaloniki: Aristoteleion University [ΑΠΘ] (unpublished).

Spencer, Frank, ed. (1997). *History of Anthropology* (2 volumes). New York and London: Garland Publishing.

Stamatopoulos, Kostas M. (2004). *The Chronicle of Tatoi (1800–2003)* [*Το Χρονικό του Τατοΐου (1800–2003)*] (2 volumes). Athens: Kapon [Καπόν].

Stathaki-Koumari, Rodoula (1993). "I. Koumaris—Personal Memories" ["I. Κούμαρης. Προσωπικές Αναμνήσεις"]. *Athropology-Bulletin of the Anthropological Society of the Friends of the Anthropological Museum* [*Ανθρωπολογία, Δελτίο της Ανθρωπολογικής Εταιρείας των Φίλων του Ανθρωπολογικού Μουσείου*], 1 (March): 9–13.

Stathatos, Nikos (1948). "Social Issues: The Premarital Health Certificate" ["Κοινωνικά ζητήματα. Το προ του γάμου πιστοποιητικόν υγείας"]. *Έθνος* (13 October): 1.

Stavropoulos, A.K. and Spyros Marketos (1986). "Clon Stéphanos and Greek Medical Geography" ["Ο Κλων Στέφανος *και η* ελληνική ιατρική γεωγραφία"]. *Materia Medica Greca* 14, no. 5: 473–5.

Stavropoulos, Alexandre M. (1970). *Bilan analytique et clinique du Centre expérimental de consultations prémaritales et conjugales de la Société hellénique d'eugénisme à Athènes.* Louvain: Université Catholique de Louvain.

Stein, Ludwig (1897). *Die sociale Frage im Lichte der Philosophie: Vorlesungen über Socialphilosophie und ihre Geschichte.* Stuttgart: Enke.

Stepan, Nancy Leys (1991). *"The Hour of Eugenics": Race, Gender, and Nation in Latin America.* Ithaca, NY, and London: Cornell University Press.

Stocking, George W. (1971). "What's in a Name? The Origins of the Royal Anthropological Institute (1837–71)". *Man—The Journal of the Royal Anthropological Institute* (New Series) 6: 309–90.

—— (1982). *Race, Culture, and Evolution: Essays in the History of Anthropology.* Chicago and London: The University of Chicago Press [2nd edition; first edition 1968].

—— (1986). "Anthropology and the Science of the Irrational: Malinowski's Encounter with Freudian Psychoanalysis". In *Malinowski, Rivers, Benedict and Others* [History of Anthropology vol. 4], edited by George W. Stocking Jr, 13–49. Madison, Wisconsin: University of Wisconsin Press.

——, ed. (1988). *Bones, Bodies, Behavior: Essays on Biological Anthropology.* Madison: University of Wisconsin Press.

—— (1995). *After Tylor: British Social Anthropology 1888–1951.* Madison: University of Wisconsin Press.

——, ed. (1996). *Volksgeist as Method and Ethic: Essays on Boasian Ethnography and the German Anthropological Tradition.* Madison: University of Wisconsin Press.

—— (2001). *Delimiting Anthropology: Occasional Inquiries and Reflections.* Madison: University of Wisconsin Press.

Stoler, Ann Laura (1995). *Race and the Education of Desire: Foucault's History of Sexuality and the Colonial Order of Things.* Durham, NC: Duke University Press.

Stone, Dan (2001). "Race in British Eugenics". *European History Quarterly* 31, no. 3: 397–425.

Striftou, Aikaterina (1935). "Jugendkunde und pädagogische Psychologie in Griechenland. Aus der Tätigkeit des Pädagogischen Instituts der Universität zu Athen. I. Wissenschaftliche Arbeiten des Laboratoriums für experimentale Pädagogik der Universität zu Athen". *Zeitschrift für Jugendkunde* 5: 126–9.

Sundhaussen, Holm (2002). "Was ist Südosteuropa und warum beschäftigen wir uns (nicht) damit?" *Südosteuropa-Mitteilungen* 42, no. 5–6: 92–105.

—— (2003). "Der Balkan: Ein Plädoyer für Differenz". *Geschichte und Gesellschaft* 29: 642–58.

Taguieff, Pierre-André (2000). *Die Macht des Vorurteils: Der Rassismus und sein Double.* Hamburg: Hamburger Edition [French edition: *La force du préjuge.* Paris: Éditions La Découverte, 1988].

Teschler-Nicola, Maria (2007). "*Volksdeutsche* and Racial Anthropology in Interwar Vienna: The 'Marienfeld Project'". In *"Blood and Homeland": Eugenics and Racial Nationalism in*

Central and Southeast Europe (1900–1940), edited by Marius Turda and Paul Weindling, 55–82. Budapest and New York: CEU Press.
Theodorou, Vassiliki and Despina Karakatsani (2011). "Eugenics and 'Puericulture': Medical Attempts to Improve the 'Biological Capital' in Interwar Greece". In *Health, Hygiene and Eugenics in Southeastern Europe to 1945*, edited by Christian Promitzer, Sevasti Trubeta and Marius Turda, 299–323. New York and Budapest: CEU Press.
Thumb, Albert (1914–1915). "The Modern Greek and his Ancestry". *Bulletin of the John Rylands Library Manchester* 2 (Oct. 1914–Dec. 1915) [repr. Nendeln, Liechtenstein: Kraus, 1967]: 22–47.
Todorova, Maria (1999). *Die Erfindung des Balkans. Europas bequemes Vorurteil.* Darmstadt: Primus Verlag [orig.: *Imagining the Balkans.* New York (et al.): Oxford University Press, 1997].
Topinard, Paul (1878). *Anthropology* (Preface by Professor Paul Broca). London: Chapman and Hall [et al.].
—— (1950). "On 'Race' in Anthropology". In *This is Race*, edited by Earl W.Count, 171–7. New York: Henry Schuman.
Trubeta, Sevasti (1999). *Die Konstitution von Minderheiten und die Ethnisierung sozialer und politischer Konflikte. Eine Untersuchung am Beispiel der im griechischen Thrakien ansässigen "Moslemischen Minderheit".* Frankfurt am Main [et al.]: Peter Lang.
—— (2004). "Das nationalsozialistische Bild von südosteuropäischen Roma am Schnittpunkt des Zigeuner- und des Südosteuropastereotyps". *Zeitschrift für Balkanologie* (Wiesbaden) 40, no. 1: 92–105.
—— (2006). "Hybridity in Post-Colonial Discourse and in Southeast-European Studies". In *Marginal Linguistic Identities: Studies in Slavic Contact and Borderland Varieties*, edited by Dieter Stern and Christian Voss, 31–43. Wiesbaden: Harrassowitz.
—— (2007). "Anthropological Discourse and Eugenics in Interwar Greece". In *"Blood and Homeland": Eugenics and Racial Nationalism in Central and Southeast Europe (1900–1940)*, edited by Marius Turda and Paul Weindling, 123–42. Budapest and New York: CEU Press.
——, ed. (2008). *The Roma in the Modern Greek State* [*Οι Ρομά στο σύγχρονο ελληνικό κράτος. Συμβιώσεις, Αναιρέσεις, Απουσίες*]. Athens: Kritiki [Κριτική].
—— (2010). "The 'Strong Nucleus' of the Greek Race: Racial Nationalism and Anthropological Science". *Focaal—European Journal of Anthropology* 58: 63–78.
Tsakalatos, Athanasios (1909). *John Lamark and his Work* [*Ιωάννης Λαμάρκ και το έργον αυτού*]. Athens: P.D. Sakellariou [Τύποις Π Δ Σακελλαρίου] [second edition: 1911].
—— (1926). "Review: Biologische Probleme der Rassehygiene und die Kulturvölker, Von Dr. med. Stavros Zurukzoglu. Monchen [sic] 1925". *Ιατρική. Ελληνική Ιατρική Επιθεώρησις* 3, no. 9: 267–8.
—— (1931). *On Lamarkianism: John Lamark, his Life and Works* [*Περί του Λαμαρκισμού. Ιωάννης Λαμάρκ ο βίος και το έργον αυτού: διάλεξις γενομένη εν τω αμφιθεάτρω του Πανεπιστημίου κατά την συνεδρίασιν της 22 Ιανουαρίου 1931*]. Athens: P.D. Sakellarios [Τύποις Εταιρείας, Π Δ Σακελλάριος].
Turda, Marius (2007a). "The Nation as Object: Race, Blood and Biopolitics in Interwar Romania". *Slavic Review* 66, no. 3: 413–41.
—— (2007b). "From Craniology to Serology: Racial Anthropology in Interwar Hungary and Romania". *Journal of the History of Behavioral Sciences* 43, no. 3: 361–77.
—— (2010). *Modernism and Eugenics.* New York: Palgrave.
Turda, Marius and Tudor Georgescu, eds (2010). *Race Ascendant: Framing Physical Anthropology in Central and Southeastern Europe, Focaal* (Special Issue) 58.
Tzavaras, Nikos (1998). "A Perversive Use of Biological Terms" ["Μια διαστροφική κατάχρηση των βιολογικών εννοιών"]. In *Six Essays on Racism* [*Έξι κείμενα για το ρατσισμό*], edited by Initiative of Citizens against Racism [Κίνηση Πολιτών Κατά του Ρατσισμού], 70–83. Athens: Paraskinio [Παρασκήνιο].
Tzavaras, Thanasis (2003). "The History of Psychology in Greece" ["Η ιστορία της ψυχολογίας στην Ελλάδα"]. *Ιστορικά* 39: 547–50.

University of Athens (Museum of the History of Education) [Πανεπιστήμιο Αθηνών. Μουσείο Ιστορίας της Παιδείας], ed. (1999). *Athens as an Educational City: From the 19th to 20th Century* [*Αι Αθήναι ως εκπαιδευτική πόλις. Από τον 19ο προς τον 20ό αιώνα*]. Athens: Communality of Athens—Cultural Section [Δήμος Αθηναίων, Πολιτισμικός Οργανισμός].

Urry, James (1993). *Before Social Anthropology: Essays on the History of British Anthropology*. Chur (Switzerland): Harwood Academic Publishers.

Uzarczyk, Kamila (2007). "'Moses als Eugeniker?' The Reception of Eugenics in Jewish Medical Circles in Interwar Poland". In *"Blood and Homeland": Eugenics and Racial Nationalism in Central and Southeast Europe (1900–1940)*, edited by Marius Turda and Paul Weindling, 283–98. Budapest and New York: CEU Press.

Valenčius, Conevery Bolton (2000). "Histories of Medical Geography". In *Medical Geography in Historical Perspective* (Medical History, Supplement no. 20), edited by Nicolaas A. Rupke, 3–28. London: The Wellcome Trust Centre for the History of Medicine at University College London.

Vallois, H.V. (1951). "UNESCO: On Race". *Man* 51: 15–16.

Valsamakis, N. (1875). "The Rich and the Poor" ["Πλούσιοι και πένητες"]. *Οικονομική Επιθεώρησις* 3, no. 30: 254.

Van der Kiste, John (1994). *Kings of the Hellenes, the Greek Kings, 1863–1974*. Stroud: Sutton.

Varon, Ontet (1994). "The 'Silence' of the Written Sources" ["Η 'σιωπή' των γραπτών πηγών"]. *Σύγχρονα Θέματα* 17, no. 52–3: 79–84.

Varvaresos, K. (1912). *Population Theory* [*Η θεωρία του πληθυσμού*]. Athens: Raftanis [Τυπογραφείον της Β Αυλής Α Ραφτάνη].

Vathis, Marinos Sp. (1925). "For Implementation: The Marriage among Syphilis-Diseased Individuals" ["Δια την πράξιν. Ο γάμος των συφιλιδικών"]. *Ιατρική—Ελληνική Ιατρική Επιθεώρησις* 3, no. 7: 178–9.

Veloudis, Georg (1970). "Jacob Philipp Fallmerayer und die Entstehung des neugriechischen Historismus". *Südost-Forschungen* 29: 43–90.

Vlavianos, Georgios (1939). *Mental Hygiene and Mental Eugenics* [*Ψυχική Υγιεινή και Ψυχική Ευγονική*] (offprint from *Academic Medicine*, Issue February 1939). Athens.

Vlavianos, Simonidis (1931). "The Fight Against Alcoholism" ["Ανταλκοολικός Αγών"]. *Ιατρική Εφημερίς* (6 September): 2.

—— (1933). "Hitler's New Law on the Sterilisation of those Affected by Hereditarily Transmitted Diseases" ["Ο νέος νόμος του Χίτλερ δια την στείρωσιν των μεταδοτικών νόσων κληρονομικώς"]. *Ιατρική Εφημερίς* (31 December): 4.

—— (1934). "Against Alcoholism. Alcoholism in Greece" ["Κατά του Αλκοολισμού. Ο Αλκοολισμός εν Ελλάδι"]. *Ιατρική Εφημερίς* (18 November): 1–4.

Voss, Christian (2003). "The Situation of the Slavic-Speaking Minority in Greek Macedonia—Ethnic Revival, Cross-Border Cohesion, or Language Death?" In *Minorities in Greece—Historical Issues and New Perspectives*, Special Issue of *History and Culture of South Eastern Europe: An Annual Journal*, edited by Sevasti Trubeta and Christian Voss, 5: 173–87.

Weber, Max (2006). *Wissenschaft als Beruf.* Stuttgart: Philipp Reclam jun [first edition 1919].

Wecker, Regina et al., eds (2009). *Wie Nationalsozialistisch ist die Eugenik? / What is National Socialist about Eugenics?* Vienna, Cologne and Weimar: Böhlau.

Wedekind, Michael (2007). "Wissenschaftsmilieus und Ethnopolitik im Rumänien der 1930/40-er Jahre". In *Herausforderung Bevölkerung. Zu Entwicklungen des modernen Denken über die Bevölkerung vor, im und nach dem "Dritten Reich"*, edited by Ursula Ferdinand and Josef Ehmer, 233–64. Wiesbaden: VS Verlag für Sozialwissenschaften.

Weindling, Paul (1989). "The 'Sonderweg' of German Eugenics: Nationalism and Scientific Internationalism". *The British Journal for the History of Science* 22, no. 3: 321–33.

—— (2011). "Racial Expertise and German Eugenic Strategies for Southeastern Europe". In *Health, Hygiene and Eugenics in Southeastern Europe to 1945*, edited by Christian Promitzer, Sevasti Trubeta and Marius Turda, 27–54. New York and Budapest: CEU Press.

Weingart, Peter (1983). "Verwissenschaftlichung der Gesellschaft—Politisierung der Wissenshcaft". *Zeitschrift der Soziologie* 12, no. 3: 225–41.
—— (1999). "Science and Political Culture: Eugenics in Comparative Perspective". *Scandinavian Journal of History* 24: 163–77.
—— (2001). *Die Stunde der Wahrheit? Zum Verhältnis der Wissenschaft zu Politik, Wirtschaft und Medien in der Wissensgesellschaft.* Weilerswist: Velbrück Wiss.
—— (2003). *Wissenschaftssoziologie.* Bielefeld: Transcript.
—— (2006). "Knowledge and Inequality". In *Inequalities of the World: New Theoretical Frameworks, Multiple Empirical Approaches,* edited by Göran Thernborn, 163–90. London, New York: Verso.
Weingart, Peter, Sabine Maasen and Ullica Segerstråle (1997). "Shifting Boundaries between the Biological and the Social: The Social and Political Contexts". In *Human by Nature: Between Biology and the Social Sciences,* edited by Peter Weingart et al., 65–7. Mahwah, NJ, and London: Lawrence Erlbaum Associates, Publishers.
Weingart, Peter et al., eds (1997). *Human by Nature: Between Biology and the Social Sciences.* Mahwah, NJ, and London: Lawrence Erlbaum Associates, Publishers.
Weisbach, Augustin (1881). "Die Schädelform der Griechen". *Mittheilungen der Anthropologischen Gesellschaft in Wien* 11, no. 2 [new series vol. 1]: 72–97.
—— (1882). "Die Schädelform der Griechen". *Mitteilungen der Anthropologischen Gesellschaft in Wien* 11 [new series 1 (1)]: 72–97.
Weiss, Sheila Faith (2004). *Humangenetik und Politik als wechselseitige Ressourcen. Das Kaiser-Wilhelm-Institut für Anthropologie, menschliche Erblehre und Eugenik im "Dritten Reich".* Berlin: Max-Plank-Gesellschaft zur Förderung der Wissenschaften e.V.
West, Cornel (2001). *Race Matters.* New York: Vintage Books.
Willke, Helmut (2002). *Dystopia. Studien zur Krisis des Wissens in der modernen Gesellschaft.* Frankfurt am Main: Suhrkamp.
Wilson, Edward O. (1975). *Sociobiology: The New Synthesis.* Cambridge, MA: Harvard University Press.
Wittich, Dieter and Horst Poldrack (1990). *Der Londoner Kongress zur Wissenschaftsgeschichte 1931 und das Problem der Determination von Erkenntnisentwicklung.* Berlin: Akademie-Verlag.
Wundt, Wilhelm (1900–21). *Völkerpsychologie, Eine Untersuchung d. Entwicklungsgesetze von Sprache, Mythus u. Sitte* (10 volumes). Leipzig: W. Engelmann.
Xirotiris, Nikolaos I. (1971). "Own Observations on the Frequency of Blood Groups among Pomaks" ["Ιδίαι παρατηρήσεις επί της κατανομής των συχνοτήτων των ομάδων αίματος εις τους Πομάκους"]. PhD, Thessaloniki [ΑΠΘ] (unpublished).
Yearley, Steven (2005). *Making Sense of Science: Understanding the Social Study of Science.* London: Sage Publications.
Yeomans, Rory (2007). "Of 'Yugoslav Barbarians' and Croatian Gentlemen Scholars: National Ideology and Racial Anthropology in Interwar Yugoslavia". In *"Blood and Homeland": Eugenics and Racial Nationalism in Central and Southeast Europe, (1900–1940),* edited by Marius Turda and Paul Weindling, 83–122. Budapest and New York: CEU Press.
Young, Robert (1995). *Colonial Desire: Hybridity in Theory, Culture and Race.* London: Routledge.
Zallonis, I.A. (1870). *Memorandum to Parliament about Impediments to Marriage—From a Medical Point of View* [*Υπόμνημα προς την Βουλήν Περί Κωλυμάτων Γάμου—Υπό Ιατρικήν Έποψιν*]. Ermoupolis: [no publsher].
Zampelios, Spiridon (1852). *Greek Folk Songs* [*Άσματα Δημοτικά της Ελλάδος*]. Corfu: Ermis [Τυπογραφείον Ερμής].
Zanella, Ines Caroline (2004). *Kolonialismus in Bildern: Bilder als herrschaftssicherndes Instrument mit Beispielen aus den Welt- und Kolonialausstellungen.* Frankfurt am Main [et al.]: Peter Lang.
Zarifi, Maria (2002). "Das deutsch-griechische Forschungsinstitut für Biologie in Piräus, 1942–1944". In *Autarkie und Ostexpansion. Pflanzenzucht und Agrarforschung im Nationalsozialismus,* edited by Susanne Heim, 206–32. Göttingen: Wallstein Verlag.

—— (2010). *Science, Culture and Politics: Germany's Cultural Policy and Scientific Relations with Greece 1933–1945*. Saarbrücken: VDM Verlag Dr Müller.

Zeepvat, Charlotte (2004). *The Camera and the Tsars: A Romanov Family Album*. Stroud: Sutton.

Zelepos, Ioannis (2002). *Die Ethnisierung griechischer Identität (1870–1912). Staat und Privatakteure vor dem Hintergrund der "Megali Idea"*. Munich: Oldenbourg.

Ziegler, Béatrice (2002). "Verhinderte Wissenschaft? Universität und Eugenik in der Schweiz am Beispiel der Laufbahn von Stavros Zurukzoglu in Bern". Lecture delivered at the Conference, "Psychiatry and Eugenics in the 19th and 20th Centuries: Switzerland in the European–American Context", Monte Verità (17–22 February) (unpublished).

Ziegler, Theobald (1891). *Die soziale Frage, eine sittliche Frage*. Stuttgart: Göschen [second edition].

Zikopoulos, Theodor (1886). *Zur Lungenparenchym-Expectoration bei Phthisis tuberculosa*. Würzburg, University, Diss., 1886, Würzburg: Thein (Stürtz).

Zimmermann, Andrew (2001). *Anthropology and Antihumanism in Imperial Germany*. Chicago and London: The University of Chicago Press.

Zimmermann, Michael (1996). *Rassenutopien und Genozid. Die nationalsozialistische "Lösung der Zigeunerfrage"*. Hamburg: Hans Christians Verlag.

Zoïa, Geneviève (1990). "L'anthropologie en Grèce". *Terrain* 14: 143–51.

Zurukzoglu, Stavros (1925a). *Biologische Probleme der Rassenhygiene und der Kulturvölker*. Munich: J.F. Bergmann.

—— (1925b). "On Eugenics; I" ["Περί ευγονίας (1)"]. *Ιατρική* 3, no. 2: 66–9.

—— (1925c). "On Eugenics; II" ["Περί ευγονίας (2)"]. *Ιατρική* 3, no. 3: 93–4.

—— (1925d). "The Problem of Degeneration" ["Το πρόβλημα του εκφυλισμού"]. *Ιατρική* 3, no. 4: 122–7.

—— (1925e). "Eugenics" ["Η ευγονία"]. *Υγεία* 2, no. 11: 221–4.

—— (1925f). "The Three Problems of Eugenics" ["Τα τρία προβλήματα της ευγονίας"]. *Υγεία* 2, no. 12: 246–8.

—— (1925g). "Eugenics" ["Η ευγονία"]. *Υγεία* 2, no. 13: 265–7.

——, ed. (1935). *Die Alkoholfrage in der Schweiz*. Basel: Benno Schwabe Verlag.

——, ed. (1938). *Verhütung erbkranken Nachwuchses. Eine kritische Betrachtung und Würdigung*. Basel: Benno Schwabe & Co-Verlag.

—— (1956a). *Die alkoholbedingte Kriminalität in der schweizerischen Armee während des Aktivdienstes 1939–1945*. Berne: Die Eidgenössische Kommission gegen den Alkoholismus.

—— (1956b). "Probleme der Erbhygiene". In *Rassenfrage-Heute*, edited by Gesellschaft für Christich-jüdische Zusammenarbeit München e.V, 71–80. Munich: Süddeutscher Verlag (2nd edition 1963).

—— (1963). *Ist Trunksucht eine Krankheit? Problem der Beitragspflicht der Krankenkassen*. Benno: Schwabe.

INDEX OF NAMES

Amantos, Konstantinos 93, 196, 216, 217
Andriotis, Nikolaos P. 93
Apostolidès, N. 38, 39
Apostolidis, Petros 262–263
Aravantinos, Anastasios 109–110
Ardavanis-Limperatos, Spiridon 132, 232

Balis, Georgios 243
Baumann, Hans 138
Bees, Nikos 124–127; also *Jews*
Bier, August 47
Boas, Franz 43
Bonaparte, Maria 117
Bordier, Arthur 32
Brabrook, Edward William 64
Brinton, Garrison Daniel 63
Broca, Paul 74

Caminopetros or Kaminopetros, Ioannis 110; also *thalassaemia*
Carathéodory, Constantin 157
Charitakis, Kostis 129, 228
Chatzivasileiou, Grigorios 213, 214, 228
Cooley, Thomas Benton 109
Coon, Carleton Stevens 172, 174
Coray, Adamantios, or Adamantius Coray(s), Adamantios Korae or Koraēs, 34–35; also *medical geography*

Danezis, Ioannis 267, 273
De Quatrefages de Bréau, Jean Louis Armand 42
Debetz Frantsevich Georgy (or Georgii Frantsevich Debets) 91, 197
Demertzis, Konstantinos 260
Deniker Jegorowitsch, Joseph 39
Diamantopoulos, G. 30
Doxiadis, Apostolos 190, 217–222, 231
Doxiadis, Spiros 273
Dragoumis, Filippos 93
Drakoulidis, Nikolaos 94, 224, 231, 233–234, 241–244 [310]; also *Aggelos Doxas* 224
Droysen, Johann Gustav 152
Duke Pavel Alexandrovich Romanov 67

Eickstedt, Egon von 128, 169
Eleftheriadis, Dimosthenis 260–261
Exarchopoulos, Nikolaos 101–103

Fallmerayer, Jakob Filipp 154, 156, 160, 172
Fessas, Faidon 269
Finke, Leonhard Ludwig 33
Fischer, Eugen 128–130, 163, 168–170, 221, 249
Folkmar, Daniel 64
Fraas, Karl Nikolaus 42
Frazer, James George 118–119
Freud, Sigmund 118
Fürst, Carl Magnus 69

Gall, Franz Joseph 113, 114
Gardikas, Konstantinos G. 93, 99–101, 205, 237
Gedeon, Sofia or Sophie 102
Gianniris, Michail 260
Gieseler, Wilhelm 129
Glinos, Dimitris 94, 159
Gobineau, Arthur de 159–160, 167, 253
Graham Bell, Alexander 108
Grosse, Ernst 169
Günther, Hans Friedrich Karl 128
Gütt, Arthur 108

Haeckel, Ernst 80
Hampe, Roland 125, 184
Hauswirth, Alfred 205
Heldreich, Theodor von 29, 42, 58
Hirsch, August 36

Jensen, Arthur R. 105

Kakouri, Katerina 111
Karavidas, Konstantinos 157–158
Kardamatis, Ioannis 159
King Constantine 67
King George I 67
King George II 67
King Otto 41, 42, 239
Kiriakidis, Stilpon 93
Kitikas, Dimitris 69
Kordatos, Ioannis 198
Kosmetatos, Georgios 108, 109, 110, 214
Koukoules, Faidon 73, 93
Koumaris, Ioannis 12–16, 19–21, 45–50, 55–56, 66–75, 77, 79, 80–86, 91, 94–96, 98, 101, 109–112, 114, 118–199, 122, 124–126, 128–130, 136–137, 139–140, 163–170,

172–189, 191, 196–197, 199, 201, 205, 247–248, 256, 258, 269, 281, 286–290
also *Koumaris John* 170, 173, 187
also *Kumaris, J.* 47
also *Mistras, Ion* 48
mixophobia, mixophobic 170, 177, 188
and racism, racist (views) 19, 142, 179, 185–186, 189, 269, 290
autobiography 21, 45, 47–50, 67–68, 84–85, 100, 186, 188, 274
on racial purity, purification 15–16, 163, 165–168, 177–183, 188
racial hygiene conception by 16, 18, 188, 189, 246–248, 256
Krikos, Alexandros 101, 122–124
Kröger, Tim 137
Krumbacher, Karl 160 [316]
Kyrle, André Forster Georg 91

Lampadarios, Emmanouil 101–103
Legrain, Paul-Maurice 225
Lenz, Fritz 128–130, 205, 213
Livas, Xerxis 141, 192, 195, 199
Logothetopoulos or Logothetopulos, Konstantinos 277, 278
Lombroso, [Cesare] 132n
Louros, Konstantinos 93
Louros, Nikolaos 18, 94, 264–278
Luschan, Felix von 34, 96, 97, 175, 176
Luschka, Hubert von 30

Maridakis, Georgios 243
Marinatos, Spiridon 112, 131, 200
Markovits/Markovich Adalbert 69, 90
Martin, Rudolf 80–81
Megas, Georgios 116
Melissinos, Konstantinos 131
Menardos, Simos 93
Mengele, Josef Dr 130
Menghin, Oswald 129
Mermigkas, Konstantinos 46
Metaxas, Ioannis 71, 126, 195, 244, 245, 258, 260, 271
Miliarakis, Spiridon 29–30
Mitsopoulos, Maximos 84
Moiseidis, Moisis M. 94, 205, 212, 214, 224–225, 229, 233–236
Mollison, Theodor 129
Moutousis, Konstantinos 18, 206, 232, 235, 236, 247, 252, 256, 257
Mponis, Konstantinos 271
Mühlmann, Wilhelm 80, 81, 89, 97

Nikoklis, Nikolaos G. 155, 156
Nikolaidis, Rigas 28

Oikonomopoulos, Nikolaos 228, 231
Oikonomos, Aristeidis 226, 228
Oikonomos, Konstantinos 153, 154, 156
Ornstein, Bernhard 34, 38, 39, 42

Panagiotou, Panagiotis P. 107, 108, 214–215
Pantazis, Georgios 87, 135
Papadiamantopoulos, A. 27
Papanastasiou, Alexandros 90, 92, 96, 97, 244
Papanoutsos, Evangelos 271, 272
Paparrigopoulos, Konstantinos 152–156
Papastratigakis, Konstantinos 260
Papillaut, George 91
Petrovich Bogdanov, Anatoly 43
Piniatoglou, Lazaros 115, 117–119, 127
Ploetz, Alfred 205, 246, 248, 250
Poincaré, Leon 32
Poulianos, Aris 16, 93, 94, 196–201
Preyer, William T. 29
Prince George of Greece 117
Prince Peter of Greece and Denmark 14, 91, 95, 98, 115–120
Princess Alexandra 67

Queen Olga Konstantinovna Romanova 67

Rasputin Yefimovich, Grigori 67
Reche, Otto 129
Reinach, Salomon 69, 91
Reinerth, Hans 128
Ripley, William Zebina 159
Rivouar, Ioannis 112–115
Rohlfs Brothers (Gerhard and Heinrich) 32–33
Romanov Pavlovich, Dmitri 67
Rosenberg, Alfred Ernst 172
Rutot, August 69, 91

Sakellariou, Georgios 101, 103–105, 225, 237, 238
Sanger, Margaret 220
Savvas, Konstantinos 93
Schwidetzky-Roesing, Ilse 91, 177, 179
Skalieris, Georgios 157
Skaltsounis, Ioannis 27, 29
Sklavounos, Georgios 45, 47, 87, 93, 126–127, 133–136, 249
also Sclavunos, G. 47
Sklavounos, Themistoklis 87, 95
Skliros, Georgios 161–162
Skoufos, Theodoros 84
Stathaki-Koumari, Rodoula 48, 49, 115

Stefanis, Konstantinos 272, 273
Steinmetz, Sebald Rudolf 213
Stéphanos, Clon 12–13, 15, 21, 31–39, 41, 43–44, 55–57, 59–63, 65–69, 81–82, 90, 97, 139, 164–165, 171, 174, 196, 287
conception of anthropology 13, 37, 59–60, 62–63, 65–66, 97, 139, 163, 165, 287; also *medical geography*
Svolos, Alexandros 244, 263

Theodorakopoulos, Ioannis 271
Topinard, Paul 163–164
Trichopoulos, Dimitris 272–273
Tsakalotos, Athanasios 249, 250
Tsikopoulos, Theodoros or Theodor Zikopoulos 28, 29, 324
Tsouderou, Virginia 270
Tzonis, Konstantinos 133, 135–136

Valaoras, Vasileios 267, 270
Varvaresos, Kiriakos 94, 243–244 [327]
Vegleris, Faidon 183, 245, 271
Venizelos, Eleftherios 157, 217, 243, 244
Verschuer, Otmar Freiherr von 128–130
Virchow, Rudolf Ludwig Karl 6, 39, 113
Vlavianos, Georgios 224, 238
Vlavianos, Simonidis 232, 234, 236, 237, 254

Weinert, Hans W. 128, 129
Weisbach, Augustin 38
Weismann, Friedrich Leopold August 108

Xirotiris, Nikolaos 93, 192, 193

Zachariadis, Nikos 198
Zallonis, Ioannis 240
Zampelios, Spiridon 152, 154–155
Zepos, Panagiotis 245n, 271
Zochios, Ioannis 28
Zurukzoglu, Stavros 18, 205n, 206, 229, 230, 246–256, 286; also *racial hygiene*

INDEX OF SUBJECTS

Albanians 153, 155, 156, 165, 174–175, 176n, 190, 193–194, 198
alcoholism 224, 249, 254, 256
Alexandra (maternity hospital Athens) 266–267, 274; see also *Louros, Nikolaos*
Anthropological Laboratory and Museum (Athens) 2, 12, 13, 61n
Anthropological Museum in Athens 10, 12–13, 19, 21, 40–46, 48–49, 56–59, 60n, 61–62, 66, 68–74, 76–77, 79, 84, 86–92, 95, 111–112, 114–115, 119, 122n, 139, 164, 165n, 174n, 287
  and colonialism 66,73–76, 115, 139, 287
  as national heterotopia 13, 61–62, 77
  building of 6, 8, 69, 72, 90, 92
  open to the public 57, 66, 67–68, 77
Anthropological Museum in Budapest 43
anthropology
  and psychoanalysis 117–118, 224n
  as fetish 14, 98, 110, 113, 114, 140
  as third biology 80
  German 6, 14, 169
  Greek 11, 12, 13, 19, 51, 66, 77, 114, 119, 141, 288, 290
  in the broad sense 2, 26, 79, 283
  see also *physical anthropology; racial anthropology; cultural anthropology; social anthropology*
antiquity cult 13, 16, 66, 118, 119, 139, 165, 171, 206, 281
antiquity Hellenic 35, 36, 37, 38, 39, 65, 77, 119, 122n, 149, 155, 160, 163, 171, 172, 175, 201
Aretaieion hospital in Athens 47
Asia Minor 34, 39n, 157–158, 176, 195, 198
refugees from 190, 191, 192n, 198, 281
atavism 39

bacteriology 249, 260n
Balkan Wars 211
Berliner Gesellschaft für Anthropologie 34n, 39, 58
Berliner Gesellschaft für Anthropologie, Ethnologie und Urgeschichte 58
biological capital 207, 222
biologism 3, 4, 279, 286, 288
  authoritative biologism 279, 283–285
  biologistic classism 285
  biosociology 4
bio-power 207, 222–223, 228, 258, 279, 281, 282–284
birth control (generally) 207, 210, 233
  eugenic 214, 218, 220, 231, 233, 239, 246
    against 261
    by Greek state institutions 239–246
botany 29, 80

cancer 224
Chams (Čames) 194
Civil War, Greek 14, 19, 56, 77, 99, 104, 121, 128, 131–137, 182, 197, 290
class, social classes 148, 162, 171, 213–214, 218, 220, 233n, 251–252, 269, 279n, 281, 283–286
  middle class 42, 208, 220–222, 230, 252, 285
  lower classes 103, 213, 217–218, 220–222, 225–226, 269, 275, 281, 285
  poorest classes 212
  upper classes 42, 213, 220, 252, 285
  weak classes 210
  working classes 208–209, 231, 252, 254
Communist Party of France 75n
Communist Party of Greece 132, 136, 185, 195, 198
Communist Review [*KOMEΠ*] 185, 186
communists 17, 182, 195, 197, 209, 210, 290
  anti-communists 127, 182, 195
criminal biology 132n, 249
criminology 93, 99, 100, 205, 206n, 237n, 238n
  positivist criminology 100
cultural anthropology 5, 65, 73, 115, 116

Darwinism 27n, 29, 39, 250
degeneration 122, 211, 212–214, 218, 220, 223, 225, 236, 250, 261, 270
demography 31, 37, 39, 60, 65, 82, 210, 228, 258
dictatorship 126, 131n, 195, 200, 236, 244n, 245, 260n, 271, 274
disease, diseases 17, 32, 33, 36, 37, 101, 106, 108, 213, 222, 223, 224, 229, 230, 232, 234, 239, 241, 243, 247, 253, 270, 273
  contagious, infectious 228, 240, 249, 262

hereditary 213, 240, 255, 261
venereal 224, 239, 241, 254, 262–263
Dromokaiteion (hospital in Athens) 260

Ethnological Museum in Athens 70–76
also *National Historical Museum* 73
ethnology 13, 31, 48, 63, 65, 67, 70, 73, 73, 74, 77, 81, 112, 115, 116, 119, 120, 157
eugenics 4, 7, 16, 17, 18, 19, 21, 82, 94, 112, 188, 203, 204, 207, 212, 214, 216, 217, 219, 220, 229, 230–236, 241–243, 247–250, 253, 256, 257–258, 264–267, 272, 279, 280, 281, 283, 285, 286
against and beyond 18, 258–263
and Greek Orthodox Church 236, 241–242
antiquity 205–207
German 18, 246, 256; see also *racial hygiene*
Jewish 206–207, 250n
liberal 17, 258, 285
negative 221
new 266
positive 213, 215
post war, reformist 263–273, 286
primitive 207
soft 266
euthanasia 205, 271, 272
Exposition Coloniale Internationale in Paris 74

Fabianists 222n, 263
fili
Greek 14, 15, 16, 18, 19, 81, 82, 96, 109, 115, 121–123, 136, 145–147, 149–192, 196, 203, 204, 212, 213, 218, 220, 222, 236, 244, 252
and bio-power 279–286
fluid constancy of 167, 179, 187
strong nucleus of 15, 152, 165,167–168, 171–172, 174, 178, 190, 195
see also *race Greek*; *Koumaris Ioannis*
First Panhellenic Congress for Hygiene (Athens) 233
folklore (studies) 63, 65, 81, 115, 116, 118
Friedrich-Wilhelm-Universität Berlin 130n

Geneoktonie 251, 252
Gesellschaft für Rassenhygiene 254
Greek Archaeological Society 59, 91n, 92, 117n
Greek blood 16, 179, 180, 181, 185, 188, 189
infection, contamination of 16, 179–181, 189
see also *racial purity*
Greek blood (newspaper) 119n, 127
Greek blood (resistance group) 127n
Greek Society of Eugenics 20, 245, 263, 264, 266–268, 271–274
Greek Society of Eugenics and Human Genetics 245n, 264, 267, 269n, 272n, 273

health 30, 31n, 32–33, 100n, 109, 185, 222, 223, 226–230, 233, 239, 244, 281, 282, 285
health card 109, 214
health certification, prenuptial (premarital) 184, 185, 219, 222, 233, 234, 235, 240, 241, 242n, 244, 245n, 254, 260, 261, 268, 270, 271n
health insurance 102, 109, 213, 227, 235, 256, 270, 275
healthcare 209, 213, 227, 228, 229, 231, 239n, 243, 268, 272
Hellene [*Ελληνίς*] (journal) 231n, 235n
Hellenism 15, 37, 149, 151–158, 160–161, 167–168, 171, 192
assimilatory power of 152, 162, 165, 167–168, 171
Historical and Ethnological Society (Greece) 39n, 70–73, 93
hygiene 6, 17, 21, 30, 31, 32, 36, 189, 206, 213n, 222, 229, 231, 248, 249, 253, 256, 261
and eugenics 17, 223–232, 236, 247, 253, 257
as obsession 260
public hygiene 227n, 228, 233, 255, 256, 257
social hygiene 213, 228, 229, 230, 232n, 237, 249, 250, 254–258

Institute for Forensic Medicine, University of Athens 132n
intelligence tests 99, 104–105
Internationale Hygiene-Ausstellung Dresden 219n, 220n

Jews 124–128, 168, 169, 177, 184, 281

Kaiser-Wilhelm-Gesellschaft 251, 274, 275, 277
Kaiser-Wilhelm-Institut für Anthropologie, menschliche Erblehre und Eugenik 20, 81
Koumaris Award 90, 140, 192, 196, 199

La difessa della Razza (journal) 186
laography 65
leprocy 224, 239, 240, 243, 246
Hansen disease 184, 239, 245, 265

malaria 159, 213, 224, 231, 254
medical microbiology 32
Military Academy for Hygiene 260n
migrants 37 77, 168, 227
  Greek emigrants 180n, 181
medical geography 12, 30, 31–35, 39
medical faculty (University of Athens) 12–13, 19, 41–42, 44, 45–46, 51, 57, 58, 68, 69n, 85n, 86, 87, 92, 117, 141, 256–257, 258n, 289
minorities 126, 168, 189, 190, 191, 194–195, 281, 290, 292
modernization 16–17, 145, 149, 157, 207–208, 217, 221, 225, 237, 246, 259, 285, 288
monstrosity, teratogenicity 107
  monsters 58
Muslims 158, 192, 194

National Council of Greek Women 230, 231, 260
National Socialism 15, 130, 137, 159, 168, 175, 177, 251, 271
NSDAP 277, 278

Orientalistic prejudices 15, 171, 172
Ottoman Empire 15, 30, 36–38, 72, 149, 150–158, 175, 189–191, 193, 226n
  legacy 153, 171–172, 175
  territory, territories 37, 157, 190, 198, 209, 213
Ottonian University in Athens 2, 42
overpopulation 17, 207–208, 210, 213, 217, 220
  post-war discourse on 268, 269

paedology 100–103
Palais des Invalides 72
Parnassus philological club 72, 92
  and hygiene 230, 231, 238
pauperism 225–228
physical anthropology 2, 3, 5, 7, 10, 13, 21, 34, 37, 39, 41, 44, 46, 48, 51, 55, 56, 67, 81, 83, 116, 139, 159, 169, 192, 197, 289
  alliance with archaeology 13, 62, 63, 64, 65, 70, 139
  university chair 10, 12, 19, 21, 41, 44–48, 56, 79–138, 140, 141, 188, 256, 258
population question 211–212, 218, 223, 233, 252, 268
poverty 17, 207, 208, 214, 222–223, 225, 227–228, 232, 259
  as social pathology 207, 223, 225
Pomaks 192, 291
prevention 186, 204, 225, 233, 247, 255, 260
promiscuity 17, 223–224
  and anti-communism 182
prostitution, prostitutes 214, 224

Quadrivio (journal) 178, 185, 186

race 15, 16, 18, 81, 91, 97, 112, 127, 136, 141, 142, 163, 165–192
  Anterior-Asiatic (*Vorderasiatisch*) 173–177, 191, 198
  by Foucault 183, 281–283
  German 124, 130
  Jewish 127, 187
  Nordic 15, 171–177, 178, 179
  Vitalrasse (vital race) 250, 251, 286
  see also *fili, Greek*
racial anthropology 6, 13, 77, 120, 139, 288
racial hybridization 162, 165, 168, 171
racial hygiene 16, 18, 188, 189, 205, 246–258, 264, 265, 281
racial intuition 170, 291, 292
racial nationalism 19, 287, 289, 290
  hidden legacy of 287, 289, 290
racial purity 15, 126, 160, 162–170, 172, 177, 178, 179, 181, 182, 183
racial types 68, 163, 164, 166, 168, 173, 176, 177, 178
  Mediterranean 173, 174, 198
Roma 125, 168, 192, 265, 291

Saint-Simonists 226
Slavophones 93, 192, 194, 195, 198
Slavs 93, 141, 153–156, 165, 190, 192–196, 198
social anthropology 14, 98, 115–120; see also *Prince Peter*
social biology 4, 219, 260, 261
social disease 17, 222, 223–232, 233, 257, 262, 263, 270
social question 13, 17, 142, 204, 208–210, 212, 226
Sociétê d'Anthropologie de Paris 39
sterilization 17, 203, 232, 233, 234, 235, 236, 238, 241, 244, 248, 254, 255, 261, 263
  voluntary 234–237, 254–255
  forced, compulsory 235–236, 238, 241, 254, 255, 265
  of criminals 237, 238
syphilis 103, 184, 224, 237, 240, 242, 243, 245, 262

thalassaemia 109, 110, 266, 269
trachoma 103, 184, 224, 237, 240, 242, 243, 245, 262
tuberculosis 213, 224, 231, 234, 237, 240, 241, 243, 254, 262
Turks 141, 180, 190–193, 195, 291

UNESCO's *First Statement on Race* 105, 178, 187, 188
University of Athens 2, 10, 12, 19, 21, 28n, 29, 38, 41–42, 44, 46–47, 49, 69, 76n, 79, 83, 93, 95, 100–101, 116n, 120, 135, 188, 206, 226, 232, 235, 238, 247, 252, 256, 260n, 276n

Vlachs, Aromanians 193, 198, 193

welfare institutions 16–17, 145, 204, 207–209, 217, 226, 232, 235, 246, 248, 269, 285
World War I 38, 48, 80, 217
World War II 4, 14, 18–19, 56, 85, 92, 94–95, 99, 110–111, 116, 118, 121–122, 127–128, 130, 136–138, 177, 187, 191, 194–195, 214n, 244, 251, 255, 258–259, 261, 263–264, 266, 268, 272–273, 280, 288, 292

Zeughaus 72
Ziel und Weg (journal) 178, 184, 186,

www.ingramcontent.com/pod-product-compliance
Lightning Source LLC
LaVergne TN
LVHW020651300926
849161LV00004B/890

* 9 7 8 9 0 0 4 2 5 7 6 6 5 *